Sensor Systems for Gesture Recognition II

Sensor Systems for Gesture Recognition II

Editor

Giovanni Saggio

MDPI • Basel • Beijing • Wuhan • Barcelona • Belgrade • Manchester • Tokyo • Cluj • Tianjin

Editor
Giovanni Saggio
Electronic Engineering
University of Rome "Tor
Vergata"
Rome
Italy

Editorial Office
MDPI
St. Alban-Anlage 66
4052 Basel, Switzerland

This is a reprint of articles from the Special Issue published online in the open access journal *Sensors* (ISSN 1424-8220) (available at: www.mdpi.com/journal/sensors/special_issues/Gesture_Sensor).

For citation purposes, cite each article independently as indicated on the article page online and as indicated below:

LastName, A.A.; LastName, B.B.; LastName, C.C. Article Title. *Journal Name* **Year**, *Volume Number*, Page Range.

ISBN 978-3-0365-8547-5 (Hbk)
ISBN 978-3-0365-8546-8 (PDF)

Cover image courtesy of Giovanni Saggio

Contents

About the Editor

Giovanni Saggio

Giovanni Saggio (GS) graduated with an M.Sc. degree in Electronic Engineering from the University of Rome "Tor Vergata" in 1991, and a Ph.D. degree in Microelectronic and Telecommunication Engineering in 1996. GS is an Associate Professor with tenure at the Electronic Engineering Department of the University of Rome "Tor Vergata" and is a coordinator of the Hiteg (Health Involved Technical Engineering Group) Lab. GS participated in 15 scientific projects, being the coordinator in 10 projects, and founded 3 Spinoffs (Captiks Srl, Seeti Srl, and Voicewise Srl). To date, GS has authored 9 books, authored/co-authored 255 scientific publications, authored/co-authored 12 patents, and was Guest Editor for 3 Special Issues related to Electronics.

Preface to "Sensor Systems for Gesture Recognition II"

Gesture recognition (GR) aims to interpret human gestures with impacts in a number of different application fields.

This Special Issue is devoted to describing and examining up-to-date technologies to measure gestures, algorithms for interpreting data, and applications related to GR.

The technologies involve camera-based systems (e.g. an optical motion capture system), and wearable sensors (e.g. an accelerometer, gyroscope, inertial measurement unit (IMU), magnetic inertial measurement unit (MIMU), electromyography (EMG), surface electromyography (sEMG), force myography (FMG), and data/sensory glove).

Data interpretations are detailed here by means of certain metrics (e.g. Euclidean distance) or of a number of classifiers (e.g. artificial neural network (ANN), grasshopper extreme learner (KTGEL), reinforcement learning (RL), deep Q-network (DQN), Random Forest (RF)).

The adopted applications are for medical purposes (e.g. rehabilitation training, control of electric prostheses, gait behavior recognition, cerebral palsy evaluation, and performance in surgical skill assessment), for social matters (e.g. emotion recognition and judgment, hand signs, sign language recognition, and activity recognition), for sports activity analysis (e.g. football kicks), for machine interaction (e.g. human–computer interaction and visual object tracking), and for animal-related application (e.g. detecting fatigue).

This Special Issue is addressed to all the researchers, professionals, and designers interested in GR and to all the users driven by curiosity and passion.

The Guest Editor expresses acknowledgment and thanks to all the involved authors.

Giovanni Saggio
Editor

sensors

MDPI

Article

Benchmarking Dataset of Signals from a Commercial MEMS Magnetic–Angular Rate–Gravity (MARG) Sensor Manipulated in Regions with and without Geomagnetic Distortion

Pontakorn Sonchan, Neeranut Ratchatanantakit, Nonnarit O-larnnithipong, Malek Adjouadi and Armando Barreto *

Electrical and Computer Engineering Department, Florida International University, Miami, FL 33174, USA; psonc001@fiu.edu (P.S.); nratc001@fiu.edu (N.R.); nolarnni@fiu.edu (N.O.-l.); adjouadi@fiu.edu (M.A.)

* Correspondence: barretoa@fiu.edu

Abstract: In this paper, we present the FIU MARG Dataset (FIUMARGDB) of signals from the tri-axial accelerometer, gyroscope, and magnetometer contained in a low-cost miniature magnetic–angular rate–gravity (MARG) sensor module (also known as magnetic inertial measurement unit, MIMU) for the evaluation of MARG orientation estimation algorithms. The dataset contains 30 files resulting from different volunteer subjects executing manipulations of the MARG in areas with and without magnetic distortion. Each file also contains reference ("ground truth") MARG orientations (as quaternions) determined by an optical motion capture system during the recording of the MARG signals. The creation of FIUMARGDB responds to the increasing need for the objective comparison of the performance of MARG orientation estimation algorithms, using the same inputs (accelerometer, gyroscope, and magnetometer signals) recorded under varied circumstances, as MARG modules hold great promise for human motion tracking applications. This dataset specifically addresses the need to study and manage the degradation of orientation estimates that occur when MARGs operate in regions with known magnetic field distortions. To our knowledge, no other dataset with these characteristics is currently available. FIUMARGDB can be accessed through the URL indicated in the conclusions section. It is our hope that the availability of this dataset will lead to the development of orientation estimation algorithms that are more resilient to magnetic distortions, for the benefit of fields as diverse as human–computer interaction, kinesiology, motor rehabilitation, etc.

Keywords: MARG; MIMU; orientation estimation; sensor fusion algorithm; dataset; orientation algorithm benchmarking

Citation: Sonchan, P.; Ratchatanantakit, N.; O-larnnithipong, N.; Adjouadi, M.; Barreto, A. Benchmarking Dataset of Signals from a Commercial MEMS Magnetic–Angular Rate–Gravity (MARG) Sensor Manipulated in Regions with and without Geomagnetic Distortion. *Sensors* **2023**, *23*, 3786. https://doi.org/10.3390/s23083786

Academic Editor: Giovanni Saggio

Received: 28 February 2023
Revised: 29 March 2023
Accepted: 5 April 2023
Published: 7 April 2023

1. Introduction

1.1. Need for a MEMS MARG Benchmarking Dataset

One of the earliest developments of a micro-machined accelerometer was reported by Roylance et al. in 1979 [1]. However, it would be about 15 years before these devices were embedded into end-user products and manufactured in large volumes [2]. By 2009, miniature gyroscopes, also developed as micro electro mechanical systems (MEMs) would become commercially available as well [3]. Both of these sensors were then packaged commercially as "6-degrees-of freedom" MEMS inertial measurement units (IMUs), sparking significant interest due to the advantages that these sensor modules could have in many prospective applications. These modules were small in size and low in weight, power consumption and heat dissipation, while simultaneously offering measurements of acceleration along three orthogonal axes and rotational speed about those same axes. In principle, these were the same measurements required in aeronautical and maritime inertial navigation systems that use the "strapdown configuration". The strapdown approach estimates orientation by integration of the gyroscope measurements and utilizes it for the appropriate projection ("resolution") of accelerometer readings to an inertial coordinate frame with one of its axes parallel to the gravitational acceleration [4,5].

For the strapdown approach, the orientation estimate must be very accurate, so that the acceleration of gravity can be correctly discounted from the resolved accelerometer readings and then double-integrated to yield a position estimate. Unfortunately, the use of MEMS IMUs has shown that the quality of the acceleration and rotational speed measurements from these miniature sensors is significantly poorer than that of their navigational or strategic counterparts [6–8]. In response, researchers have focused efforts on the development of orientation estimation approaches that can be resilient to the imperfections of signals from the MEMS accelerometers and gyroscopes. In particular, the offset frequently found in the outputs from MEMS gyroscopes is highly damaging to any process that uses those signals for orientation estimation. This is because rotational speeds must be iteratively accumulated (integrated) in one form or another to keep a running tally of total rotation from a starting orientation, i.e., the quantification of the current orientation of a rigid body with respect to its initial orientation. The offset in the outputs of MEMS gyroscopes varies from "run-to-run" and it even experiences significant "in-run" fluctuations [7]. This makes complete and permanent cancellation of gyroscope offsets extremely difficult to achieve and gives rise to orientation "drift" errors when not properly counteracted.

Accordingly, many new MEMS orientation estimation approaches strive to implement some form of information fusion, so that the accelerometer readings may be used to enhance (often correct) initial orientation estimates obtained from gyroscopic readings. In this spirit, the "6-degrees-of-freedom" IMUs were augmented with tri-axial magnetometers incorporated first on the same module enclosure and later as additional components in the same chip as the accelerometer and gyroscope. These are the sensor modules that we prefer to designate as "magnetic–angular rate–gravity" (MARG) modules, although they are sometimes identified also as "9-degrees-of-freedom" IMUs or magnetic and inertial measurement units (MIMUs). For the last several decades, researchers have proposed multiple approaches to solve the problem of real-time MARG orientation estimatation, attempting to fulfill the promise that MEMS IMUs first seemed to have in the earlier XXI century. In their 2021 paper, Nazarahari and Rouhani summarized the breadth and depth of the different approaches proposed over the last 40 years [9], finding that there may not be a single existing algorithm that outperforms the rest under all conditions. Instead, they found that certain algorithms seem to perform better when specific considerations (e.g., latency, computational complexity, characteristics of the environment in which the MARG is used, etc.) are given priority. Accordingly, when they identify the future research challenges in the field, they state "... we suggest that test platforms and benchmarking studies are required to identify the most effective SFAs (Sensor Fusion Algorithms), as well as techniques that could improve the accuracy and robustness of SFAs." [9].

Useful application of MEMS MARG modules in areas such as human–computer interactions and human motion studies requires significant accuracy and robustness in the MARG orientation estimation algorithms used. For example, if a physical hand-held controller with a single MEMS MARG attached to it is used as a ray-casting pointer for three-dimensional virtual environments (e.g., [10]), orientation estimation errors between the actual orientation of the hand-held controller and the orientation of the virtual ray created will be detrimental to the task when pointing to objects at increasing virtual depths, as a virtual object will span a smaller amount of degrees of visual field when it is simulated at increasing virtual depth (away from the user). Conversely, any orientation error will be projected to a longer error distance when the ray intersects deeper planes in the virtual environment. In more complex uses of MARG modules for human–computer interactions, such as their utilization in instrumented gloves for real-time hand tracking and gesture identification (e.g., [11]), each finger is frequently modeled as a kinematic chain. Therefore, the position of the fingertip is computed as a composition of the lengths and orientations of all three finger segments (proximal phalanx, intermediate phalanx, and distal phalanx), as estimated by the MARGs attached to them. If each MARG orientation estimate contains errors, the estimated position of the fingertip may be significantly flawed, resulting in an unacceptable internal representation of the user's hand in the computer. Furthermore, developing techniques that will allow a user to "grab" virtual objects with their virtual hand will demand a high level of accuracy in the virtual representations of position and orientation of all hand segments.

In principle, MARG module orientation estimation is possible based on gyroscope data, particularly when periodically corrected by information from the accelerometer and the magnetometer. However, a critical challenge that emerges is the risk of applying a correction when the assumptions made for the utilization of secondary sensor (accelerometer or magnetometer) data are not met [12]. If such inappropriate corrections are fully applied, significant error may be introduced in the current and future orientation estimates.

For magnetometer corrections, the assumption is, typically, that the geomagnetic vector has the same magnitude and the same direction throughout the complete operational space in which the MARG will be used. Indeed, the geomagnetic vector, considered in isolation, would be constant in magnitude and direction within the reduced environments (e.g., a room or even a city) in which these types of MARG applications will take place. However, in the modern built environment, it is essentially unavoidable to find ourselves near large ferromagnetic objects, such as metal furniture and even structural elements embedded in our dwellings, laboratories, and buildings. These objects could have high magnetic permeability that may cause the "bending" of the magnetic field lines, yielding localized distortions of the magnetic field in their neighborhood. As a result, possible local geomagnetic distortions cannot be overlooked in the evaluation of MARG orientation estimation algorithms.

Accordingly, we believe that there is a need for benchmarking datasets that challenge the MEMS MARG orientation algorithms in different ways. Specifically, the study of the resilience of MARG orientation algorithms that use magnetometer readings of local distortions of the geomagnetic field has attracted a lot of attention in recent years and should be fostered by the creation of calibrated (i.e., containing ground truth orientations) datasets collected in spatial regions where the geomagnetic field is both normal *and* distorted. The impact of magnetic field disruptions on MARG orientation algorithms within fields such as biomechanics research and human–computer interaction became particularly clear after the publication of the 2009 paper by deVries et al. [13], who advised researchers to "'Map' your laboratory on ferromagnetic characteristics ... " and "Preferably use IMUs well away from floors, walls, and ceilings". These limiting concerns were echoed more recently (2019) by Picerno in his comprehensive survey of techniques for studying joint kinematics by using inertial and magnetic sensors [14]: "Unfortunately, the presence of ferromagnetic disturbances distorts the sensing of the local magnetic north. This negatively affects the reliability of the estimated sensor's orientation and may, thus, compromise the usability of such application in the clinical settings, which are normally characterized by ferromagnetic materials and related interferences."

The perceived need for a calibrated dataset that may be used for benchmarking the performance of MARG orientation algorithms in regions with and without magnetic distortions has prompted us to develop the FIU MARG Dataset (FIUMARGDB), which we introduce in this paper.

1.2. Related Datasets and Studies

As it became clear that different MARG orientation estimators exhibited performance advantages for different scenarios, and that their "tuning", i.e., the assignment of specific numerical values to operational parameters of the estimator, was critical, a more urgent need for testing common data emerged. One of the earliest data repositories created specifically for MARG evaluation was "RepoIMU", developed by Szczesna et al. from The Silesian University of Technology, Gliwice, Poland, in 2016 [15]. The paper that presented the dataset expresses the authors' belief that, at the time, "A similar repository was not found." While the authors note three contemporary datasets (one for the study of pedestrian navigation [16], one for human activity recognition [17], and one that combines readings from five Xsens inertial measurement units and data from a Kinect system [18]), they make a compelling case to conclude that none of those datasets would be truly appropriate for benchmarking the performance of MARG sensor modules. The third dataset references is, perhaps, the closest to the kind of dataset needed for MARG orientation algorithm benchmarking, but includes data from the Kinect system as "ground truth", which is known to compare poorly in accuracy and stability to data available from a multicamera motion capture system.

The RepoIMU repository includes recordings from IMU sensor(s) and a Vicon Nexus optical motion capture system under two scenarios: (a) movement of a "wand" (T-stick) with one Xsens MTi-G-28 A53 G35 MARG (in this case, the Vicon system used six markers) and (b) movement of three custom-built IMUs mounted on each of the three sections of an articulated pendulum (in this case, the Vicon system used eight markers). Each wand experiment was reported in a single file. Each pendulum experiment resulted in three separate files (one for each of the IMU modules). The files are comma-separated value (CSV) files with headers in the first row. Each file includes records (rows) comprising a timestamp (in milliseconds) and tri-axial accelerometer, gyroscope, and magnetometer data from the corresponding IMU, as well as quaternion orientation estimates for the corresponding segment. The repository contains a total of 95 recordings, and is available in GitHub [19].

The paper implies that the IMU readings and the Vicon system readings were initially recorded to separate files, as a detailed description of the "data synchronization" process that needed to be performed is included. In fact, the paper mentions that the Vicon system was operated always at a sampling frequency of 100 Hz, whereas the sampling frequency for the IMU signals was different for different tests (as shown in Table 2 of the RepoIMU paper, which shows values of 90 or 166 Hz). In addition, the IMUs used are not necessarily the same type as the low-cost miniature MARG boards that are available at the time of writing (e.g., 3-Space™ Nano IC, from Yost Labs, 3.8 mm × 5.2 mm × 1.1 mm, under USD 20 in large quantities [20]). The Xsens MTi g modules are more complex, combining a MEMS IMU, a GPS, and a barometer, with dimensions of 58 × 58 × 33 mm [21]. The characteristics of the custom IMUs used in the pendulum recordings are not necessarily similar to the characteristics of contemporary low-cost miniature MARG chips. Furthermore, the RepoIMU dataset did not control or insert any form of known magnetic field interference as part of the test scenarios.

In the last 2 years, the need for an objective and detailed comparison of the performance of the many available MEMS MARG estimators has prompted researchers to apply some of the leading estimators to the same sets of MARG data, which they have collected for that purpose. In some cases, researchers have made the data available to other interested parties. That is the case in the study of the accuracy of ten orientation estimation algorithms performed by Caruso et al. (Politecnico di Torino, Bio Robotics Institute in Pisa, University of Berlin and University of Sassari) [22]. The aim of their work was to analyze the accuracy of ten orientation estimators (called sensor fusion algorithms, SFAs), across a matrix of three rotation rates (slow, medium, and fast) by three (pairs of) commercial MARG modules: Xsens-MTx, APDM-Opal, and Shimmer-Shimmer3. For all recordings, the six MARGs were attached to a wooden board, which also had eight reflective spherical markers on it, so that the reference orientation of the whole board (as a rigid body) could be determined by a Vicon T20—Nexus 2.7 optical motion tracking system involving 12 cameras. In this case, also the MARG data and the optical motion capture data seem to have been initially recorded in separate files (since the optical motion capture data "were first processed in Nexus 2.7" and the paper describes a two-step process of synchronization). Both the Matlab algorithm implementations and the three Matlab data (*. mat) files with MARG and Vicon data were made available through IEEE Dataport [23].

The main emphasis of this research effort was the implementation, "tuning" (i.e., search for best parameters for each algorithm), and comparison of the algorithms. The creation of the dataset was a means to that end. All three of the MARG modules used were more complex than the low-cost miniature MARG boards most readily available (such as the Yost Labs 3-Space™ Nano IC). The three recording scenarios did not control or insert any form of known magnetic field interference as part of the test environment.

Another MARG dataset, which was also developed in order to perform a comparative assessment of orientation estimation algorithms, is the one used for a broad study comparing the performance 36 orientation estimators (SFAs) performed by Nazarahari and Rouhani from the University of Alberta, Canada [24]. They applied the algorithms to data from three Xsens MTws MIMUs attached to rigid plastic plates equipped with four retro-reflective markers and fixed to the subject's foot, shank, and thigh. The reflective markers were tracked by an eight-camera Vicon motion capture system. The recordings involved the participation of nine able-bodied participants. Each participant performed actions in two phases; phase I (DataShort.mat records) included standing and brief episodes of walking, turning, jumping, and hopping in order to explore various motion patterns and intensities (each complete trial lasted 137 $\pm$ 7 s.) and phase II (DataLong.mat records) included standing and longer intervals of walking and turning to explore highly dynamic long-duration tasks (each complete trial lasted 393 $\pm$ 3 s.)

The Xsens MTws are closer in size to the miniature MEMS MARG chips (e.g., Yost Lab's 3-Space™ Nano IC), but they are commercialized in a different price range. In this effort, the main focus was on the implementation, tuning, and comparison of the algorithms, but the associated data (from Subject 2) remain available (as Matlab data files) on the website of the Neuromuscular Control and Biomechanics Laboratory of the University of Alberta [25], and include the ground truth foot, shank, and thigh orientation results (as quaternions) from the Vicon system. In a paper that describes the comparison of the algorithms, no mention is made of the introduction or control of known magnetic distortions, other than the decision to implement the SFA algorithms using the foot-worn MIMU only because it "was close to the ground surface and experienced the highest magnetic disturbance compared to shank/thigh MIMUs, according to De Vries et al.".

Previous studies and datasets have not specifically established contrasting recording conditions that would expose the MARG to environments with and without magnetic disturbances. There have been studies where those contrasting magnetic conditions were studied, but their authors have not made the corresponding MARG datafiles accessible to other interested parties. Roetenberg et al. [26,27] studied short MARG records where an Xsense MT9 MARG was rotated in alternating locations that they characterized as "free space" and "close to 3.75 kg of metal" [26]. Subsequently, they performed three types of tests [27]. They first studied the magnetic disruption effect on signals from a static Xsens MT9 MARG as "an iron cylinder of 3.75 kg was placed near the sensor module for 10 min without moving the sensor". A second series of 10 quasi-static tests included rotations of + and -90° performed along the three axes. "After these rotations, the iron cylinder was placed at 5 cm of the module and a new sequence of rotations was performed in opposite directions. The iron was then taken away and the sensor was rotated 90° along the x axis and -90° back." In the third experiment, three 10 cm carbon fiber sticks with optical markers in their ends were attached at orthogonal directions on the MARG and the assembly was attached to a 50 cm long stick "and moved by hand near a large iron tool case". This allowed recording of the MARG signals while an orientation reference (ground truth) was obtained by a Vicon 370 3-D optical tracking system with six cameras. "The movements consisted of small and large rotations along multiple axes at different velocities and different distances from the ferromagnetic case."

In these studies, the data were utilized internally by Roetenberg's research group to develop and evaluate their magnetic disturbance compensation approach, in which varying weight is assigned to the contributions of the magnetometer signals in the correction stage of a Kalman filter. The emphases of these studies and publications were on the crafting of the enhanced Kalman filter orientation estimator and no mention was made of availability of the MARG data to external parties.

The growing interest in comparing the performance of diverse MEMS MARG orientation algorithms and the lack of available datasets that include recording situations that deliberately recorded MARG signals in regions with and without geomagnetic field disruptions has prompted us to develop the FIUMARGDB dataset. Our intent was to collect recordings partially taken in magnetically disrupted regions that were specifically and purposely set up. We asked volunteers to execute a fixed sequence of pre-specified rotations and translations moving a low-cost, commercially available MEMS MARG module, and we have recorded from multiple volunteers so as to capture the different movement idiosyncrasies that different users of the MARG-based human–computer interaction device could exhibit.

2. Materials and Methods

In this section, we briefly describe the setup used for the recording of the files, the MARG module used to record the signals, and the optical motion capture system that was used to simultaneously produce estimates of MARG orientation and position that can be used as "ground truth" for benchmarking the results of multiple MARG orientation algorithms. We also describe in detail the sequence of translations and rotations that the subjects were instructed to execute.

2.1. Recording Environment

As the goal was to obtain data recordings where the MARG would be operating in both magnetically undistorted environments and magnetically distorted regions, our initial concern was to set up an area for the recordings that would not (originally) have magnetic distortions. To this end, the three locations in which the MARG would be operated were defined within a region in our laboratory in which we had previously repeatedly mapped the magnetic field at intervals of 1 foot (25.4 cm) in all three orthogonal directions [28]. The three locations where the MARG would operate during the recordings, (H), (A), and (B), were defined in the portion of the previously mapped space where the magnetic field vectors had been found to have the same orientation and magnitude, away from any large ferromagnetic objects. All the necessary supports were made from wood and glued together (avoiding the use of metallic fasteners).

A 3′ (91.44 cm) by 2′ (60.96 cm) poster presentation cardboard was placed horizontally to provide the subjects with a visual plane of reference (although the subjects were instructed to hold the MARG above this reference plane, never allowing the MARG to touch the cardboard, except at the "home location" at the beginning and end of the recording). The reference plane was at an approximate height of 1 m above the floor of the laboratory.

Paper labels with the letters "H", "A", and "B" were pasted on the horizontal reference plane to guide the movements executed by the subjects. These three locations were arranged as a capital "L" that had been mirrored along its vertical stroke, with (A) located at the intersection of the two strokes. (H) was located about 30 cm in the approximate direction north (For repeatability, the (H), (A), and (B) locations in our setup were placed on lines that run parallel to the grid defined by the tiles in the floor of our laboratory. That grid is only approximately oriented south–north and east–west.) from (A) and (B) was located about 55 cm in the approximate direction west of (A). The relative distances between locations (H), (A), and (B) are displayed in Figure 1.

Since the series of manipulations instructed to the subjects requires them to start by picking up the MARG from location (H), where it would be resting on a "cradle", the inertial frame of reference for orientation purposes would naturally be the same as the body frame of reference at that initial moment of the recording (which we will refer here as "startup"). As the MARG we used (see below) adopts a left-handed orthogonal set of axes, that would be also the one naturally used for the (fixed) inertial frame of reference. Those axes are as described in Table 1.

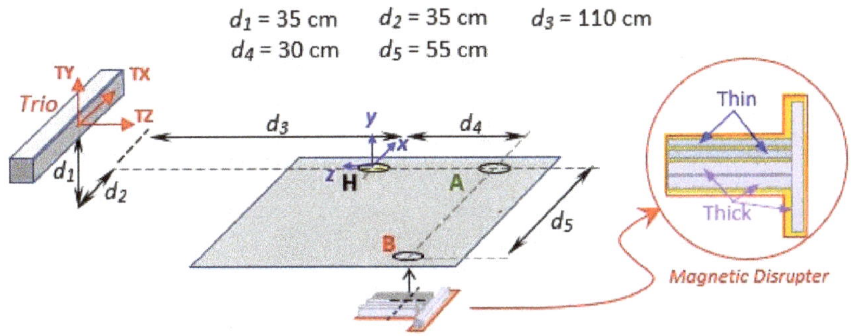

Figure 1. Relative position of locations (H), (A), and (B) with respect to the origin of coordinates for the optical motion capture Trio system. The Trio system uses a right-handed coordinate system (TX, TY, TZ). This is used only for position coordinates. The MARG uses a left-handed coordinate system (x, y, z). (The "Thin" and "Thick" identifiers for the steel bars are explained in Section 2.2.3).

Table 1. Direction of inertial frame axes (for orientations) with respect to labels (H), (A), and (B).

AXIS	AXIS DIRECTION
x AXIS	Parallel to the (B) to (A) direction, positive towards (A)
y AXIS	Parallel to the floor-to-ceiling direction, positive towards the ceiling
z AXIS	Parallel to the (A) to (H) direction, positive towards (H)

2.2. *MARG Module, Optical Motion Tracking System, and Magnetic Disrupter Used*

2.2.1. MARG Module Used

The MARG sensor used for the recordings was a 3-Space™ Wireless 2.4 GHz Attitude and Heading Reference System (AHRS)/inertial measurement unit (IMU) from Yost Labs, Portsmouth, OH, USA, 45662 (https://yostlabs.com/, accessed on 15 February 2023). We chose to use the 3-Space MARG because the manufacturer makes it available in a wide spectrum of versions, all built surrounding the same basic sensor with different types of enclosures and communication alternatives [29]. This will accommodate widely varying user needs in such a way that no superfluous features need to be purchased. The 3-Space family spans a wide range, from the Nano IC model, a low-cost single surface-mount integrated circuit, to 2.4 GHz wireless or Bluetooth versions and even a watertight USB/RS232 module version. The 3-Space sensor has been validated with calibrated movements performed by an industrial robot and found to be appropriate for a prospective application in a study performing joint angle analyses of surgeons performing laparoscopic surgery [30].

The version of the 3-Space MARG we used was contained in a 60 mm × 35 mm × 15 mm plastic enclosure. The MARG exchanges data and commands with the host personal computer through a matching receiver ("dongle") connected to a USB port in the host. For our recordings, the MARG enclosure was firmly attached to the center of an OptiTrack (plastic) "hand rigid body". Three M4 12.7 mm (diameter) reflective spheres were attached to three of the six available prongs of the hand rigid body in such a way that the MARG was located approximately at the center of the triangle defined by the three reflective spherical markers. A lightweight wooden handle was added to the plastic "hand rigid body" so that volunteers could more easily manipulate the MARG. Figure 2 shows the complete wand assembly manipulated by the volunteer subjects.

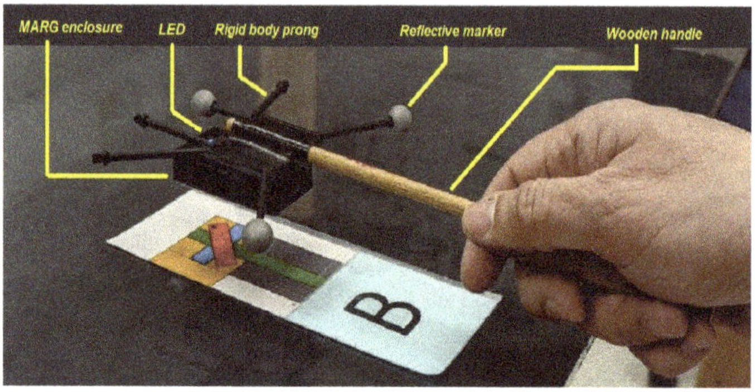

Figure 2. Wand assembly manipulated by the volunteer subjects, shown in Pose 1 <Default Pose> over location (B).

As mentioned before, the default coordinate axis frame for the MARG we used is an orthogonal (Cartesian) left-handed system, where the positive Z (body) axis points to the LED located at one edge of the module's largest face (ordinarily lit in blue). Since orientation estimates commonly represent the accumulation of 3D rotations from an initial orientation, the default inertial frame will be considered to match the body frame at startup, and therefore all orientations are indicated as rotations from that initial orientation to the body orientation at any time during the recording. In other words, the orientations of the (body reference frame of the) MARG are referenced to the body frame orientation at startup, which becomes the default (fixed) inertial frame orientation. In all the recording runs, an effort was made to place the MARG in such a way that the initial body frame axes (and therefore the fixed inertial frame axes) matched the directions described in Section 2.1.

The full set of specifications for the 3-Space MARG we used can be consulted in the "3-Space Sensor Miniature Attitude and Heading Reference System With Pedestrian Tracking User's Manual" (https://yostlabs.com/wp/wp-content/uploads/pdf/3-Space-Sensor-Users-Manual-3.pdf, accessed on 15 February 2023) [29].

It is important to note that, following the native frame of reference of the MARG used, the inertial reference frame used for orientations is as described in Section 2.1. However, the coordinate axis for positions is completely independent of the MARG and will be defined, instead, in accordance with the standards for the optical motion tracking system, described next. This is particularly relevant, as the MARG we used generates internal orientation estimates (as quaternions) utilizing some selectable orientation estimation algorithms. In our files, we also recorded the quaternion orientations generated internally by the MARG using a Kalman filter.

2.2.2. Optical Motion Capture System Used

The optical position and orientation tracking system used during the recordings was the V120:Trio system from OptiTrack (OptiTrack is a company of NaturalPoint, Inc., Corvallis, OR, USA, 97339). This is a system that includes three cameras with infrared filters mounted in a rectangular prism enclosure (58.42 cm $\times$ 4.06 cm $\times$ 5.08 cm). Each camera is surrounded by a ring with 26 infrared LEDs illuminating the field of view of the cameras. Since the cameras are mounted in the enclosure at the factory, the positional relationships of the camera are pre-calibrated and do not require user intervention. Furthermore, position tracking of an infrared reflective marker (or set of markers defining a rigid body), as well as the orientation of a set of markers defining a rigid body, can be set up simply in the accompanying Motive software (Ver. 2.3.2). A complete list of the technical specifications of the V120:Trio system can be found at (https://optitrack.com/cameras/v120-trio/specs.html, accessed on 15 February 2023) [31].

The longest dimension of the V120:Trio system was placed parallel to the (B)-to-(A) line of our setup, about 140 cm to the north of the (B)-to-(A) line and leveled approximately 35 cm above the reference plane of our setup (poster presentation cardboard).

The OptiTrack Motive software identifies the position of a particular marker within each of the three images captured by the cameras. With precise knowledge of the camera characteristics and the positional relationships between the cameras, the Motive software calculates the three-dimensional px, py, pz position of the spherical marker in a right-handed orthogonal coordinate system which has its origin at the location of the central camara of the Trio device, with its TZ axis coinciding with the optical axis of that camera (pointing towards the scene viewed by the camera). The TX axis for the positions reported by the Trio system runs parallel to the longest dimension of the enclosure. Therefore, with respect to the (H), (A), and (B) labels of our setup, the directions of the Trio axes are specified in Table 2, with the origin of the positional coordinates at the location of the central camera of Trio.

Table 2. Direction of V120:TRIO axes (for positions) with respect to labels (H), (A), and (B).

AXIS	AXIS DIRECTION
TX AXIS	Parallel to the (B) to (A) direction, positive towards (A)
TY AXIS	Parallel to the floor-to-ceiling direction, positive towards the ceiling
TZ AXIS	Parallel to the (H) to (A) direction, positive towards (A)

If at least three markers can be detected in the images of the three Trio cameras, they can be designated in Motive as a "Rigid Body", and the software can then track the position of the geometric center of the triangle and the orientation of a 3D vector from the center of the triangle to one designated marker. For our recordings, the designated marker was selected so that the 3D vector that Trio tracks for orientation runs parallel to the Z axis of the body frame of the MARG. Therefore, at every sampling interval, the Trio system reported the three coordinates, px, py, and pz, of the center of the triangle (according to the axes TX, TY, and TZ) in meters and the orientation of the triangular rigid body as a quaternion. In order to adapt the orientation estimate from Trio to the conventions established for the MARG orientation estimates (for example, the internally generated quaternion that used a Kalman filter for orientation estimations), the following manipulations were applied to define the recorded q_{Trio} quaternion (cam_qx, cam_qy, cam_qz, cam_qw) from the quaternion originally calculated by the Trio system stored in rbData (it must be noted that we used the following ordering of the quaternion components: the 1st, 2nd, and 3rd components are the qx, qy, and qz vector components, respectively. The 4th component is qw, the scalar component).

$$cam\_qx = rbData.\ qz \times (-1) \tag{1}$$

$$cam\_qy = rbData.\ qw \tag{2}$$

$$cam\_qz = rbData.\ qx \times (-1) \tag{3}$$

$$cam\_qw = rbData.\ qy \tag{4}$$

The goal of operating the V120:Trio system while the MARG was recording accelerations, rotational speeds, and magnetic field components was to have, for each sampling instant of the MARG sensor data, an independent measurement of the orientation of the MARG (orientation of the rigid body defined by the three spherical reflectors) and its position (the position of the center of the triangle defined by the three spherical reflectors). Accordingly, the orientation calculated by the V120:Trio system (to be referred as the "Trio orientation", q_{Trio}) can act as a "ground truth" for orientation, given its resilience to movement characteristics and magnetic disruptions. Then, the MARG signals can be processed by a variety of MARG orientation algorithms, and their results can be compared to the Trio orientations to assess which algorithm, and under which conditions, yields orientation estimates that more closely resemble the ones from Trio.

The placement of the Trio device was chosen to obtain a good compromise between closeness of the cameras to the markers at any point during the recordings and certainty that the three markers would always remain within the "field of view" of all three cameras. The difficulty of finding an ideal placement for optical motion capture (OMC) systems such as the Trio device is recognized in the motion analysis community. For example, Hindle, Keogh, and Lorimer acknowledge "maintaining a line of sight to each marker throughout the movement is a major challenge when using 3D OMC as markers often become displaced and/or occluded" [32]. We tried a number of combinations of the three distances, d1, d2, and d3, in Figure 1, arriving at the values for these distances shown in the figure. Nonetheless, there can be rare instances during the manipulations performed by the subjects in which the line of sight from any of the three cameras to either one of the spherical markers is obstructed by the MARG holder, or even by another marker. In those cases, the Trio system cannot provide an orientation estimate and repeats the values of quaternion components calculated for the last valid estimation. These events (isTracked = 0) occur in individual sampling instants or in short intervals lasting a few sampling instants and do not distort the overall progression of the quaternion components significantly. Nonetheless, to identify those rare instances, the files also include a flag variable for each sampling instant that is "1" if all the markers were detectable and "0" when at least one of the markers was not detectable. This "isTracked" flag would allow interested users to process the dataset files, applying the interpolation approach of their choice to overwrite the repeated quaternion component values present when "IsTracked" has a value of 0. We have included the Matlab function qTrioFixed = TrioInterp(qTrio, isTracked) in the repository, which performs linear interpolation on the individual quaternion components during the intervals in which "Is Tracked" has a value of 0.

In our data collection setup, both the Trio system and the MARG were connected to the same personal computer host. Our recording software was set to request samples from both systems every 8.3 milliseconds (i.e., at a rate of 120 Hz), and both pieces of information were written simultaneously to a single hard disk file, avoiding the need for after-the-fact synchronization of two different files from each experimental run.

2.2.3. Magnetic Disrupters

One of the priorities in the creation of the files for the dataset was the inclusion, in each experimental run, of both intervals where the MARG would be operating in a magnetically undistorted environment and intervals in which the same device would be subjected to the same type of manipulations but in a region of space known to have distortion of the geomagnetic field. To fulfill the assumption of a uniform, undistorted geomagnetic field in the neighborhood of locations (H) and (A), we defined all three locations for the experiment in a region of space away from furniture that would comprise large ferromagnetic objects. Then, in order to introduce a purposeful magnetic distortion in the neighborhood of location (B), we placed five bars of M35 high speed steel (HSS) from Accusize Industrial Tools, Richmond Hill, Ontario, Canada, under (B), just below the poster presentation cardboard that provided a visual reference plane for the subjects. All five of the steel prisms had a length of 6 inches (15.24 cm), but three of them ("thick") were 0.5″ × 0.5″ (1.27 cm × 1.27 cm) in cross-section, whereas two of them ("thin") were 0.25″ × 0.25″ (0.635 cm × 0.635 cm) in cross-section. Both of the "thin" bars and two of the "thick" bars were aligned north-to-south, and the remaining "thick" bar was placed with a west-to-east orientation at the south end of the other four bars, as shown in Figure 1.

2.3. Sequence Instructed to the Subjects

With the recording environment set up as described in the previous sections, we asked each of the volunteer subjects to grab the assembly containing the MARG module by the wooden handle and we instructed them to perform a prescribed series of translations and rotations. This series of movements was first demonstrated by one of the experimenters who performed the movements following the slides of a PowerPoint slide show that was being displayed on a computer monitor in front of the subject location, so that the correct sequence of movements would not depend exclusively on the memorization by the subject. Then, the subjects were asked to perform the same sequence of movements with the MARG while they were also shown the PowerPoint slides.

The sequence of movements, listed in Table 1, starts and ends with the MARG resting at the "home location" (H), such that the two largest faces of the MARG enclosure are parallel to the floor and the LED in the enclosure is on the edge that is opposite to the subject, as shown in Figure 2. That initial orientation of the MARG is called the <Default Pose>. After execution of every translation or rotation, the subjects were asked to hold the corresponding "pose" for a count of one to five. It was recommended that, in each one of the poses, the subjects tried to hold the lowest point of the complete MARG holding assembly just below the height of two cigarette packs stacked one top of each other (measured as 44.1 mm). (Two actual cigarette packs were placed on the poster board, away from locations (H), (A), and (B), to serve as a visual reference for the subjects.) Depending on the specific pose, the lowest point of the assembly could correspond to one of the reflective spheres, one of the plastic prongs of the plastic rigid body, etc.

In the <Default Pose>, the axes of the body frame of the MARG are oriented in the same way as the axes of the inertial reference frame, described in Table 1. According to the movement sequence defined in Table 3, only sequence steps 12 to 19, corresponding to Poses 6, 7, 8, 9, and 10, take place in the neighborhood of location (B), i.e., under the influence of a distorted magnetic field (steps 2 and 20 are transitions moving in and then out of the distorted magnetic field). The rotations in each of the steps are specified with respect to the fixed inertial frame axes (following the left-hand frame convention).

Table 3. Sequence of steps in each recording run.

Sequence Step	Location	Rotation	Resulting Pose
1	H	(Initial location and pose for the task)	1 <Default Pose>
2	(to) A	After translation H to A, yields	1
3	A	+90° Z Axis, yields	2
4	A	−90° Z Axis, yields	1
5	A	+90° X Axis, yields	3
6	A	−90° X Axis, yields	1
7	A	+90° Y Axis, yields	4
8	A	−90° Y Axis, yields	1
9	A	−45° Y Axis and + 90° X Axis, yields	5
10	A	+45° Y Axis and − 90° X Axis, yields	1
11	(to) B	Just translation A to B	6 (same orientation as 1)
12	B	+90° Z Axis, yields	7
13	B	−90° Z Axis, yields	6
14	B	+90° X Axis, yields	8
15	B	−90° X Axis, yields	6
16	B	+90° Y Axis, yields	9
17	B	−90° Y Axis, yields	6
18	B	−45° Y Axis and + 90° X Axis, yields	10
19	B	+ 45° Y Axis and − 90° X Axis, yields	6
20	(to) H	Just translation back to H	1

The volunteer experimental subjects were recruited from the student body, faculty, and staff of Florida International University. Each subject was given a small enticement (32 GB USB memory or hand-held multimeter) for his/her participation. The experimental procedure was approved by the FIU Internal Review Board (IRB). All subjects were 18 years or older (ages 27.4 ± 7.3 years), without known motion impairments. Each subject held the MARG assembly with their dominant hand. We placed emphasis on including recordings from multiple different human subjects because the database was developed for MARG use in human–computer interaction applications. Therefore, we sought to capture as much as possible of the variability in speeds, trajectories, and stability of poses that can reasonably be found in application of MARG modules within hand-held devices for human–computer communication. This is also the reason why the instructions to the subjects were not exhaustive, leaving room for the idiosyncrasies of movement from each individual. This means that we expected variability in the timing and "accuracy" with which each volunteer subject held the MARG in the instructed poses, which is what would also naturally occur in the ordinary use of a three-dimensional computer interface device (e.g., WiiMote, Nintendo Switch Joy-Con, etc.). This "inaccuracy" on the part of each of the subjects does not impact the intended benchmarking use of the dataset, as the estimation of orientations by new algorithms (even in imperfectly executed poses) will be compared to a "ground truth" estimation of the actual pose held, provided by the Trio system. Our reasons for proposing the orientation estimates from the Trio system as a trustworthy ground truth are detailed in the discussion section (Section 4.1).

A video recording of the sequence of manipulations listed in Table 3, annotated with the identifiers of the poses (P1, ..., P10) as they occur, is available as part of the Supplementary Materials for this paper.

2.4. Verification of the Magnetic Disruption Established near Location B

We verified that the magnetic field near (B) was disrupted (changed) by the presence of the steel bars measuring the field in the X, Y, and Z directions (according to the blue coordinate axes in Figure 1) both at Location (A) and Location (B). We recorded a file ("LONGRUN.csv", available in the dataset in folder "Extra_files_1") in which the MARG started at (H) in the Default Pose (i.e., Pose 1, with the axes of the MARG body frame in the directions indicated by the blue arrows in Figure 1), was translated without rotation to (A), was held there for more than 5 min, was translated without rotation to (B), was held there for more than 5 min, and then it was taken back to (H) without rotation. When we examined the mean and standard deviation in Gauss of the magnetometer readings over 500 consecutive samples, first in (A) and then in (B), we found:

$$
\begin{aligned}
&MagnetoXYZinA \\
&= [(-0.0568, 0.00013), (-0.2264, 0.00010), (0.2177, 0.00014)] \\
&MagnetoXYZinB \\
&= [(-0.0287, 0.00110), (0.0717, 0.00120), (-0.2707, 0.00130)]
\end{aligned}
$$

Here, we can see that all three average magnetometer readings have changed substantially, with the Y and Z components even changing sign, which confirms the magnetic disruption in (B). We also observed that the standard deviations were small (more than one order of magnitude smaller than the averages), which confirms that the readings were essentially constant while the MARG was held in (A) and while the MARG was held in (B). That is, the magnetic disruption at (B) is constant, without variations through time.

3. Results

The result of our data collection effort is the compilation of the FIUMARGDB dataset, which contains the simultaneous signals from the MARG sensors and the Trio system for the sequence described in Table 3, executed by 30 volunteer subjects. The length of each record varies, as different subjects executed the sequence of steps in Table 3 at slightly different paces. Amongst the 30 records in the FIUMARGDB dataset, the minimum record lasts 51.50 s and the longest record lasts 153.96 s, with an average of 100.46 s and a standard deviation of 27.21 s. The repository also includes some additional recordings (e.g., the "LONGRUN.csv" record described in the previous paragraph and others), which are found inside folders labeled "Extra_Files_1" to "Extra_Files_4".

3.1. File Organization

Each of the files in FIUMARGDB is a comma-separated value (CSV) ASCII file, where each row contains data collected at a different sampling instant from both the MARG module and the Trio system. The only exception is the very first line in the file, which contains the column headers, also separated by commas.

Table 4 provides the most important aspects of the organization of each file in FIUMARGDB.

Table 4. Organization of the files in FIUMARGDB.

Entity (Units)	Column	Data (Header)
Timestamp (ms)	1	Timestamp
Trio position (m)	2	pos_x
	3	pos_y
	4	pos_z
Trio orientation (normalized unit quaternion)	5	cam_qx
	6	cam_qy
	7	cam_qz
	8	cam_qw
Kalman filter orientation (normalized unit quaternion)	9	ss_qx
	10	ss_qy
	11	ss_qz
	12	ss_qw
Gyroscope readings (rad/s)	13	gyro_x
	14	gyro_y
	15	gyro_z
Accelerometer readings (g)	16	acc_x
	17	acc_y
	18	acc_z
Magnetometer readings (Gauss)	19	mag_x
	20	mag_y
	21	mag_z
Confidence Factor	22	stillness
isTracked	23	isTracked

The "Confidence Factor" (also described as a "stillness" measure) is a variable computed within the MARG, described as "a value indicating how much the sensor is being moved at the moment. This value will return 1 if the sensor is completely stationary, and will return 0 if it is in motion. This command can also return values in between indicating how much motion the sensor is experiencing." [29]. Similarly, "isTracked" is a flag that normally takes on the value of 1, indicating successful operation of the Trio system, but may take on the alternative value of 0 if one of more of reflective markers tracked by Trio is not visible.

As shown in Table 4, every file in the FIUMARGDB dataset provides all the elements needed to compare the performance of any given MARG orientation estimation algorithm to the orientation estimates from the Trio system as "ground truth". Readings from the accelerometer, the gyroscope, and the magnetometer can be fed to the algorithm under study and its output, expressed as a quaternion for each sampling instant, can be compared to the Trio quaternion provided in the file. Entries in columns 9–12 provide the components of the orientation quaternion calculated in real time by the onboard Kalman filter in the MARG, as an example of orientation estimate result.

3.2. *Visualization of the Contents of a Representative File*

In this subsection, we present the data contained in a representative file (rec03.csv) from the FIUMARGDB dataset. Furthermore, we emphasize in the visualizations how the level of resilience of a given MARG orientation algorithm to magnetic disturbances might be gauged. The FIUMARGDB repository also includes the Matlab functions used to create Figures 3–5.

Figure 3 shows the information obtained from the MARG. This figure was created in Matlab after the contents of the CSV file were read into the workspace. Here, we display the evolution through time of the three accelerometer channel values (acc_x, acc_y, and acc_z), the three gyroscope channel values (gyro_x, gyro_y, and gyro_z), and the three magnetometer channel values (mag_x, mag_y, and mag_z) in the top three subplots, respectively. Finally, the bottom subplot displays the evolution of the four components of the quaternion orientation calculated by the Trio system (cam_qx, cam_qy, cam_qz, and cam_qw). It is in this plot that the timing of the poses can best be recognized. Poses were identified by their numbers (underscored), using blue font for Poses 1–5 held at location (A) and red font for Poses 6–10 held at location (B), where the magnetic field was distorted. It can be seen that while the accelerometer and gyroscope signals during the 2nd part of the record are very similar to those in the 1st part (since the sequence of rotations executed in (B) was the same as those executed in (A)), the magnetometer signals for Poses 6–10 are clearly distorted with respect to those observed for Poses 1–5, as expected.

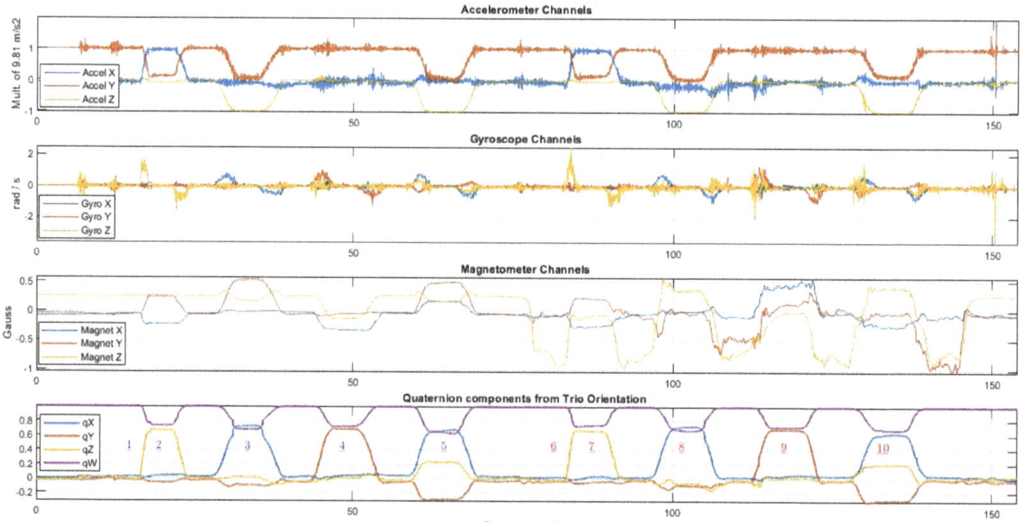

Figure 3. From top to bottom: plot of the three accelerometer channels; plot of the three gyroscope channels; plot of the three magnetometer channels; and temporal evolution of the four components of the rigid body orientation estimate from the Trio system as a quaternion. (Underlined numbers indicate the poses).

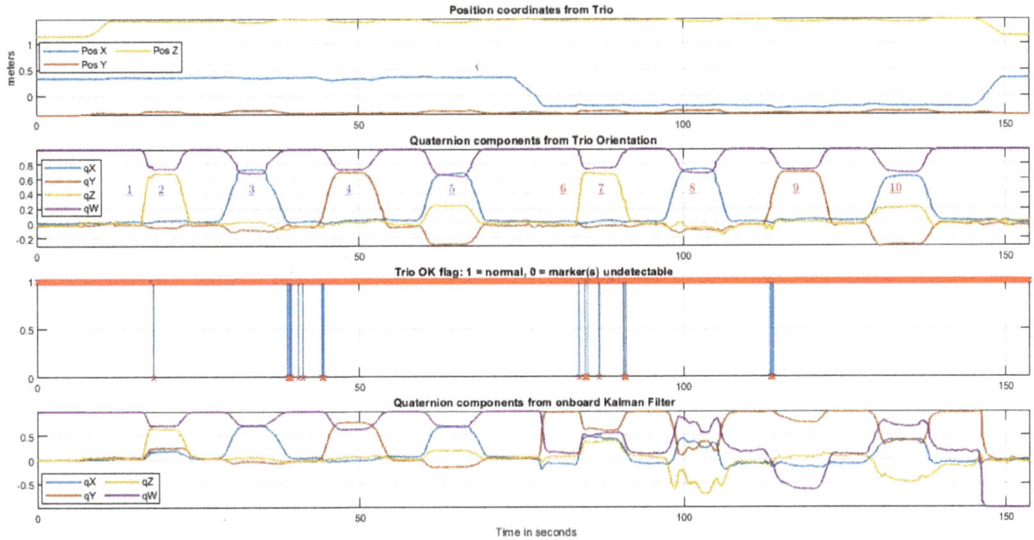

Figure 4. From top to bottom: plot of the three position coordinates with respect to the origin of the Trio frame (TX, TY, TZ); temporal evolution of the four components of the rigid body orientation estimate from the Trio system as a quaternion (converted to the MARGs left-hand frame); plot of the isTracked flag through the recording; and temporal evolution of the four components of onboard Kalman filter quaternion orientation estimate. (Underlined numbers indicate the poses).

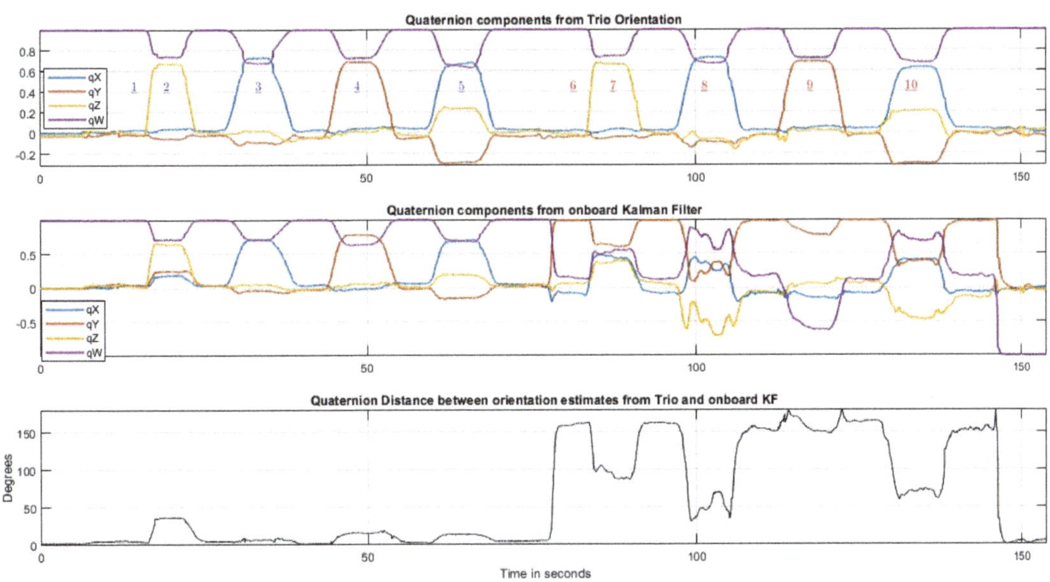

Figure 5. From top to bottom: temporal evolution of the four components of the rigid body quaternion estimate from the Trio system as a quaternion (converted to the MARGs left-hand frame); temporal evolution of the four components of onboard Kalman filter quaternion estimate; and temporal evolution of the "quaternion distance" [33] (angle) between the two orientation quaternions plotted above. (Underlined numbers indicate the poses).

Figure 4 shows all the information generated by the Trio system. The top subplots display the evolution through time of the position coordinates (pos_x, pos_y, and pos_z). The second subplot displays (again) the time evolution of the components of the quaternion calculated by the Trio system (cam_qx, cam_qy, cam_qz, and cam_qw), with the same labeling of poses used in Figure 3. The next subplot displays the values of the "isTracked" flag. This flag recorded a value of 1 (all reflective markers were observable by the Trio system) most of the time, with only a few occurrences of the value 0, which identifies the few sampling instants in which not all three of the reflective markers were observable. These occurrences were rare during the recordings. For example, in the file visualized in Figures 3 and 4, isTracked was 1 for 98.65% of the samples. This figure also includes, at the bottom, the time evolution of the four components of the orientation quaternion calculated by the onboard Kalman filter implemented in the Yost Labs 3–Space MARG.

Figure 5 shows an example of the kind of assessment of MARG orientation estimators that can be performed using the files in FIUMARGDB. At the top, the quaternion components from the Trio system are shown. These orientations, expressed as quaternions, can be taken as the orientation "ground truth". The middle subplot in the figure shows the evolution of the quaternion components generated by the onboard Kalman filter. It can be observed that during the first part of the recording both estimates are very similar. However, when the MARG was translated to location (B), where the magnetic field was distorted (Poses 6–10), the altered magnetometer readings that can be seen in Figure 3 negatively impact the performance of the Kalman filter orientation estimation. As a result, the Kalman filter components take on erroneous values, whereas the estimation process in the Trio system is not affected and produces a very similar sequence of orientation estimates for Poses 6–10 as that produced for Poses 1–5 in location (A).

From the series of four-valued quaternions from Trio and the series of four-valued quaternions from the Kalman filter, is possible to derive a series of "quaternion distance" measurements throughout the complete record. This can be obtained through the "dist" command (angular distance in radians), which, according to Matlab, "returns the angular distance in radians between two quaternions" [33]. This instantaneous error measure is similar to the one used in [24].

The bottom subplot in Figure 5 displays the evolution through time of the "quaternion distance" (already converted to degrees) between the orientation estimates from the Trio system and the onboard Kalman filter. It is easy to recognize that the distance increases significantly during the interval of the recording in which the MARG was at location (B). Therefore, if the orientation estimate from the Trio system is considered the "ground truth", this last graph can be interpreted as indicating that the performance of the Kalman filter in MARG orientation estimation degraded significantly while the MARG was in the magnetically distorted environment.

Figure 6 shows the root mean square (RMS) value of the quaternion distance in degrees (Trio vs. Kalman filter) for each of the recordings in the FIUMARGDB dataset. The red trace is the RMS value incurred only while the MARG was in the neighborhood of location (B), as identified by negative values in the position coordinate TX ("at B"). The blue trace is for the RMS value computed in the remainder of the recording run ("not at B"). While the RMS of the quaternion distance "at B" varies from record to record, we can see that it is typically much higher than the RMS "not at B". The average and standard deviation values are 118.0371° and 39.4526°, respectively, for at B and 11.8065° and 5.2525°, respectively, for not at B, which suggests that most of the quaternion distance resulted from the lack of resilience of the orientation estimation algorithm under magnetically distorted conditions.

It should be pointed out that for our recordings we configured the AHRS filter onboard the Yost 3-Space module to execute the simple implementation of a Kalman filter, just as a basic item for comparison. The Yost 3-Space module is also capable of implementing faster orientation filters, such as "Q-COMP (quaternion complementary) filter" and "Q-GRAD (quaternion gradient descent) filter", instead of the Kalman filter algorithm [29].

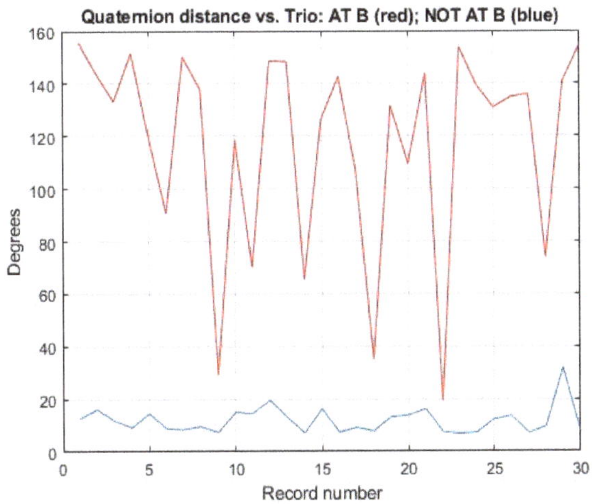

Figure 6. RMS values of quaternion distance (Trio vs. onboard Kalman filter) for the recordings in the FIUMARGDB. The red trace represents the RMS value calculated only in the interval when the MARG was in the neighborhood of Location B (at B). The blue trace represents the RMS value of the remainder of the record (not at B).

In the repository, within the folder "Extra_Files_2", we provide three recordings from runs in which the MARG was configured to implement the Q-COMP filter, and three recordings from runs in which the MARG was configured to implement the Q-GRAD filter. For each of those groups of three records, and for the first three records from our main dataset (rec01.csv, rec02.csv, and rec03.csv), we computed the root mean square (RMS) value of the quaternion distance (e.g., bottom plot in Figure 5 for the Kalman filter orientations from rec03.csv) over the complete run. Table 5 shows the means and standard deviations we found for each type of filter (the values for Kalman filter, KF, refer to RMS computed from files rec01.csv, rec02.csv, and rec03.csv).

Table 5. RMS values of quaternion distance using various onboard orientation filters.

	Q-GRAD	Kalman Filter	Q-COMP
Q distance mean (°)	83.8225	96.8412	100.6647
Q distance std. dev. (°)	8.5298	6.5633	2.1149

We can see in Table 5 that Q-GRAD recorded a slightly lower RMS value of the quaternion distance and Q-COMP recorded a slightly higher RMS value of quaternion distance with respect to the Kalman filter used for all the recordings (30) in our main dataset. Appendix B shows representative plots of quaternion components and quaternion distance (with respect to the Trio orientations) for Q-GRAD and Q-COMP, which can be compared to the lower two plots in Figure 5. In the Q-GRAD plots, we can see that the orientation estimates deteriorate more gradually than for Q-COMP. This might be the reason for the lower RMS value of the quaternion distance displayed by Q-GRAD.

4. Discussion

The main objective of our development of FIUMARGDB was to create a series of MARG data files (i.e., readings from the MARGs tri-axial accelerometer, tri-axial gyroscope, and tri-axial magnetometer) that would be accompanied by the corresponding series of orientation measurements (obtained by the Trio optical motion capture system) which could be considered "ground truth" values of MARG orientation. In particular, we sought to create such combined MARG–ground truth orientation recordings in the following context:

- The MARG signals should come from a low-cost, commercially available MEMS MARG module, as it is for these modules that the signal processing requirements are most challenging but the potential rewards are most promising.
- The environmental context and movements carried out by the MARG module should be similar to those a MARG module may experience in its application in body movement (e.g., hand and fingers movement) tracking for the purpose of human–computer interaction (as that is our area of work [12,34,35]).

Accordingly, we used the 3-Space Attitude and Heading Reference System (AHRS)/ inertial measurement unit (IMU) from Yost Labs (https://yostlabs.com/, accessed on 15 February 2023). Yost Labs offers the same basic MARG system in a variety of packages and with different forms of communication to a host. The basic versions can be affordable (particularly if purchased in medium or large quantities). However, we employed the 3-Space MARG version that communicates wirelessly to a PC host to avoid the disruptive tethering effect that a wired connection from the sensor could have.

4.1. Discussion of the Main Set of Recordings

To explore the types of circumstances in which the MARG may operate as part of a human–computer interaction system, we asked a number of volunteer subjects to perform the same pre-defined sequence of translations and rotations with the MARG at two locations. The magnetic field at the first location, (A), was assumed to be the undisturbed local geomagnetic field. In contrast, magnetic disrupters (described in Section 2.2.3) were placed under the second location, (B), so that the magnetic field sensed by the MARG would be distorted by design. We aimed at capturing the different movement characteristics that could be expected from diverse human operators (e.g., speeds, specific trajectories, continuity of motion, etc.) by recording the same sequence of actions executed by several volunteers.

The generation of independently obtained orientation estimates was achieved by affixing the 3-Space MARG module to a plastic "rigid body hand (emulator)" with three reflective markers, as recommended by the manufacturers of the V120:Trio tracking system, in order to generate orientation measurements of the rigid body defined by the markers.

The appropriateness of using the orientations calculated from the Trio system as "ground truth" to assess the effectiveness of MARG orientation estimation algorithms stems from the documented position tracking accuracy of this type of optical motion capture (OMC) system, and from the fact that the orientation computation procedures from marker positions are not iterative (as opposed to most MARG orientation estimation approaches). The manufacturer of the VT120:Trio system makes the general statement (https://optitrack.com/applications/movement-sciences/#accuracy, accessed on 15 February 2023) that "OptiTrack systems typically produce less than 0.2 mm of measurement error, even across large tracking areas—even of those 10,000 sq ft or more. In smaller measurement areas, OptiTrack systems regularly produce errors of 0.1 mm or less", which references the 2017 study by Aurand et al. [36], where they assessed the position tracking accuracy of an OptiTrack system that employed 42 cameras for tracking a large volume of 135 m^3. Aurand et al. concluded that "the OMC system demonstrated submillimeter mean accuracy at every location in the capture volume, and error was found to be less than 200 μm in 97% of the capture volume (using all 42 cameras)", also indicating that the errors were found to be less than 200 μm in 91% of the capture volume if only 21 cameras were involved. While their study dealt with a much larger capture volume that necessitates the involvement of larger numbers of cameras, they also commented that the errors measured in a study by Eichelberger et al. [37], which "measured inter-marker distance using 6–10 Vicon cameras within a 13.2 m^3 (5.5 m × 1.2 m × 2 m) capture volume ... were of the same order of magnitude as the current study". This level of position tracking accuracy in the same type of optical motion capture systems as the V120:Trio supports the use of its orientation estimates as "ground truth".

While the orthogonal Cartesian axis frames used by the Trio system and by the 3-Space MARG are not the same (the former is right-handed and the latter is left-handed), care was exercised to set the initial orientation of the MARG so that each of its axes would be precisely parallel to one of the Trio axes (e.g., to make x parallel to TX and y parallel to TY in Figure 1). If, nonetheless, for a particular recording, the initial direction of the MARG x axis was inadvertently misaligned with respect to TX by a small rotation, this "rotational offset" may be compensated for by applying a compensatory rotation (encoded as a quaternion, q_C) to the results of a MARG-based orientation estimate, such as, for example, the Kalman filter estimate, q_{Kalman}:

$$q_{KalmanComp} = q_C \otimes q_{Kalman} \tag{5}$$

where q_{Kalman} is the Kalman filter quaternion at any given sampling instant (whose components in the file are ss_qx, ss_qy, ss_qz, and ss_qw according to Table 4), q_C is the (constant) "compensatory quaternion", ⊗ indicates the quaternion product, and $q_{\text{KalmanComp}}$ is the compensated quaternion for that sampling instant. Appendix A describes how the values of the components of q_C can be obtained from an FIUMARGDB file and the matrix equation needed for the compensation of each MARG-based orientation quaternion.

4.2. Supplementary Recordings: Reverse Location Itinerary and Alternative Disrupter Placements

Beyond the main set of recordings that constitute our dataset, we have sought to provide users with two critical variations of the manipulations of the MARG and the location of the magnetic disrupter.

4.2.1. Reverse Itinerary Recordings

We have included (in the folder Extra_Files_3) seven recordings in which the setup is the same as in the main recordings of the dataset, except that the locations are visited in a reverse circuit. That is, at the beginning of the recording, the MARG is picked up from (H) and it is translated, without rotation, to location (B), where there is a disrupted magnetic environment, *first*. At (B), the same sequence of poses as usual (Poses 6, 7, (6), 8, (6), 9, (6), 10, (6)) are held *and then* the MARG is translated, without rotation, to location (A), where the usual poses (Poses 1, 2, (1), 3, (1), 4, (1), 5, (1)) are held. Finally, the MARG is returned to the home location (H). These recordings may be helpful in assessing the capability of a given orientation estimator to "recover" and provide correct orientation estimates in (A) if the estimates generated first in (B) were erroneous. These recordings, with names that start with "rer" (last r meaning "reverse") instead of "rec", can also be visualized with the same Matlab functions and scripts as provided for the standard recordings (e.g., "rec03.csv").

4.2.2. Alternative Positioning of the Magnetic Disruptor

We also provide five sets of three recordings each in which the magnetic disruptor cluster shown in Figure 1 was not placed directly under location (B), but instead it was placed under locations 15 cm to the east of (B) and/or shifted by −15 cm, 0 cm, or +15 cm from south to north with respect to the original location under (B). Figure 7 shows the alternative locations of the magnetic disrupter xDy (where x = 1 or 2 and y = 0, 1, or 2). The 15 recordings are included in the folder Extra_Files_4 of the dataset, and their names follow the convention xDy-n.csv, where n is 1, 2, or 3. For completeness, this folder also contains three files from the main group of files, which can be used for comparison purposes: rec01.csv = 1D1-1.csv, rec02.csv = 1D1-2.csv, and rec03.csv = 1D1-3.csv. For a first assessment of the impact of magnetic disrupter placement on MARG orientation estimate at Location (B), we evaluated the RMS value of the quaternion difference (Kalman filter vs. Trio estimates) just while the MARG was at location (B) for each of the 18 files. The itinerary for all these files was the "standard itinerary", i.e., the one followed during the recording of all the files in the main dataset: Location (H)–Location (A)–Location (B)–Location (H). (We identified that the interval in which the MARG was in the neighborhood of location (B) with the interval of the recording in which the coordinate TX of the position reported by Trio was negative.) Figure 7 shows vertical bars with the average RMS quaternion difference (Trio vs. Kalman filter) observed at location (B) when the magnetic disrupter was under each of the six locations.

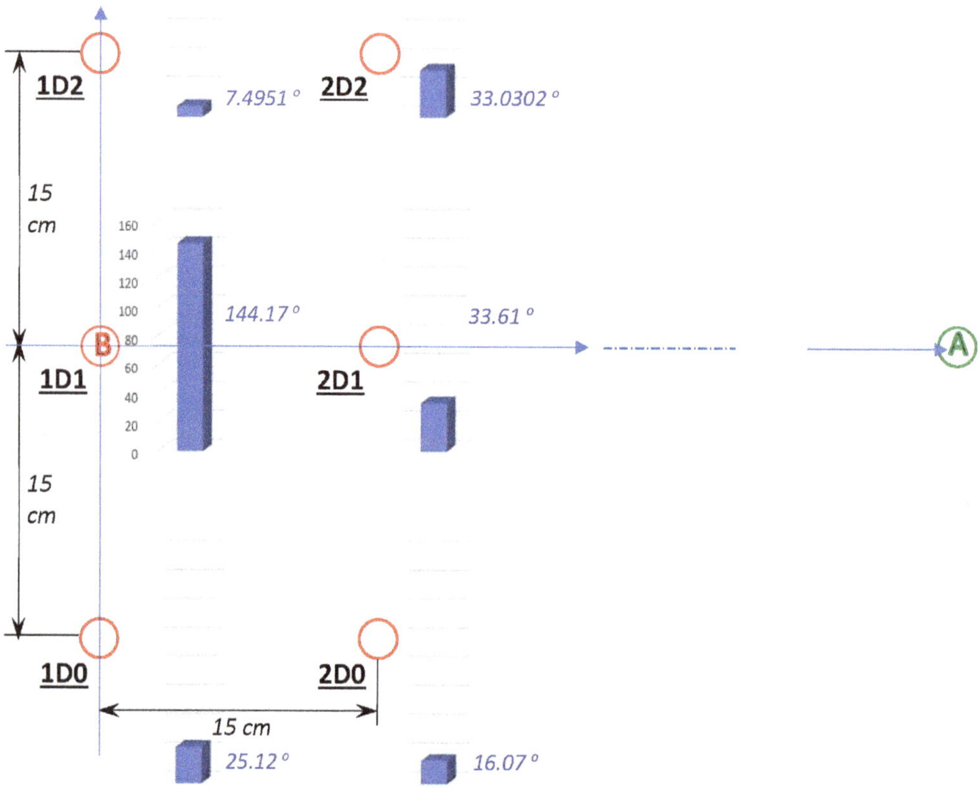

Figure 7. Alternative locations of the magnetic disrupter and corresponding effects in Location (B). (The circled A and B letters represent Locations (A) and (B), respectively.) The red circles are the locations of the magnetic disrupter labeled as XDY, where 1D1 is the standard placement at (B). The heights of the blue prisms represent the RMS value of Kalman filter estimation quaternion distance with respect to the Trio orientation estimation (in degrees) while the MARG was hovering over location (B) and Poses 6–10 were held according to the standard sequence.

The additional 15 recordings may be used by orientation estimation algorithm designers to expose their algorithms to magnetic distortions that have different degrees of strength and are centered at locations around the point of MARG test. The heights of the blue prisms in Figure 7 followed a configuration that was partially expected. Except for the placement of the disrupter right below the area where the MARG operated (placement 1D1), the impact on the performance of the onboard Kalman filter was greatest when the disrupter was placed 15 cm to the east (2D1) or to the south (1D0) of the testing point where the MARG was operated (Location B). There was a smaller impact when the disrupter was placed 21.21 cm in a southeast direction from B (2D0). However, the impacts on the performance at (B) when the disrupter was placed at 1D2 and 2D2 did not match the results for placement in 1D0 and 2D0. This may be, however, a result of the asymmetrical configuration of the disrupter, which has a "thick" steel bar across its south boundary, without a corresponding bar across its north boundary. In any case, Figure 7 confirms that the additional 15 recordings in folder Extra_Files_4 will offer a wider variety of magnetic disruptions that algorithm designers can use for testing. Ultimately, a truly robust orientation estimation algorithm should not degrade under any of the conditions represented in the 15 additional files and it should also be resilient to avoid degradation under the stronger magnetic disruptions involved in the standard files (rec01.csv through rec30.csv) of our dataset.

5. Conclusions

This paper has presented the FIUMARGDB dataset of MARG signals accompanied by "ground truth" orientation estimation quaternions. This dataset is meant to facilitate the benchmarking of orientation estimation algorithms that use signals from the accelerometers, gyroscopes, and magnetometers in a MARG module to compute the orientation of the module, typically as a quaternion. Benchmarking has becoming increasingly important because it has been found that different orientation estimation algorithms may be more severely affected in their performance under certain circumstances. FIUMARGDB was created specifically to expose orientation estimation algorithms to operating environments with and without magnetic distortion.

FIUMARGDB was developed for the benchmarking of orientation algorithms in a context that might be similar to the one experienced by MARG modules used in human–computer interaction devices. Accordingly, all the records were obtained while the MARG module was moved by a human subject with their dominant hand. This defines the range and speed of MARG rotations and translations that were recorded. Similarly, we included recordings created with the participation of multiple human volunteers to capture the corresponding possible variations in speed, trajectory, continuity of movement, and steadiness of poses held.

We selected a basic, low-cost MARG module for the recordings, since the required processing of signals from this type of MARG might be the most challenging. At the same time, very small and low-cost MARG modules will be the best suited type for developing some human–computer interaction devices, such as an instrumented glove to track the orientation and configuration of the hand of a computer user (which may require the inclusion of many MARG modules in the glove). Similar priorities may apply to MARG usage in the fields of human motion studies and motor rehabilitation.

The FIUMARGDB dataset and Matlab programs for its use can be accessed through this URL: https://github.com/LABDSP/FIUMARGDB_marg_signals_and_reference_orientations.git (accessed on 15 February 2023).

Supplementary Materials: The following supporting information can be downloaded at: https://www.mdpi.com/article/10.3390/s23083786/s1, Video: FIUMARGDB_sequence.mp4.

Author Contributions: Conceptualization, P.S., N.R., N.O.-l., and A.B.; methodology, P.S., N.R., N.O.-l., and A.B.; software, P.S., N.R., N.O.-l., and A.B.; formal analysis, P.S., N.R., N.O.-l., and A.B.; investigation, P.S., N.R., N.O.-l., M.A., and A.B.; resources, A.B. and M.A.; data curation, N.R. and P.S.; writing—original draft preparation, P.S. and A.B.; writing—review and editing, N.O.-l., M.A., and A.B.; visualization, P.S., N.R., and N.O.-l.; supervision, M.A. and A.B.; project administration, M.A.; funding acquisition, M.A. and A.B. All authors have read and agreed to the published version of the manuscript.

Funding: This research was funded by The National Science Foundation (NSF), grant number CNS-1920182, and Dr. Neeranut Ratchatanantakit was supported by the FIU Dissertation Year Fellowship (DYF) Program.

Institutional Review Board Statement: The study was conducted in accordance with the Declaration of Helsinki and approved by the Institutional Review Board of Florida International University (with protocol approval number IRB-23-0088) for studies involving humans.

Informed Consent Statement: Informed consent was obtained from all subjects involved in the study.

Data Availability Statement: The record files of the FIUMARGB dataset described in the paper, as well as Matlab programs for their use, can be accessed through this URL: https://github.com/LABDSP/FIUMARGDB_marg_signals_and_reference_orientations.git (accessed on 15 February 2023).

Conflicts of Interest: The authors declare no conflict of interest.

Appendix A

In reference to Figure 1, where the blue coordinate frame (x, y, z) corresponds to the initial orientation (at startup) of the MARG and the red coordinate frame (TX, TY, TZ) corresponds to the Trio system, efforts were made in the recording of the files to set the initial orientation of the MARG so that x and TX would be parallel, and y and TY would be parallel too. Most MARG orientation algorithms will adopt the blue frame as the "inertial frame", reporting future MARG orientations with respect to it. Under those circumstances, the conversion equations shown in Section 2.2.2 will adjust the quaternion results of Trio orientation estimation to be directly comparable to the quaternion MARG orientation estimates.

If, on the other hand, a misalignment of the red and blue frames exists at startup, such that, for example, the x axis is rotated with respect to the TX axis, the MARG orientation estimates will likely be referenced to that inertial frame, which has a constant "rotation offset" with respect to the estimates generated by the Trio system.

Vince [38] showed that a quaternion, q_3, that represents the compounding of the rotations represented by q_1 first, followed by q_2 afterwards, is obtained simply through this quaternion product:

$$q_3 = q_2 \otimes q_1 \tag{A1}$$

Therefore, the "rotation offset" can be removed by obtaining a new, compensated orientation estimate, e.g., $q_{\text{KalmanComp}}$, which is the composition of the original MARG-based estimate, e.g., q_{Kalman}, modified by an appropriate compensation quaternion, q_C:

$$q_{KalmanComp} = q_C \otimes q_{Kalman} \tag{A2}$$

To figure out the values needed in the components of q_C, we focus on the fact that Equation (A2) must be true throughout the complete record, but, in particular, at startup, when the MARG-based algorithms would indicate a MARG orientation encoded by the quaternion q_{Kalman} = [ss_qx, ss_qy, ss_qz, ss_qw] = [0, 0, 0, 1], which indicates "no rotation", i.e., there is no rotation (yet) mediating between the MARG body frame and the inertial frame. At startup, the Trio orientation in general will be q_{Trio0} = [cam_qx0, cam_qy0,cam_qz0,cam_qw0].

(Note: Since the equations in Appendix A are based on those by Vince, the scalar part of all quaternions in this appendix are placed first, as opposed to the convention used in the body of the paper, where the scalar is shown as the last (fourth) component of the quaternions.) Vince [38] indicated that for two quaternions

$$q_a = [s_a, \boldsymbol{a}] = [s_a, x_a\boldsymbol{i} + y_a\boldsymbol{j} + z_a\boldsymbol{k}] \tag{A3}$$

$$q_b = [s_b, \boldsymbol{b}] = [s_b, x_b\boldsymbol{i} + y_b\boldsymbol{j} + z_b\boldsymbol{k}] \tag{A4}$$

the product can be computed through this matrix product (note s_a and s_b are the scalar components of quaternions q_a and q_b, respectively):

$$\begin{aligned} q_a \otimes q_b &= [s_a, \boldsymbol{a}] \otimes [s_b, \boldsymbol{b}] \\ &= [s_a s_b - \boldsymbol{a}\cdot\boldsymbol{b}, s_a\boldsymbol{b} + s_b\boldsymbol{a} + \boldsymbol{a}\times\boldsymbol{b}] \\ &= \begin{bmatrix} s_a & -x_a & -y_a & -z_a \\ x_a & s_a & -z_a & y_a \\ y_a & z_a & s_a & -x_a \\ z_a & -y_a & x_a & s_a \end{bmatrix} \begin{bmatrix} s_b \\ x_b \\ y_b \\ z_b \end{bmatrix} \end{aligned} \tag{A5}$$

Therefore, substituting q_C = [qC_x, qC_y, qC_z, qC_w], the compensating quaternion we want to find for q_a, and the initial MARG-based orientation quaternion, [0, 0, 0, 1], for q_b, their product should yield the initial q_{Trio0} quaternion:

$$\begin{bmatrix} qC\_w & -qC\_x & -qC\_y & -qC\_z \\ qC\_x & qC\_w & -qC\_z & qC\_y \\ qC\_y & qC\_z & qC\_w & -qC\_x \\ qC\_z & -qC\_y & qC\_x & qC\_w \end{bmatrix} \begin{bmatrix} 1 \\ 0 \\ 0 \\ 0 \end{bmatrix} = \begin{bmatrix} cam\_qw0 \\ cam\_qx0 \\ cam\_qy0 \\ cam\_qz0 \end{bmatrix} \quad (A6)$$

That is:

qC_w = cam_qw0; qC_x = cam_qx0; qC_y = cam_qy0; qC_z = cam_qz0.

In other words, the components of the "compensating quaternion", q_C, are just the components of the first quaternion generated by the Trio system at startup. Once those values are read from the first row of any given file in FIUMARGDB, the compensation of each MARG estimated quaternion (for example the values ss_qx, ss_qy, ss_qz, and ss_qw from the Kalman filter) only need to be pre-multiplied by the corresponding 4-by-4 matrix to yield the four values of the compensated quaternion:

$$\begin{bmatrix} qC\_w & -qC\_x & -qC\_y & -qC\_z \\ qC\_x & qC\_w & -qC\_z & qC\_y \\ qC\_y & qC\_z & qC\_w & -qC\_x \\ qC\_z & -qC\_y & qC\_x & qC\_w \end{bmatrix} \begin{bmatrix} ss\_qw \\ ss\_qx \\ ss\_qy \\ ss\_qz \end{bmatrix} = \begin{bmatrix} ss\_qwComp \\ ss\_qxComp \\ ss\_qxComp \\ ss\_qwComp \end{bmatrix} \quad (A7)$$

where the elements in the last column vector to the right in the above equation can be used to substitute the original elements of the quaternion encoding the original (uncompensated) result of the MARG-based orientation estimation.

It can be noticed that, for the file visualized in Figure 4, the initial Trio quaternion has components cam_qx0 = 0.0032, cam_qy0 = −0.0280, cam_qz0 = −0.0222, and cam_qw0 = 0.9993. Therefore, the compensating matrix in Equation (A7) will be very similar to a 4-by-4 identity matrix, which means that the differences between the compensated quaternion and the original quaternion will probably be negligible.

Appendix B

We recorded six additional files with the standard setup and performing the same standard sequence of steps indicated in Table 3 but choosing different onboard orientation filters. A summary of the performance of those filters, in comparison with the performance of the Kalman filter chosen for the basic set of records in our dataset, is provided in Table 5.

Here, we present plots of the evolution of the quaternion components and the evolution of the quaternion distance (with respect to the Trio orientation estimation, acting as a ground truth) for the two alternative orientation filters. These plots also display the numerical value of RMS of the quaternion distance for the complete record (in degrees).

Figure A1 shows the results for a representative file in which the Q-COMP filter was chosen (file qcomp2.csv). In this case, the RMS value was 100.66°. The traces in this figure can be compared with the lower two subplots of Figure 5, which were obtained from file rec03.csv, in which the Kalman filter was selected. The RMS value for that case was 90.03°.

Figure A2 shows the results for a representative file in which the Q-GRAD filter was chosen (file qgrad1.csv). In this case, the RMS value was 86.20°. This slightly lower level of error may result from the more gradual deterioration of the orientation estimate that seems to take place when the Q-GRAD filter was used. However, these observations seem to confirm that the orientation estimates from the three types of filters degrade noticeably when the MARG is operating in a magnetically distorted area.

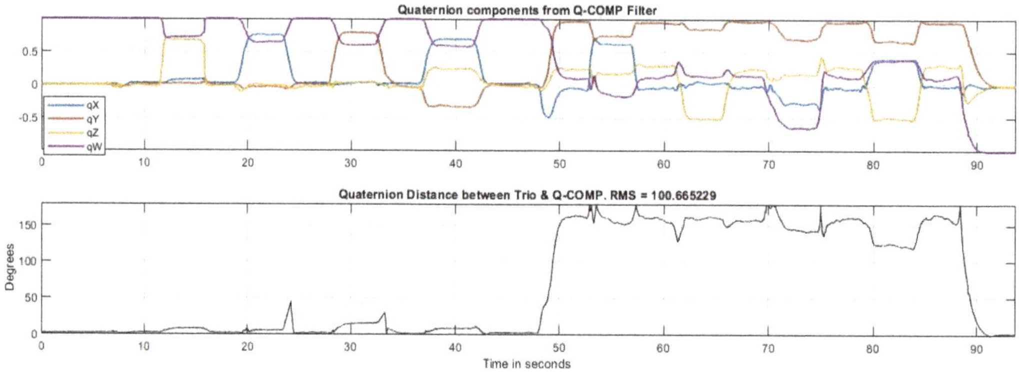

Figure A1. (**Top**) evolution of the quaternion components generated by the Q-COMP onboard filter. (**Bottom**) evolution of the quaternion distance between the Trio quaternion result and the Q-COMP quaternion result.

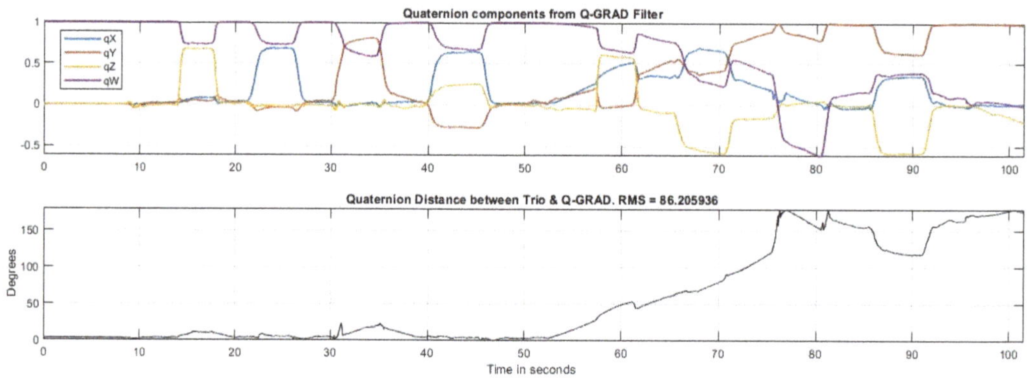

Figure A2. (**Top**) evolution of the quaternion components generated by the Q-GRAD onboard filter. (**Bottom**) evolution of the quaternion distance between the Trio quaternion result and the Q-GRAD quaternion result.

References

1. Roylance, L.M.; Angell, J.B. A batch-fabricated silicon accelerometer. *IEEE Trans. Electron Devices* **1979**, *26*, 1911–1917. [CrossRef]
2. Lee, I.; Yoon, G.H.; Park, J.; Seok, S.; Chun, K.; Lee, K.-I. Development and analysis of the vertical capacitive accelerometer. *Sens. Actuators A Phys.* **2005**, *119*, 8–18. [CrossRef]
3. Johnson, R.C. 3-Axis MEMs gyro chip debuts. *EE Times*. 26 October 2009. Available online: https://www.eetimes.com/3-axis-mems-gyro-chip-debuts/ (accessed on 15 February 2023).
4. Titterton, D.H.; Weston, J.L.; Institution of Electrical Engineers. *Strapdown Inertial Navigation Technology*; Institution of Electrical Engineers: Stevenage, UK, 2004.
5. Savage, P.G. *Strapdown Analytics*; Strapdown Associates: Maple Plain, MI, USA, 2000.
6. Woodman, O.J. *An Introduction to Inertial Navigation*; Technical Report No. 696, UCAM-CL-TR-696; University of Cambridge: Cambridge, UK, 2007; ISSN 1476-2980.
7. Aggarwal, P. *MEMS-Based Integrated Navigation*; Artech House: Boston, MA, USA; London, UK, 2010.
8. Foxlin, E. Motion Tracking Requirements and Technologies. In *Handbook of Virtual Environments, Design, Implementation, and Applications*; Stanney, K.M., Ed.; Lawrence Earlbaum Associates: Mahwah, NJ, USA, 2002.
9. Nazarahari, M.; Rouhani, H. 40 years of sensor fusion for orientation tracking via magnetic and inertial measurement units: Methods, lessons learned, and future challenges. *Inf. Fusion* **2021**, *68*, 67–84. [CrossRef]
10. Ro, H.; Byun, J.-H.; Park, Y.J.; Lee, N.K.; Han, T.-D. AR Pointer: Advanced Ray-Casting Interface Using Laser Pointer Metaphor for Object Manipulation in 3D Augmented Reality Environment. *Appl. Sci.* **2019**, *9*, 3078. [CrossRef]

11. Kortier, H.G.; Sluiter, V.I.; Roetenberg, D.; Veltink, P.H. Assessment of hand kinematics using inertial and magnetic sensors. *J. Neuroeng. Rehabil.* **2014**, *11*, 70. [CrossRef] [PubMed]
12. Ratchatanantakit, N.; O-larnnithipong, N.; Sonchan, P.; Adjouadi, M.; Barreto, A. A sensor fusion approach to MARG module orientation estimation for a real-time hand tracking application. *Inf. Fusion* **2023**, *90*, 298–315. [CrossRef]
13. de Vries, W.H.K.; Veeger, H.E.J.; Baten, C.T.M.; van der Helm, F.C.T. Magnetic distortion in motion labs, implications for validating inertial magnetic sensors. *Gait Posture* **2009**, *29*, 535–541. [CrossRef] [PubMed]
14. Picerno, P. 25 years of lower limb joint kinematics by using inertial and magnetic sensors: A review of methodological approaches. *Gait Posture* **2017**, *51*, 239–246. [CrossRef] [PubMed]
15. Szczęsna, A.; Skurowski, P.; Pruszowski, P.; Pęszor, D.; Paszkuta, M.; Wojciechowski, K. Reference Data Set for Accuracy Evaluation of Orientation Estimation Algorithms for Inertial Motion Capture Systems. In Proceedings of the Computer Vision and Graphics, Cham, Switzerland, 10 September 2016; pp. 509–520.
16. Angermann, M.; Robertson, P.; Kemptner, T.; Khider, M. A high precision reference data set for pedestrian navigation using foot-mounted inertial sensors. In Proceedings of the 2010 International Conference on Indoor Positioning and Indoor Navigation, Zurich, Switzerland, 15–17 September 2010; pp. 1–6.
17. Chavarriaga, R.; Sagha, H.; Calatroni, A.; Digumarti, S.T.; Tröster, G.; Millán, J.d.R.; Roggen, D. The Opportunity challenge: A benchmark database for on-body sensor-based activity recognition. *Pattern Recognit. Lett.* **2013**, *34*, 2033–2042. [CrossRef]
18. Banos, O.; Calatroni, A.; Damas, M.; Pomares, H.; Rojas, I.; Sagha, H.; Mill'n, J.d.R.; Troster, G.; Chavarriaga, R.; Roggen, D. Kinect=IMU? Learning MIMO Signal Mappings to Automatically Translate Activity Recognition Systems across Sensor Modalities. In Proceedings of the 2012 16th International Symposium on Wearable Computers, Newcastle, UK, 18–22 June 2012; pp. 92–99.
19. Szczęsna, A. RepoIMU: Reference Data Set for Accuracy Evaluation of Orientation Estimation Algorithms for Inertial Motion. Capture Systems. 2016. Available online: https://github.com/agnieszkaszczesna/RepoIMU (accessed on 15 February 2023).
20. YostLabs. 3-Space Nano IC-Product Description Page. Available online: https://yostlabs.com/product/3-space-nano/ (accessed on 15 February 2023).
21. Xsens. MTi-G Miniature AHRS with Integrated GPS. Available online: https://studylib.net/doc/18864299/xsens2020503-brochure-mti (accessed on 15 February 2023).
22. Caruso, M.; Sabatini, A.M.; Knaflitz, M.; Gazzoni, M.; Croce, U.D.; Cereatti, A. Orientation Estimation Through Magneto-Inertial Sensor Fusion: A Heuristic Approach for Suboptimal Parameters Tuning. *IEEE Sens. J.* **2021**, *21*, 3408–3419. [CrossRef]
23. Caruso, M. MIMU_OPTICAL_SASSARI_DATASET. 2021. Available online: https://ieee-dataport.org/documents/mimuopticalsassaridataset (accessed on 15 February 2023).
24. Nazarahari, M.; Rouhani, H. Sensor fusion algorithms for orientation tracking via magnetic and inertial measurement units: An experimental comparison survey. *Inf. Fusion* **2021**, *76*, 8–23. [CrossRef]
25. Nazarahari, M. Sensor Fusion Algorithm for MIMU Data. 2021. Available online: https://www.ncbl.ualberta.ca/sensor-fusion (accessed on 15 February 2023).
26. Roetenberg, D.; Luinge, H.; Veltink, P. Inertial and magnetic sensing of human movement near ferromagnetic materials. In Proceedings of the Second IEEE and ACM International Symposium on Mixed and Augmented Reality, Washington, DC, USA, 7–10 October 2003; pp. 268–269.
27. Roetenberg, D.; Luinge, H.J.; Baten, C.T.M.; Veltink, P.H. Compensation of magnetic disturbances improves inertial and magnetic sensing of human body segment orientation. *IEEE Trans. Neural Syst. Rehabil. Eng.* **2005**, *13*, 395–405. [CrossRef] [PubMed]
28. Ratchatanantakit, N.; O-larnnithipong, N.; Barreto, A.; Tangnimitchok, S. Consistency Study of 3D Magnetic Vectors in an Office Environment for IMU-based Hand Tracking Input Development. In Proceedings of the Human-Computer Interaction. Recognition and Interaction Technologies, Cham, Switzerland, 26–31 July 2019; pp. 377–387.
29. YostLabs. 3-Space Sensor Miniature Attitude & Heading Reference System With Pedestrian Tracking User's Manual. 2017.
30. Hislop, J.; Isaksson, M.; McCormick, J.; Hensman, C. Validation of 3-Space Wireless Inertial Measurement Units Using an Industrial Robot. *Sensors* **2021**, *21*, 6858. [CrossRef] [PubMed]
31. OptiTrack. Specifications of the V120:Trio Motion Capture System. Available online: https://optitrack.com/cameras/v120-trio/specs.html (accessed on 15 February 2023).
32. Hindle, B.R.; Keogh, J.W.L.; Lorimer, A.V. Inertial-Based Human Motion Capture: A Technical Summary of Current Processing Methodologies for Spatiotemporal and Kinematic Measures. *Appl. Bionics Biomech.* **2021**, *2021*, 6628320. [CrossRef] [PubMed]
33. Mathworks. dist: Angular Distance in Radians. Available online: https://www.mathworks.com/help/nav/ref/quaternion.dist.html (accessed on 15 February 2023).
34. O-larnnithipong, N.; Barreto, A. Gyroscope drift correction algorithm for inertial measurement unit used in hand motion tracking. In Proceedings of the 2016 IEEE SENSORS, Orlando, FL, USA, 30 October–3 November 2016; pp. 1–3.
35. O-larnnithipong, N.; Barreto, A.B.; Ratchatanantakit, N.; Tangnimitchok, S.; Ortega, F.R. Real-Time Implementation of Orientation Correction Algorithm for 3D Hand Motion Tracking Interface. In *Universal Access in Human-Computer Interaction. Methods, Technologies, and Users, Proceedings of the 12th International Conference, UAHCI 2018, Las Vegas, NV, USA, 15–20 July 2018*; Springer International Publishing: Cham, Switzerland, 2018; pp. 228–242.
36. Aurand, A.M.; Dufour, J.S.; Marras, W.S. Accuracy map of an optical motion capture system with 42 or 21 cameras in a large measurement volume. *J. Biomech.* **2017**, *58*, 237–240. [CrossRef] [PubMed]

37. Eichelberger, P.; Ferraro, M.; Minder, U.; Denton, T.; Blasimann, A.; Krause, F.; Baur, H. Analysis of accuracy in optical motion capture–A protocol for laboratory setup evaluation. *J. Biomech.* **2016**, *49*, 2085–2088. [CrossRef] [PubMed]
38. Vince, J. *Quaternions for Computer Graphics*; Springer: London, UK; New York, NY, USA, 2011.

sensors

MDPI

Article

A System of Emotion Recognition and Judgment and Its Application in Adaptive Interactive Game

Wenqian Lin [1,*], Chao Li [2] and Yunjian Zhang [3]

1 School of Media and Design, Hangzhou Dianzi University, Hangzhou 310018, China
2 College of Computer Science and Technology, Zhejiang University, Hangzhou 310027, China
3 College of Control Science and Technology, Zhejiang University, Hangzhou 310027, China
* Correspondence: jiangnanshui253@126.com

Abstract: A system of emotion recognition and judgment (SERJ) based on a set of optimal signal features is established, and an emotion adaptive interactive game (EAIG) is designed. The change in a player's emotion can be detected with the SERJ during the process of playing the game. A total of 10 subjects were selected to test the EAIG and SERJ. The results show that the SERJ and designed EAIG are effective. The game adapted itself by judging the corresponding special events triggered by a player's emotion and, as a result, enhanced the player's game experience. It was found that, in the process of playing the game, a player's perception of the change in emotion was different, and the test experience of a player had an effect on the test results. A SERJ that is based on a set of optimal signal features is better than a SERJ that is based on the conventional machine learning-based method.

Keywords: emotion judgment system; adaptive interactive game; set of optimal signal features; sensor

Citation: Lin, W.; Li, C.; Zhang, Y. A System of Emotion Recognition and Judgment and Its Application in Adaptive Interactive Game. *Sensors* **2023**, *23*, 3250. https://doi.org/10.3390/s23063250

Academic Editor: Giovanni Saggio

Received: 8 February 2023
Revised: 5 March 2023
Accepted: 13 March 2023
Published: 19 March 2023

1. Introduction

Emotion plays an important role in daily life and is a critical factor that affects the process of an individual's cognitive, communication, and decision-making abilities. Physiological signals, such as skin electricity, electrocardiogram, pulse wave, and facial electromyogram, can be used to recognize and judge individuals' emotions [1]. In the past few years, research has reported on the recognition and judgment of emotions based on physiological signals. The fusion of multiple emotional modalities was proposed by Khezri et al. [2] to improve the performance of an emotion judgment system. In the presented emotion recognition system, recorded signals with the formation of several classification units identified the emotions independently, and considerable improvement was obtained. A new approach for the empirical identification of affected regions, which was based on skin conductance, was put forward by Liapis et al. [3]. Their findings identified the regions in the valence–arousal rating space that might reliably indicate self-reported stress while using interactive applications. A new method recognizing the emotional state of six individuals was given by Yoo et al. [4]; the method possessed good performance accuracy and could make a distinction between one emotion and other possible emotional states. Yang et al. [5] put forward a new method, which was based on skin conductance, that classified the emotion image based on the electroencephalography signals; the new method bridged the emotion gap by building a relationship between experiential information and the expected emotion experience of the viewer, and the results showed that the method could bring about a pleasant experience. Jang et al. [6] experimentally evaluated the dependability of physiological signal changes initiated by multiple emotions by measuring six basic emotions; they indicated that the physiological signals based on heart rate, skin conductance, and blood volume pulse were more reliable than those evaluated at baseline. An algorithm was put forward by Sepulveda et al. [7] to ameliorate the emotion recognition extracted from electrocardiogram signals using wavelet transform for signal analysis, and the algorithm, when combined with wearable devices, proved to be effective for classifying emotion. In order to recognize emotion based on multimodal physiological signals, Zhang et al. [8] proposed a deep-fusion framework, which displayed higher class separability in emotion recognition, and this framework was more effective in subject-independent emotion recognition than other fusion methods. A framework of multimodal emotion classification using multichannel physiological signals was proposed by Yan et al. [9]; they pointed out that it was significant to develop adaptive decision fusion strategies in the process of emotion classification.

Although some achievements have been made in recognizing and judging emotions from physiological signals, based on the above research, there is still room for improvement in the accuracy of judgment and the universality of application, for example, in the field of interactive games.

There is an increasing interest in creating games that are based on interaction technology. Although interactive games belong to the category of human–computer interaction, they are different from general human–computer interaction in the following aspects: (1) Compared with general human–computer interaction, interactive games pay more attention to the process of interaction rather than the result of interaction; (2) the meaning and purpose of general human–computer interaction are determined by the user's purpose and task, while those in interactive games are determined by the purpose and operation form of the game itself; (3) general human–computer interaction is more stable and focuses on the durability of functions to consolidate the user experience, while interactive games focus on waking up a series of user experiences; and (4) whether in interactive content or control systems, interactive games have ample room for innovation.

For a long time, game developers have tried to apply physiological signals to the process of playing games, hoping that a player's experiential state could be captured in real time when a player realized the capture, so as to enhance the interest and intelligence of the game. As early as 1984, CalmPute designed a device called CalmPrix that operated racing games based on the skin electrical signals. In 1998, Nintendo released the physiological sensor Teris64. In 2010 and 2011, Nintendo released Wii accessories based on physiological signal sensors. In 2011, Ubisoft also announced the development of similar products. However, these products have not been widely used. One of the reasons is that the equipment is too cumbersome and complex to wear, and another reason is that it does not conform to the operation habits of players in some aspects.

Lv et al. [10] designed and evaluated a touchless motion interaction technology and developed three primitive augmented-reality games with 11 dynamic gestures. Players interacted with the augmented-reality games using hand/feet gestures in front of the camera, which triggered the interaction event to interact with the virtual object in the scene. Vachiratamporn et al. [11] analyzed the affective states of players prior to and after witnessing a scary event in a survival horror game by collecting the player-affect data through their own affect annotation tool that allowed the player to report his affect labels while watching his recorded game play and facial expressions; the results showed that players were likely to get more fearful of a scary event when they were in the suspense state, and heart rate was a good candidate for the detection of player affect. Du et al. [12] presented an approach to detect a subject's emotions using heartbeat signals and facial expressions, and the approach had high accuracy and less computational time for four emotions when playing different games. Baldassarri et al. [13] put forward two kinds of interactive games to promote communication and attention experience; one is to consider emotions to measure a user's attention, concentration, and satisfaction, and the other is to use a tangible desktop to promote cognitive planning. Kalantarian et al. [14] proposed a method of automatically extracting emotion marker frames from a game and training a new emotion classifier to get over the limited function of the existing emotion recognition platform for children with autism spectrum disorders. Sekhavat et al. [15] studied the degrees to which the expression of the manifested emotions of an opponent could affect the emotions and, consequently, the game-play behavior by performing a comprehensive user study to estimate the emotions of players in emotion-aware versus emotion-agnostic game versions.

Nacke et al. [16] divided physiological signals into directly controllable and non-directly controllable signals and then asked the subjects to rate the game experience with and without controlled signals, respectively. The results showed that the effect of the game experience with directly controllable signals was better than that without controlled signals. However, there is a lack of research on the effect of the game experience with non-directly controllable signals. In addition, in terms of interactive game design, designers often deliberately encourage players to consciously change their emotional state through equipment and testing, which affects the player's experience. In the design of this paper, the relationship between the game and the player is exchanged. Instead of letting players deliberately adapt to the game, the game is designed to automatically change the level according to the player's emotional state and to adapt to the player's emotional changes so as to increase the self-adaptability and the fun of the game and enhance the intelligence and naturalness of the interaction when players are not aware of it. Arendse et al. [17] evaluated the framework using player action data from the platforming game Super Mario Bros, and the results that were based on the presented framework were better than the existing work. Svoren et al. [18] built a dataset that consisted of demographic data assembled from ten participants playing Super Mario Bros and showed that the video game, together with facial expressions, could be used to predict the blood volume pulse of the subject. Granato et al. [19] predicted the subjects' emotions during video game sessions and indicated that the obtained results could improve the game design. Izountar et al. [20] proposed a VR-PEER adaptive exergame system and developed a virtual reality-based serious game as a case study. The test results showed that fifteen participants expressed the usefulness of the system in motor rehabilitation processes. Kandemir and Kose [21] presented improved human–computer interaction games in attention, emotion, and sensory–motor coordination, and specially designed the interface and the difficulty levels of the games for the use of people from different age groups and with different disabilities. The tested results showed that the games had a positive effect on children. Penev et al. [22] examined the engagement level and therapeutic feasibility of a mobile game platform for children with autism by designing a mobile application, GuessWhat, which delivered game-based therapy to children in home settings through a smart phone; the tested results showed that the GuessWhat mobile game was a viable approach for the efficacious treatment of autism and further support for the possibility that the game could be used in natural settings to increase access to treatment when barriers to care exist.

Therefore, this paper consists of the following three parts. (1) A system of emotion recognition and judgment is built by collecting the change in physiological signals induced by emotional change and obtaining a set of optimal signal features. (2) The test on the above system is performed by 10 subjects playing the game Super Mario. The player's emotional trend is triggered by the special events in the game. Meanwhile, the non-directly controllable physiological signals are detected to assess the effect of the game experience. (3) To illustrate the advantages of the optimal signal features, the emotional trend changes produced by the emotion recognition and judgment system based on the set of optimal signal features and based on the conventional machine learning-based methods are compared.

2. Emotion Recognition

Emotions should be evaluated and classified before recognizing and judging them. The widely used valence–arousal (V-A) model is usually used to evaluate and classify emotions. In the V-A model, V and A represent the degree of emotional pleasure and emotion arousal, respectively [23]. Based on the discrete emotion classification, the emotions of fatigue, tension, happiness, and depression are first designated using the extracted four poles of emotion classification, which then is extended from four poles into a plane where, as shown in Figure 1, the four quadrants express, respectively, the high-arousal and positive-valence (quadrantI: HAPV), high-arousal and negative-valence (II: HANV), low-arousal and negative-valence (III: LANV), and low-arousal and positive-valence (IV: LAPV).

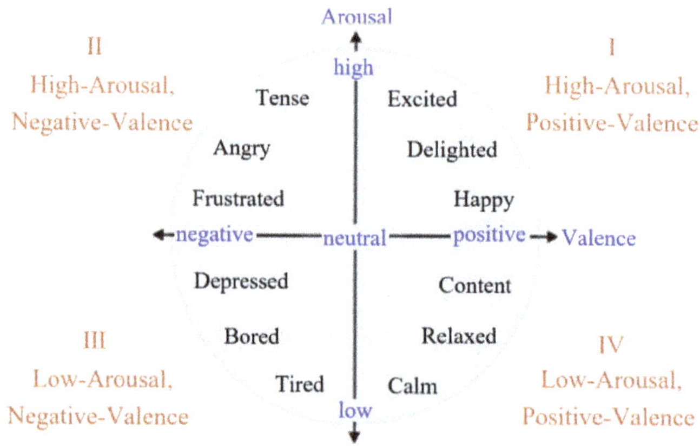

Figure 1. Valence–arousal model.

3. Emotion Judgment

The establishment of an emotion recognition and judgment system through physiological signals includes the following basic steps.

3.1. Signals of Skin Electricity and Pulse Wave

In a system of emotion recognition and judgment, individuals' physiological signals usually consist of pulse wave, skin electricity, electrocardiogram, and facial electromyogram. Which physiological signals should be used depends on the specific situation. Skin electrical signal and pulse wave signal are used in the present study. Since the skin electrical signal is easily interfered with by other signals inside the human body in processing, the noise interference generated by the hardware itself should be removed with the following formula:

$$f(t) = \text{Serial\_P\_R} - \frac{[(2014 + 2 \times \text{Serial\_P\_R}) \times 10{,}000]}{(512 - \text{Serial\_P\_R})}, \quad (1)$$

in which "Serial_P_R" is the data of the skin electrical signal, and the numbers are the debugged data based on the hardware properties.

In order to improve the performance of the computer analysis and processing, the discrete wavelet transform is used to denoise the physiological signals and decompose the signals into different frequency bands with low-pass and high-pass filtering.

The skin electrical signal is decomposed in three layers and then denoised using the wdencmp function in MATLAB; finally, all segments of the signal are normalized within the value range from 0 to 100.

The key factors to reflect the pulse wave signal are main wave, dicrotic anterior wave, dicrotic notch, and dicrotic wave. The amplitudes involved are main, dicrotic anterior, dicrotic notch, and dicrotic wave. The time refers to the time from the starting point of the waveform period to the lowest point of the dicrotic notch and to the peak c point of the main wave and the period of one waveform.

The pulse wave signal is smoothed and filtered using the Butterworth low-pass filter, which has the characteristics that the frequency response curve in the pass band is flat to the maximum without ripple and gradually drops to zero in the stop band:

$$H(u,v) = \frac{1}{1 + [D(u,v)/D_0]^{2n}} \quad (2)$$

The low-pass cut-off frequency is set to 10 Hz. The relevant parameters of the pulse wave signal are normalized after filtering.

3.2. Dimensionality Reduction in the Original Signal Feature

Dimensionality reduction in the original physiological signals is implemented with principal component analysis (PCA) because the direct fusion of the original physiological signals results in too much computation. Dimensionality reduction leads to a high efficiency and precision in the classification of emotion recognition. In the process of PCA, the weight of the original feature of the physiological signals is first calculated in the principal component, and then the weight threshold of each feature is taken as the criterion for choosing the feature. The original features with a larger weight than the threshold are chosen to form a new subset of the optimal features.

The method of Pearson correlation coefficient (PCC) is employed to determine the relation of the emotional interval, and these features are based on the subset of the optimal features. The PCC is performed on the features of four kinds of emotion trends and employed to extract the significance P of the features; then, the threshold of the optimal features correlated with four kinds of emotional trends is obtained according to the correlation coefficient and significance P of the features. In the present study, the optimal features are composed of "BpNN50", i.e., the percentage of the main pulse wave interval larger than 50 ms, "range" of the skin electrical signal, and "1dmean", i.e., mean value of the first order difference of the skin electrical signal. "BpNN50", "range", and "1dmean" are defined as follows:

$$\text{BpNN50} = \frac{\text{count}|x_{i+1} - x_i| > 50\text{ms}}{N-1}, i = 1, \cdots, N-1 \quad (3)$$

$$\begin{aligned} \text{range} &= \max(x) - \min(x) \\ \text{1dmenn} &= \frac{1}{N-1} \sum_{i=1}^{N-1} (x_{i+1} - x_i) \end{aligned} \quad (4)$$

where x is the discrete signal value, I is the ith signal, and N is the total number of signals.

3.3. Model of Emotion Judgment

According to the above description, the skin electrical signal and pulse wave signal are selected as the physiological signals to establish the emotion judgment system. The specific operation process is shown in Figure 2, where "BpNN50" is the percentage of the main pulse wave interval larger than 50 ms as shown in Equation (3); "range" is the range of the skin electrical signal and pulse wave signal as shown in Equation (4); "1dmean" is the mean value of the first order difference of the skin electrical signal (Equation (4)); $n_{\max}$ and $n_{\min}$ are n corresponding to maximum and minimum signal $x_{\max}$ and $x_{\min}$, respectively; and x_{th} is the normalized threshold of the physiological signal features:

$$x_{th} = \sum_{i=1}^{n} \left(\frac{x_n - x_{\min}}{x_{\max} - x_{\min}}\right)/n \quad (5)$$

where n is the number of signals, and x_n is the nth signal.

The range of the skin electrical signal shows a strong positive correlation between the emotional trends of LAPV and HANV, so the range of the skin electrical signal is used to judge LAPV and HANV. Since the skin electrical signal and pulse wave signal are extracted with the device worn by the fingers, considering the simplicity of the interactive device, the pulse wave and skin electrical signals are selected as the physiological signals of the emotional judgment model.

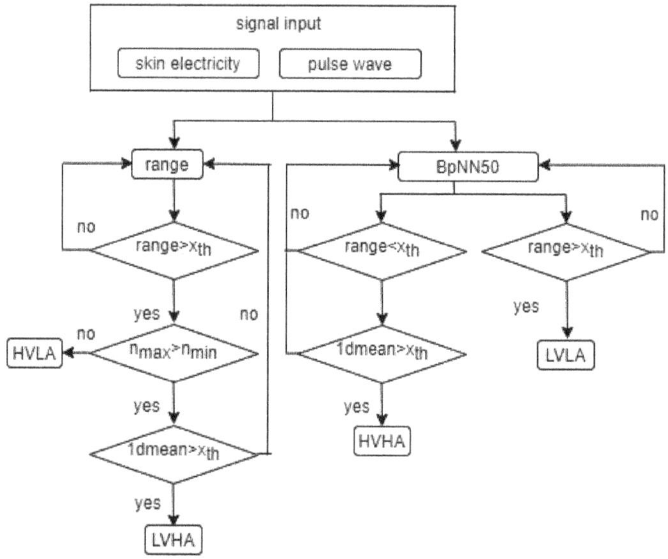

Figure 2. Model of emotion judgment.

The program continues when the range of the skin electrical signal is larger than the normalized threshold x_{th}, otherwise, returning. Then, the program continues if $n_{\max} > n_{\min}$, otherwise it is determined as LAPV. Finally, it is determined as HANV if the mean value of the first order difference of the skin electrical signal 1dmean is larger than the normalized threshold x_{th}, otherwise, returning to range.

The pulse wave signal meeting "BpNN50" goes to the next step. On the one hand, it is determined as LANV if the range of the pulse wave signal is larger than the normalized threshold x_{th}, otherwise, returning. On the other hand, the program continues when the range of the pulse wave signal is less than the normalized threshold x_{th}, otherwise, returning. Finally, it is determined as HAPV if "1dmean" > x_{th}, otherwise, returning.

4. Design of Emotion Adaptive Interactive Game

The game Super Mario is adapted through judging the corresponding special events triggered by the player's emotional trend, which enhances the player's game experience. The game Super Mario was chosen for the following two reasons: (1) The game is simple to play and easy to operate. It is suitable for players at all levels and can reduce the likelihood that the test results are affected by the participants' proficiency in the game. (2) Various events in the game that affect the subjects' emotions are independent, clear, easy to divide, and meet the requirements of our experiment for emotional arousal.

4.1. About the Game Super Mario

(1) The core mechanism of the game

The core mechanism of the Super Mario game is to use the keyboard to move on the map while acquiring resources and avoiding enemies. When the player presses the left or right direction keys of the keyboard, the character will move in the corresponding direction. When the player presses the space bar on the keyboard, the character will jump up, and this action can be used to avoid an enemy. If the player jumps up and steps on an enemy's head, he can destroy the enemy. In addition, the player can also pick up special energy items by jumping and hitting designated props to make the character bigger, move faster, jump higher, or attack an enemy with fireballs.

(2) The gameplay

The Super Mario game is easy to operate, easy to learn, moderate in difficulty, and suitable for all types of players. The content of the game is easy to understand, the rules are clear, the rewards and penalties are clear, and each game event is relatively independent.

(3) The objectives of the game and the expected player experience

The goal of the Super Mario game is to let players experience various events in the game to wake up different emotions. It is easy to cut the emotional change data caused by the event because the triggering environment and conditions of each event in this game are relatively independent. It is expected that players will generate corresponding emotional changes due to various events in the game and recover their emotions to the standard value in the gentle stage between events.

4.2. Special Events of the Game

In the process of game adaptation, all picture materials come from the materials of sharing package in the network [24] as shown in Figure 3. The package was chosen from the network because, first, these pictures can better induce the emotional trend of the subjects, and second, they can enhance the game experience of the subjects. The special events of the game are shown in Figure 4.

Figure 3. Interactive game based on Unity3D.

When the emotional trend is HANV (high-arousal and negative-valence), the players enter a state of negative-valence. At this time, the game adaptive system will trigger an event that is opposite negative-valence and that can directly affect the emotion, i.e., rewarding the player with a large number of mysterious bricks that can help the character upgrade in order to adjust the player's emotion from negative-valence to positive-valence. When the emotional trend is LAPV (low-arousal and positive-valence), the players enter a state of low-arousal. At this time, the game adaptive system will trigger an event that is opposite low-arousal and that can directly affect the emotion, i.e., making appear a large number of small monsters that can increase the difficulty of the game and raise the interest of the player in order to adjust the player's emotion from low-arousal to high-arousal. When the emotional trend is HAPV, the subjects are in an ideal entertainment state without any reaction. When the emotional trend is LANV, the subjects enter a new game scene in order to stimulate the player's interest and emotion.

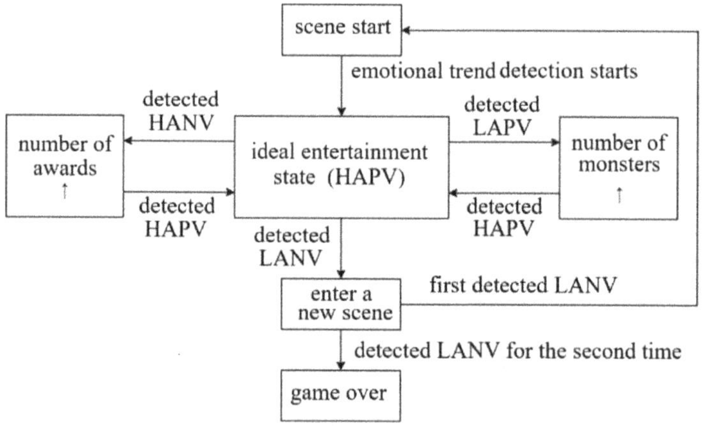

Figure 4. Special event activation conditions.

The character action is performed using the Vector2.x and Vector2.y functions of Unity2D. The regeneration and activation of the game are controlled using the instantiate function, and whether the subject's emotion remains in the type of emotional trend is judged using the while loop function.

5. Test Results and Analysis

It is easy to directly observe the variation of emotion trend from the waveform of the skin electrical signals, which are recorded with the speed function "Time.deltatime*" and the "Debug.Log()" function for saving the calculation time and accelerating calculation progress. In the process of the calculation, the time length of the calculation segment is kept consistent with the test because the skin electrical signals are varying continuously, and the calculation is carried out every 10 s.

Ten subjects aged between 24 and 30 years were selected for the test. Among the ten subjects, there were six men and four women, and half of the subjects had previously participated in similar tests.

5.1. Design of the Emotion Judgment System

In order to illustrate the performance of the emotion judgment system described in Section 3 and compare the effectiveness of the emotion judgment through the physiological signal data, two approaches of emotion judgment were designed in the test as shown in Figure 5.

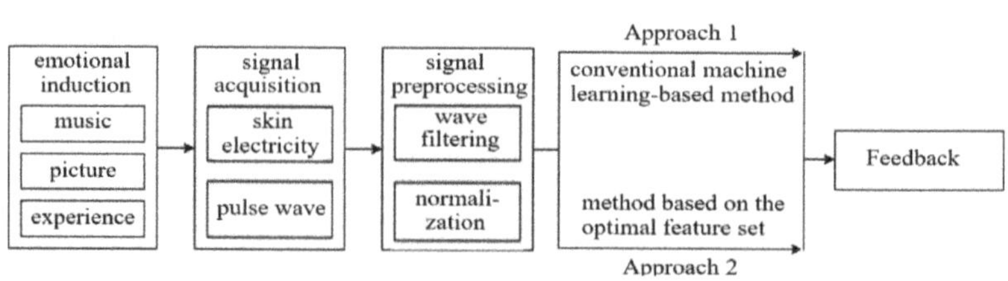

Figure 5. Two approaches of emotion judgment.

In Approach 1, the conventional machine learning-based method is used to judge the emotional trend through the physiological signal data, and the data split is set up with 70% as the training set and 30% as the validation set, i.e., introducing:

using Accord.MachineLearning.VectorMachines.Learning;

using Accord.Statistics.Kernels;
and then using the support vector machine toolkit of C# language.

In Approach 2, the features of the signal are separated into a time domain, frequency domain, and feature related with the physiological process. The time domain is determined with 18 pulse wave signals and 24 skin electrical signals. In the frequency domain, the feature mold and computation approach for the signals of skin electricity and pulse wave are similar. The feature related with the physiological process consists of seven skin electrical signals and 10 pulse wave signals. Based on the separation of features, the dimensionality reduction in the original signals is conducted to render the emotional recognition more effective and accurate. Using the PCA, the principal components are obtained, and the weight threshold of each feature of the signal on the principal component is taken as the criterion for the selection of the feature. In the end, the original features, which play the leading role, are defined as a subset of the optimal feature. Based on the subset, the PCC is employed to determine the relation of the optimal features and emotional interval. The PCC is used to compute the features of four emotion trends and extract the significance P of the features. According to the correlation coefficient and significance P, the threshold of the optimal features related to the emotional trends is defined, and the threshold is used to judge the emotional trends. Therefore, the real-time ability and interactivity of Approach 1 and 2 are compared with the tests.

In the test, the number of the subjects' emotions that was activated by special events was recorded as shown in Figure 6. Each subject underwent two rounds of tests; Approach 1 and Approach 2 as shown in Figure 5 were used to judge the emotion trend in the first and second rounds of the test, and the results are shown in Tables 1 and 2, respectively, where the significance test has been performed in the comparisons. In Tables 1 and 2, the more times the subject is activated by the emotional trend, the sharper the subject's perception to the change in the emotional trend. For example, in Figure 6, subject 3 is activated eight times by special events as shown in Table 2: two of them enter a new scene and end the game due to the state of negative-valence and low-arousal (LANV), and the other two special events are activated three times each.

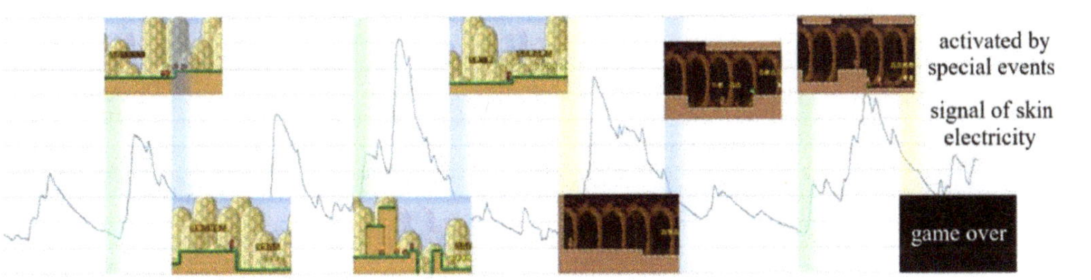

Figure 6. Records of the subjects' emotions activated by special events.

Table 1. The number of the subject's activated emotions with Approach 1.

Subjects	Participated in Test before	HANV	LAPV	HAPV	LANV
1	yes	2	1		1
2	yes	3	0		1
3	yes	2	1		2
4	yes	4	1		1
5	yes	0	2		2
	total numbers of 1~5	11	5		7
6	No	0	0		0

Table 1. *Cont.*

Subjects	Participated in Test before	HANV	LAPV	HAPV	LANV
7	No	3	1		0
8	No	0	1		2
9	No	2	3		1
10	No	2	1		1
total numbers of 6~10		7	6		4
total numbers of 1~10		18	11		11

Table 2. The number of the subject's activated emotions with Approach 2.

Subjects	Participated in Test before	HANV	LAPV	HAPV	LANV
1	yes	2	3		2
2	yes	4	1		1
3	yes	3	3		2
4	yes	3	1		2
5	yes	4	3		0
total numbers of 1~5		16	11		7
6	No	0	0		1
7	No	2	2		0
8	No	3	1		1
9	No	4	2		2
10	No	2	2		2
total numbers of 6~10		11	7		6
total numbers of 1~10		27	18		13

5.2. Result Analysis

The second round of the test results of the third subject showed that the subject had activated eight special events in total, two of which were due to the low level of pleasure and arousal to enter a new scene and end the game, and the other two special events had been activated three times respectively, i.e., the probability of activation of special events is higher compared with the first round of the test. Other subjects have similar test results, indicating that the emotion judgment system based on the set of optimal signal features is better than that based on the conventional machine learning-based method in interactivity.

Emotion recognition is delayed in most cases, especially at the beginning of the test. There may be two reasons for this. One is that the subject has just started the test and has not yet fully entered the test state. The other is that it takes a certain amount of time to achieve high arousal. However, the recognition results of several emotional trends of most subjects are basically correct, meeting the expectations of the test.

It can be observed from Tables 1 and 2 that (1) each subject is activated to varying degrees by emotional trends, showing that the emotion judgment system and design of the emotional adaptive interactive game presented in this paper are effective; (2) the number of times that each subject is activated by the emotional trend is different, indicating that the subjects' perception to the change in the emotional trend is different; (3) the number of times activated by the emotional trend for subjects who participated in the test before are basically larger than that for subjects who did not participate in the test, indicating that the test experience of the subject has an impact on the test results; and (4) the number of times activated by the emotional trend obtained with Approach 2 are larger than that obtained with Approach 1, showing that the emotion judgment system based on the set of optimal signal features is better than that based on the conventional machine learning-based method.

6. Conclusions

In order to further study the effectiveness of an emotion judgment system and the effect of a game experience with non-directly controllable signals, a system of emotion recognition and judgment is established, and an emotion adaptive interactive game is designed by adapting the game Super Mario. A total of 10 subjects were selected for the test on the interactive game and emotion judgment system; meanwhile, the results using the emotion judgment system based on a set of optimal signal features and conventional machine learning- based method are compared. The main conclusions are summarized as follows.

(1) The emotion judgment system and design of the emotional adaptive interactive game are effective. The game, which is adapted through judging the corresponding special events triggered by the player's emotional trend, can enhance the player's game experience.

(2) In the process of playing the game, the player's perception to the change in the emotional trend is different, and the test experience of the players has an impact on the test results.

(3) The emotion judgment system based on the set of optimal signal features is better than that based on the conventional machine learning-based method.

Author Contributions: Conceptualization, W.L.; methodology, W.L. and C.L.; software, W.L. and C.L.; validation, C.L. and W.L.; writing, W.L. and C.L.; resources, W.L. and Y.Z.; review, Y.Z. All authors have read and agreed to the published version of the manuscript.

Funding: This work was supported by the National Natural Science Foundation of China (Grant no. 12132015).

Institutional Review Board Statement: Not applicable.

Informed Consent Statement: Not applicable.

Data Availability Statement: Data sharing not applicable.

Conflicts of Interest: There are no conflict of interest regarding the publication of this paper.

References

1. Gouizi, K.; Reguig, F.B.; Maaoui, C. Emotion recognition from physiological signals. *J. Med. Eng. Technol.* **2011**, *35*, 300–307. [CrossRef] [PubMed]
2. Khezri, M.; Firoozabadi, M.; Sharafat, A.R. Reliable emotion recognition system based on dynamic adaptive fusion of forehead biopotentials and physiological signals. *Comput. Meth. Programs Biomed.* **2015**, *122*, 149–164. [CrossRef]
3. Liapis, A.; Katsanos, C.; Sotiropoulos, D.G.; Karousos, N.; Xenos, M. Stress in interactive applications: Analysis of the valence-space based on physiological signals and self-reported data. *Multimed. Tools Appl.* **2017**, *76*, 5051–5071. [CrossRef]
4. Yoo, G.; Seo, S.; Hong, S.; Kim, H. Emotion extraction based on multi bio-signal using back-propagation neural network. *Multimed. Tools Appl.* **2018**, *77*, 4925–4937. [CrossRef]
5. Yang, M.Q.; Lin, L.; Milekic, S. Affective image classification based on user eye movement and EEG experience information. *Hum. Comput. Interact.* **2018**, *30*, 417–432. [CrossRef]
6. Jang, E.H.; Byun, S.; Park, M.S.; Sohn, J.H. Reliability of physiological responses induced by basic emotions: A pilot study. *J. Physiol. Anthropol.* **2019**, *38*, 15. [CrossRef]
7. Sepulveda, A.; Castillo, F.; Palma, C.; Rodriguez-Fernandez, M. Emotion recognition from ECG signals using wavelet scattering and machine learning. *Appl. Sci.* **2021**, *11*, 4945. [CrossRef]
8. Zhang, X.W.; Liu, J.Y.; Shen, J.; Li, S.J.; Hou, K.C.; Hu, B.; Gao, J.; Zhang, T. Emotion recognition from multimodal physiological signals using a regularized deep fusion of kernel machine. *IEEE T. Cybern.* **2021**, *51*, 4386–4399. [CrossRef]
9. Yan, M.S.; Deng, Z.; He, B.W.; Zou, C.S.; Wu, J.; Zhu, Z.J. Emotion classification with multichannel physiological signals using hybrid feature and adaptive decision fusion. *Biomed. Signal Process. Control* **2022**, *71*, 103235. [CrossRef]
10. Lv, Z.H.; Halawani, A.; Feng, S.Z.; Rehman, S.; Li, H.B. Touch-less interactive augmented reality game on vision-based wearable device. *Pers. Ubiquitous Comput.* **2015**, *19*, 551–567. [CrossRef]
11. Vachiratamporn, V.; Legaspi, R.; Moriyama, K.; Fukui, K.; Numao, M. An analysis of player affect transitions in survival horror games. *J. Multimodal User Interfaces* **2015**, *29*, 43–54. [CrossRef]

12. Du, G.L.; Long, S.Y.; Yuan, H. Non-contact emotion recognition combining heart rate and facial expression for interactive gamming environments. *IEEE Access* **2020**, *8*, 11896–11906. [CrossRef]
13. Baldassarri, S.; Passerino, L.; Ramis, S.; Riquelme, I.; Perales, F.J. Toward emotional interactive videogames for children with autism spectrum disorder. *Univers. Access Inf. Soc.* **2021**, *20*, 239–254. [CrossRef]
14. Kalantarian, H.; Jedoui, K.; Washington, P.; Wall, D.P. A mobile game for automatic emotion-labeling of images. *IEEE Trans. Games* **2020**, *12*, 213–218. [CrossRef]
15. Sekhavat, Y.A.; Roohi, S.; Mohammadi, H.S.; Yannakakis, G.N. Play with one's feelings: A study on emotion awareness for player experience. *IEEE Trans. Games* **2022**, *14*, 3–12. [CrossRef]
16. Nacke, L.E.; Kalyn, M.; Lough, C.; Mandryk, R.L. Biofeedback game design: Using direct and indirect physiological control to enhance game interaction. In Proceedings of the 29th Annual CHI Conference on Human Factors in Computing Systems, Vancouver BC Canada, 7–12 May 2011; pp. 103–112.
17. Arendse, L.J.; Ingram, B.; Rosman, B. Real time in-game play style classification using a hybrid probabilistic supervised learning approach. *Commun. Comput. Inf. Sci.* **2022**, *1734*, 60–77.
18. Svoren, H.; Thambawita, V.; Halvorsen, P.; Jakobsen, P.; Garcia-Ceja, E.; Noori, F.M.; Hammer, H.L.; Lux, M.; Riegler, M.A.; Hicks, S.A. Toadstool: A dataset for training emotional intelligent machines playing Super Mario Bros. In Proceedings of the MMSys '20: Proceedings of the 11th ACM Multimedia Systems Conference, Istanbul, Turkey, 8–11 June 2020; Association for Computing Machinery: New York, NY, USA, 2020; pp. 309–314.
19. Granato, M.; Gadia, D.; Maggiorini, D.; Ripamonti, L.A. An empirical study of players' emotions in VR racing games based on a dataset of physiological data. *Multimed. Tools Appl.* **2020**, *79*, 33657–33686. [CrossRef]
20. Izountar, Y.; Benbelkacem, S.; Otmane, S.; Khababa, A.; Masmoudi, M.; Zenati, N. VR-PEER: A personalized exer-game platform based on emotion recognition. *Electronics* **2022**, *11*, 455. [CrossRef]
21. Kandemir, H.; Kose, H. Development of adaptive human-computer interaction games to evaluate attention. *Robotica* **2022**, *40*, 56–76. [CrossRef]
22. Penev, Y.; Dunlap, K.; Leblanc, E.; Kline, A.; Kent, J.; Ng-Thow-Hing, A.; Liu, B. A mobile game platform for improving social communication in children with autism: A feasibility study. *Appl. Clin. Inform.* **2021**, *12*, 1030–1040. [CrossRef]
23. Marimpis, A.D.; Dimitriadis, S.I.; Goebel, R. A Multiplex connectivity map of valence-arousal emotional mode. *IEEE Access* **2020**, *8*, 70928–170938. [CrossRef]
24. Available online: https://www.html5tricks.com/html5-mario.html (accessed on 23 September 2022).

sensors

MDPI

Article

Sign2Pose: A Pose-Based Approach for Gloss Prediction Using a Transformer Model

Jennifer Eunice [1], Andrew J [2], Yuichi Sei [3] and D. Jude Hemanth [1,*]

[1] Department of Electronics and Communication Engineering, Karunya Institute of Technology and Sciences, Coimbatore 641114, India
[2] Computer Science and Engineering, Manipal Institute of Technology, Manipal Academy of Higher Education, Manipal 576104, India
[3] Department of Informatics, The University of Electro-Communications, Tokyo 182-8585, Japan
* Correspondence: judehemanth@karunya.edu

Abstract: Word-level sign language recognition (WSLR) is the backbone for continuous sign language recognition (CSLR) that infers glosses from sign videos. Finding the relevant gloss from the sign sequence and detecting explicit boundaries of the glosses from sign videos is a persistent challenge. In this paper, we propose a systematic approach for gloss prediction in WLSR using the Sign2Pose Gloss prediction transformer model. The primary goal of this work is to enhance WLSR's gloss prediction accuracy with reduced time and computational overhead. The proposed approach uses hand-crafted features rather than automated feature extraction, which is computationally expensive and less accurate. A modified key frame extraction technique is proposed that uses histogram difference and Euclidean distance metrics to select and drop redundant frames. To enhance the model's generalization ability, pose vector augmentation using perspective transformation along with joint angle rotation is performed. Further, for normalization, we employed YOLOv3 (You Only Look Once) to detect the signing space and track the hand gestures of the signers in the frames. The proposed model experiments on WLASL datasets achieved the top 1% recognition accuracy of 80.9% in WLASL100 and 64.21% in WLASL300. The performance of the proposed model surpasses state-of-the-art approaches. The integration of key frame extraction, augmentation, and pose estimation improved the performance of the proposed gloss prediction model by increasing the model's precision in locating minor variations in their body posture. We observed that introducing YOLOv3 improved gloss prediction accuracy and helped prevent model overfitting. Overall, the proposed model showed 17% improved performance in the WLASL 100 dataset.

Keywords: sign language recognition; gloss prediction; transformer; pose-based approach; pose estimation; deep learning

Citation: Eunice, J.; J, A.; Sei, Y.; Hemanth, D.J. Sign2Pose: A Pose-Based Approach for Gloss Prediction Using a Transformer Model. *Sensors* **2023**, *23*, 2853. https://doi.org/10.3390/s23052853

Academic Editor: Giovanni Saggio

Received: 13 December 2022
Revised: 25 February 2023
Accepted: 28 February 2023
Published: 6 March 2023

1. Introduction

Sign language, which has its own underlying structure, grammar, syntax, and complexities, is the main mode of communication among the *Deaf Community*. To comprehend sign language, one must consider a plethora of factors involving hand movements, head, hand posture, shoulder posture, location of the lips, and facial expressions. However, in an environment where spoken language is much more prevalent, the deaf community faces challenges of communication barriers and separation from society. To alleviate communication difficulties, understanding sign language as a spoken language is becoming incredibly valuable.

The early stages of sign language research focused primarily on sign language recognition (SLR). SLR focuses on action recognition from the performed sign language sequence without paying attention to its grammatical and linguistic structures. In other words, SLR interprets the performed signs of alphabets [1], numbers [2], or symbols [3] from either static images or continuous sequences of images [4] and is categorized into Isolated SLR [5] and Dynamic SLR [6]. Continuous SLR recognizes sign postures from a continuous sequence of sign language videos which can either be an isolated word video or a continuous spoken sentence sequence, whereas isolated SLR recognizes sign postures from a single static image. Prior systems relied on hidden Markov model-based sequence recognition [7] and per-image feature extraction [8]. The effectiveness of automatic voice recognition served as inspiration for this pipeline. The design of the features that needed to be retrieved posed the biggest challenge in SLR. It was challenging to create a reliable algorithm that could extract the key linguistic elements, such as hand form [9], body movements [10], and face expression [11], even though they had already been recognized. Later, with the advancement of deep learning, manually constructed feature extraction was replaced by automatically extracted features using CNN models [2,12,13]. The overfitting, class imbalance, and exploding gradient problems caused them to perform poorly despite carrying out automatic feature extraction. Likewise, they significantly lagged in encoding the object's orientation and position. Soon, many hybrid models combined with CNN and HMM [14], CNN with DCGAN [15], CNN with LSTM [16,17], CNN with SVM [18], and CNN with hybrid segmentation [19] emerged. The outbreak of 3D CNN [17,20,21] created outstanding growth in spatio-temporal feature extraction.

Although deep learning has produced state-of-the-art results in the various challenges of SLR [16,22], to enhance the training process of the end-to-end sequence translation process, deep learning models require annotated datasets to tune CSLR models. For this to happen, the model should be trained with isolated words to increase the performance of the CSLR models. To resolve this issue, Chen et al. [23] proposed a transfer learning-based approach. This approach addressed data scarcity by gradually pretraining visual and linguistic modules from general domains into the target domains to some extent. This strategy also required annotated data to improve the model's performance. The development of better-trained sign language translation models is hampered by a lack of data. Owing to this issue, the performance of current CSLR models needs to be improved. Though various methods and architectures have been proposed to address exact interpretations of sign language through SLR and CSLR, there still lacks meaningful translation of the performed sign language. Ever since the advent of deep learning and its application in computer vision, the pairing of vision and language has received a lot of attention.

Sign language translation (SLT) [24] is the transcription of a sign language video to spoken sentence phrases, paying attention to all the rich underlying grammatical structures that allow the user to understand the underlying language model, spatial representations, and the mapping pattern between the sign and spoken language. SLT is far more complex than SLR because it considers additional visual cues such as body posture, facial expressions, and signing position. While performing sign transcription, which is literally a written version of sign performance, glosses are the intermediary representation. Glosses are words associated with a specific sign, also known as a label [25]. The structure of glosses differs from that of spoken languages. They serve as the foundation of complete sign sequence translation. For example, if a signer performs a sign sequence for the phrase "The weather is too cold today", the sign translation model suggests the relevant glosses, such as "weather", "cold", and "today". In this paper, we focus on enhancing gloss prediction accuracy in isolated SLR by reducing computational and timing complexities. The end goal of SLT is the transcription of sign language into spoken sentences. End-to-end translation and two-stage translation are the two types of SLT. End-to-end translation directly translates the sign video from the sign sequence [26], whereas two-stage SLT generates intermediate glosses from the sign video; from the glosses, the spoken sentence translation is generated while accounting for the underlying rich grammar [27].

Although end-to-end translation requires less work in terms of components and relies on naturally occurring data rather than domain knowledge and specialist linguistics, these models require a large amount of training data to achieve the aforementioned benefits. In contrast, in two-stage translation, the intermediate gloss representation settles out-of-vocabulary issues that frequently occur in end-to-end translation [28]. Therefore, understanding the importance of gloss in SLT, in this work, we concentrate on word-level gloss prediction to enable appropriate gloss prediction in the case of translating continuous sign sequences.

In this research, we propose a Sign-Pose2Gloss Prediction Transformer that eliminates the need for expensive and time-consuming labelled data to train the model. Thus, rooting out a mathematical pattern between elements is unnecessary since our model has self-supervised learning abilities. In our approach, considering the real-time challenges faced by the SLT, we suggest a novel pose-based model trained and evaluated on the large-scale word-level American sign language (WLASL) dataset for gloss prediction. In our manuscript, we use the sample gloss figures from asl-lex.org. We acknowledge them with the following citations [29,30]. The model's input will be dynamic videos made up of sign words, as shown in Figure 1, which explains how the model operates. To distinguish essential frames from redundant frames, we propose the modified histogram difference approach in conjunction with the Euclidean distance algorithm. In comparison to all other pose-based models and frameworks in use, this process made our model more accurate at predicting even similar gloss words. Additionally, we employ hand-crafted augmentation techniques, including in-plane rotation, joint angle rotation, and perspective modification to extracted pose vectors to enable our model to be considerably more adaptable to real-world applications. Furthermore, by learning the location of the target pose vectors using a bounding box, we further prevent our model from overfitting and generalization by utilizing YOLOv3 to normalize the pose vectors.

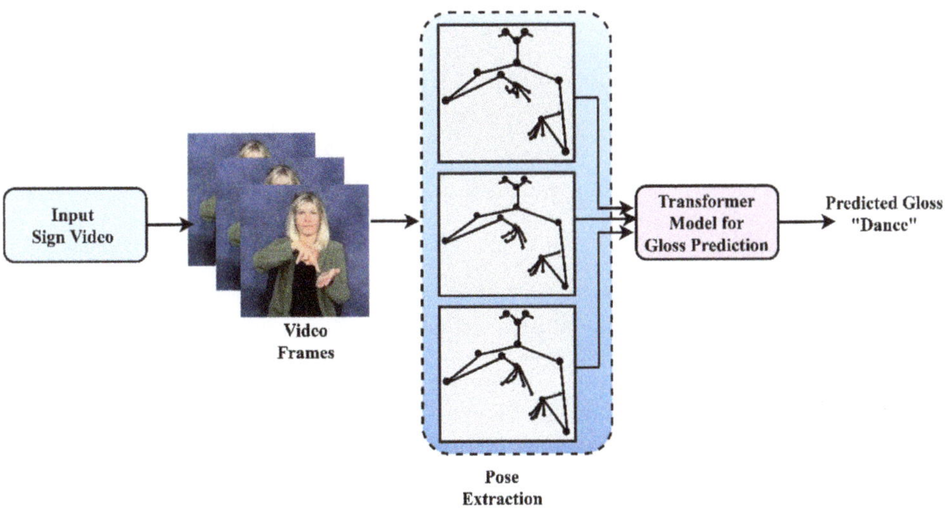

Figure 1. Overview of gloss prediction from sign poses—WLASL using a standard transformer.

On the basis of the human pose-based modelling, we emphasize isolated SLR for gloss prediction knowing that recognizing word-level sign itself is exceptionally hard and serves as a basic core element for recognizing continuous sentences in CSLR. We consider that a person's skeletal motion greatly contributes to the messages they are expressing. Inspired by the transformer architecture proposed by Ashish et al. [31] with slight modifications to the standard transformer model, we evaluate the potential of the proposed transformer model. Transformer models perform remarkably well in tasks requiring sequence processing and are relatively inexpensive computationally.

Our key contributions in pose-based word-level sign language recognition (WSLR) include:

1. We introduce a novel approach for our pose-based WLSR using a keyframe extraction technique to discard the irrelevant frames from the critical frames. To perform this

keyframe extraction, we use a modified histogram difference algorithm and Euclidean distance algorithm through which our model achieves 5% improvement compared to other existing pose-based state-of-the-art results on all the subsets of the WLASL dataset (WLASL 100, WLASL 300, WLASL1000, WLASL 2000).

2. We employ augmentation techniques that let our model fit and be adapted for any additional real-time dataset in generalizing so that it can handle the real-time scenario. For this, we adopt in-plane rotation with perspective transformation and joint rotation, which has the added benefit of enabling our model to recognize poses executed at various angles, with various hand sizes, and even at various locations.
3. We introduce a novel pose normalization approach in WLSR using YOLO v3, through which our approach has seen significant improvement of up to 20% for the exact detection of the pose vectors in the signing space.
4. To predict the glosses from the normalized pose sequence, we propose a novel method through a Sign2Pose Gloss prediction transformer, which attains the highest top 1% recognition accuracy of 80.9 in WLASL 100, 64.21 in WLASL 300, 49.46 WLASL 1000, and 38.65 WLASL 2000, surpassing all state-of-the-art outcomes from the existing pose-based models.

The body of the article is structured as follows: Section 2 outlines the prior work in the field of sign language translation and offers insight into the issues that still need to be resolved to considerably increase identification accuracy using pose-based approaches. Section 2.1 discusses the significance and impact of gloss in continuous sign language translation. The two SLT criteria are discussed in depth in Section 2.2 to clarify how two-stage SLT translation overcomes the challenges of end-to-end SLT. Section 2.3 summarizes the importance of video summarization techniques in SLT. Section 2.4 explains how key point generation and pose estimation aid in recognizing finer details in sign sequences for exact gloss prediction. The procedures and approaches for carrying out gloss prediction with our suggested model are described in Section 3. Section 4 provides a quick overview of the proposed Sign2PoseGloss prediction transformer's detailed architecture. Section 4 discusses the design and validation of the experiment. The performance assessment of our model with an architecture based on appearance and pose is carried out in Section 5, and the results are discussed. Finally, we summarize the research with future scope.

2. Related Works

Sign language translation requires visual content and gloss annotation. As discussed in Section 1, the end goal of SLT is to provide natural language spoken sentences. Sign language translation is performed after gloss recognition and further continuous sign language translation. Therefore, in this section, we present the previous literature concepts of deep learning in CSLR models to understand how the role of gloss stands as the backbone for CSLT. To facilitate bidirectional communication between the deaf community and society, building a robust model which would be capable of translating sign language into spoken sentences and vice versa is necessary. Further, we describe the techniques used to address the complexities of video processing and the issues with appearance-based methods in gloss prediction. Furthermore, this study summarizes existing techniques for keypoint extraction and pose estimation models, as well as the requirement for a systematic approach to gloss prediction with reduced processing time complexity. The related works mainly focus on the deep learning-based SLT model to analyze the state-of-the-art results. The following sections summarize the concepts and methods related to SLT.

2.1. Significance of Glosses in Vision-Based CSLT

Recognizing the exact gloss representation for the performed sign sequence plays a significant role in CSLT. The biggest challenge of a CSLT system is the insufficient annotated dataset, identifying the explicit boundaries of signed words from the extracted frames of sign video and the transcription of target sentences from the extracted gloss sequences. In the initial phase of work, Hidden Markov models [32–34] were widely used for capturing the temporal feature information. Cui et al. [35] proposed a DNN (Deep Neural Network) for temporal feature extraction and RNN for sequence learning. In his framework, he suggested an iterative training process that includes gloss annotations from video segments and an alignment proposal module that generates the aligned sentence sequences from the extracted glosses. It is evident from this approach that the iterative process of sequence learning eliminates the need for massive amounts of information to train an HMM model. Although these modalities are superior in learning temporal dependencies, the integrated approach of multiple modalities necessitates more investigation because the performed sign gestures have concurrently related streams of information. Further, Sharma et al. [36] proposed a deep transfer learning approach employed for sign sentence recognition. In their deep learning-based network model, they used a convolutional neural network along with bi-directional LSTM and connectionist temporal classification (CTC). The added advantage of using this model is it can be trained to recognize the sequence of sentences without any requirement of any prior knowledge in an end-to-end fashion. However, connectionist temporal classification faces severe overfitting during computation. To resolve this issue, Niu et al. [37] used stochastic fine-grain labelling while training the model. For extracting gloss information from sign video frames, the model should know contextual information to extract the actual context of the sign with gloss. To ensure this, Anirudh et al. [38] proposed a pose-based SLR for gloss identification with contextual information using a graph convolutional network (GCN) and BERT transformer. Though this model concentrates on both spatial and temporal fusion extraction, combining a pose-based approach with image-based features will further enhance model performance. On the other hand, Cui et al. [39] proposed a model for real-time CSLR where they used RNN to address mapping issues with relevant glosses by designing a weakly supervised detection network using a connectionist temporal and alignment proposal for continuous spoken sentence translation. Further, this method requires improvement to handle multi-modal information.

To make this easy, transfer learning is employed by initially training the deep learning network using an isolated word dataset so that the problem is addressed. Rastgoo et al. [16] adapted this transfer learning technique using a post-processing algorithm to address the limited labelled dataset issue.

2.2. End-to-End and Two-Stage Translation in SLT

With the recent advancement in neural machine translation, recent works have concentrated on designing a gloss-free model to generate textual content directly from visual domains using cross-modal mappings without any intermediate glosses. Zhao et al. [40] proposed a novel framework for sign video to spoken sentence generation using three key modules. In their model, they replaced the gloss generation module with a word existence module that checks the word existence in the input sign video. For this, they applied a CNN encoder–decoder for video feature extraction and a logistic regression classifier for word existence verification. However, in the existing proposed model, there still exist challenges in visual-to-text direct mappings. Additionally, training an SLT model is challenging for longer sentences/video sequences, and decoding a sentence from the input sign video after extracting finite dimensional features is tedious. Further, a key point normalization method to normalize the skeleton points of the signer was proposed by ref. [41] to translate sign videos into spoken sentences directly without any intermediate gloss. They applied the stochastic frame selection method for sampling and frame augmentation and transcribed sign language videos into spoken sentences using attention models. However, direct sign-to-text translation outcomes were no better. Since end-to-end translation requires a huge amount of information to train and tune the model, two-stage SLT is the better option for CSLT; however, it is time-consuming to process the input sequence.

When compared with gloss, mid-level representation drastically improves SLT performance [24]. Additionally, sign-to-gloss translation averts long-term dependencies [42], and the number of sign glosses from a particular sign video are minimal when compared with the number of frames in the video [14]. Therefore, combining gloss representation with recognition and translation of sign language, a unified architecture is proposed by Camgoz et al. [43] that jointly learns continuous sign language recognition and translation achieved by CTC, thereby improvising sequence-to-sequence learning and performance independent of ground truth timing information. The detailed summary of the existing deep learning models for two-stage SLT is discussed in Table 1.

Table 1. Summary of existing methods for gloss prediction using two-stage SLT.

Ref.	Translation Type	Technique for Gloss Prediction	Dataset	Performance Metric	Remarks
[38]	Sign2Gloss2Text	Graph convolution network (GCN) and bi-directional encoder representations from transformer (BERT)	WLASL	88.67 at top 10% accuracy on 100 gloss recognition	Image-based feature extraction enhances the performance of the model.
[44]	Sign2Gloss2Text	Human key-point estimation	KETI sign language	BLEU4—65.83 (Key points: Hand, body)	Performance would improve on improving key-point detection
[45]	Sign2Gloss2Text Gloss2Text	Spatial-temporal transformer and spatial-temporal RNN	Phoenix 2014T	BLEU4-24.00	Dataset is restricted to the weather forecast
[46]	Sign2Gloss2Text	Temporal graph convolution network (TGCN)	WLASL	62.63% at top 10 accuracy on 2000 gloss recognition	Labelling a large number of samples requires advanced deep algorithms to pave the way from word-level to sentence-level annotations
[47]	Sign2Gloss2Text	Context-aware GAN, temporal convolution layers (TCL), and BLSTM	Phoenix 2014T, CSL, and GSL signer independent	23.4%, 2.1%, and 2.26% WER, respectively	Complexity and data imbalance in GAN network
[48]	Sign2Gloss2Text	Transformer	WLASL100, WLASL300, and LSA 64	63.18%, 43.78%, and 100% recognition accuracy	Shows better outcomes on even smaller datasets
[49]	Sign2Gloss2Text	Intensity modifier	Phoenix 2014T	BLEU1-26.51	Lacks spatial and temporal information for black translation and lack of proper evaluation metrics.

In the same way, sign-to-gloss→gloss-to-text is one of the best translation protocols, where instead of training a network for text-to-text translation from scratch, they provide better translation results for gloss-to-text translation. In our approach, we propose a Sign2Gloss translation protocol network using a modified standard transformer.

2.3. Video Analysis and Summarization

Sign language translation takes time to process continuous sign sequences. As a result, incorporating video summarization or video processing techniques into SLT may improve gloss recognition accuracy in the Sign2Gloss translation protocol. Video summarization and video processing, on the other hand, are very common in video recognition and action recognition tasks [50]. The primary goal of video processing is to choose a set of frames to facilitate fast computation while processing lengthy videos. Yao et al. [51] proposed a key frame extraction technique based on multifeatured fusion for processing dance videos in order to recognize various dance motions and steps. Furthermore, a smart key-frame extraction technique was proposed by Wang et al. [52] for vehicle target recognition.

This model integrates the scale-invariant feature transform (SIFT) and the background difference algorithm, coupled with the concept of criterion factor K, to significantly divide and categorize the frames into non-mutation and mutation frames. The redundant frames are dissimilar frames that are discarded. However, because it skips a greater number of frames compared to SLT, this method is only appropriate for vehicle recognition. To resolve these missing details in frame extraction methods, Li et al. [53] proposed a new concept called sparse coding for key frame extraction with log-regularizer. This method overcomes the challenges of losing pertinent data while discarding redundant frames while performing key frame extraction. However, this method is unsuitable for complex videos because it strips away high-level semantic information from the video.

2.4. Pose-Based Methods for SLT

Human pose-based architecture is not only used for action recognition but it has also been applied to perform specific tasks in WSLR and SLT since the advancement of deep learning. Pose estimation is either performed using a probabilistic graphical model or using pictorial structures [54]. So far, human pose estimation has achieved outstanding results for static or isolated images. However, it underperformed for real-time or dynamic images such as video because of issues with tracking occlusions, motion blur during the transition, and its inability to capture the temporal dependency between the extracted video frames. The poses/skeletal holds positional information of a human body pose and can provide important cues [55]. Using the RWTH-Phoenix 2014 T dataset, a skeleton-based graph convolution network was proposed for end-to-end translation. It used only 14 key points, omitting fine-grained key points in fingers and faces, resulting in poor end-to-end translation performance. However, skeletal-based methods have gained attention in modern research methods since they are independent of background variations. Further, in skeleton-based SLR models, RGB-based skeletal methods outperforms well. To overcome this performance degradation stated in the previous work, Songyao et al. [50] proposed a skeleton-aware multi-modal ensemble for RGB frames, which has 33 key points, including key points in the nose, mouth, upper body, and hands. This framework makes use of multi-modal information and utilizes a sign language graph convolution neural network (SL-GCN) to build embedded dynamics. Further, in another work, maxim et al. [56] investigated the enhancement of recognition performance in SLR models by fine-tuning the datasets. Additionally, the author analyzed whether it is possible to use these models in a real-time environment without GPU.

Yang et al. introduced the graph convolution neural network model to deal with the temporal dependency among extracted frames. Followed by him, many others proposed various methods for pose estimation, such as the GCN-BERT method [38], key point extraction methods using open pose [57], action structured graph convolution networks [58], and MS-G3D for spatial–temporal graphical convolution networks.

The pose-based approach proposed by Youngmin et al. [57] introduced video processing and key point extraction techniques. These techniques aided in frame selection and key point extraction for precise body movement and location. Sign-to-text translation protocol was used in this pose-based approach. However, direct translation from sign language video to spoken sentence produced no good results. In addition to these methods, automatic sign language translation is possible by merging the NLP transformers and computer vision. For such tasks, the video-transformer network was proposed by Coster et al. [59]. However, these transformer networks require a huge amount of labelled data corpus to fine-tune and train ASLR models. This method is evaluated using the large-scale annotated Turkish Sign Language data corpus that eliminates the need for a large, annotated data corpus.

3. Materials and Methods

In this section, we discuss the baseline methods for our proposed Sign2Pose Gloss prediction transformer architecture that efficiently predicts gloss words from dynamic videos. Identifying the explicit boundaries of sign words from sign videos is a practical difficulty faced by CSLR/SLT systems. Though many techniques have been proposed earlier to solve the end-to-end translation model for efficient mapping of predicted words with the target sentence, the existing systems do have some snags. Intermediate gloss prediction substantially increases the translation outcomes of SLT systems. Therefore, we propose a novel method for efficient gloss prediction using a Sign2Pose Gloss prediction transformer that significantly identifies the intermediate gloss for the given input video sequence. Initially, the system is validated using the WLASL [60] dataset for word-level gloss videos using a sign-to-gloss network translation protocol. As stated in Section 2.1, the proposed model can enhance the efficiency of a two-stage SLT system, reducing the need for a large, annotated vocabulary. Furthermore, it is not required to tune and train a model from scratch when using this proposed word-level gloss prediction transformer; it can be employed as a pre-trained gloss network in two-stage SLT. The methods are subdivided into four phases, namely (i) key-frame extraction, (ii) pose estimation from key-frame, (iii) pre-processing, and (iv) pose normalization. This section briefly elaborates on the key components and steps of the proposed Sign2Pose Gloss prediction transformer.

3.1. Dataset Description

We have a wide variety of gestural corpora because sign language is not a universal language. For example, the Chalearn synthetic hand [61] dataset contains realistic 3D human male hand gestures, InterHand 2.6 M [62] is a 3D representation of interacting hands, and the TheRusLan [63] data corpus contains 22,200 audio samples with text annotations. The AUTSL [64] data corpus is a large corpus multi-modal Turkish Sign Language dataset with 226 signers and 38,336 isolated sign video samples. MS-ASL [62] data corpus is a massive corpus with 1000 signs performed by 200 signers in real-world environments. Because we plan to concentrate on the entire body posture to capture the most precise details in the body posture for gloss prediction, we opt for the word-level American Sign Language dataset for our experiments. We train our model using the large-scale signer-dependent word-level American sign language (WLASL) publicly available benchmark dataset [46]. The gloss videos for the above-mentioned dataset are collected from multifarious public websites that hold the gloss annotations for dialects in ASL along with the details of metainformation such as bounding box, temporal boundary, and signer diversity. The average length of videos in the dataset is around 2.41 s. The sign instances are performed by 119 native American signers. Initially, the collected data from multiple public resources planned for tutoring SL led to diversity in signing dialects, styles, and backgrounds suitable for real-time sign language classification. This dataset is categorized into 4 subsets based on different vocabulary sizes such as WLASL 100, WLASL 300, WLASL 1000, and WLASL 2000. These four subsets are grouped by choosing the top K glosses where K = {100, 300, 1000, 2000}. A detailed description of the dataset is briefed in Table 2.

Table 2. WLASL dataset description.

Categories	Content	Type	Glosses	Samples	Mean (Avg. Instances/Class)	Signers
WLASL 100	Video with Aligned Sign/Sentence with text and Gloss	RGB	100	2038	20.38	97
WLASL 300			300	5117	17.1	109
WLASL 1000			1000	13,168	13.16	116
WLASL 2000			2000	21,083	10.54	119

3.2. Key Frame Extraction Technique

The dynamic sign word video has multiple video frames with multiple repeated gestures and transition phases between the successive gestures after extraction. This method retains the best representations of shots from extracted frames and discards the redundant frames. Thus, processing all such extracted frames requires a high-power computational system and takes a huge amount of computational time. Therefore, we propose a key frame extraction method using a modified histogram difference algorithm method for discarding unnecessary frames to efficiently boost the performance of the proposed Sign2Pose Gloss prediction transformer for dynamic sign word videos and overcome the timing overhead and computational complexities. The main objective of this method is to decide the specific key frames from actual frames for each signed word that are notable in terms of different gesture positions and thereby disposing of the unwanted poses or gesture positions and transition phases.

We divide this key frame extraction into two phases. In our first phase, we extract the frames from the given input video in a successive manner and then calculate the threshold with mean and standard deviation after applying the modified histogram difference algorithm. The distance between the current and the difference frame is calculated using the Euclidean distance algorithm. After measuring the distance between the frames, the mean and standard deviation is calculated. In our next phase, the threshold value denoted as "T_h" is set, and the measured distances are compared with the threshold.

Let us denote the input video as "I", and the frames are represented as "f". Initially, the frames extracted from the input video are RGB frames. Then, RGB frames are converted to grayscale frames to compute the absolute difference between the frames using the absolute difference algorithm. Therefore, let the RGB frames be denoted as "f_{RGB}", grayscale frames are denoted as "f_{GRAY}", histogram difference is assumed as "H_{diff}", and 'ℕ' denotes the number of bins in the histogram.

$$H_{diff}(t) = \left| \sum_{j=0}^{\mathbb{N}} H_{(t-1)}(j) - H_{(t)}(j) \right|. \tag{1}$$

After computing the difference, we apply mean and standard deviation where "μ" is used as a symbol for mean calculation, and "σ" denotes standard deviation. The distance between the successive frames is calculated using the Euclidean difference algorithm "E_d".

$$E_d(p,q) = \sqrt{(p_1 - q_1)^2 + (p_2 - q_2)^2}. \tag{2}$$

Let "p" and "q" be the two points in a frame, and let the coordinates of "p" be (p_1, p_2) and "q" be (q_1, q_2). For "n" dimensions, the formula can be more generalized as follows:

$$E_d = \sqrt{\sum_{i=1}^{n} (p_i - q_i)^2}, \tag{3}$$

where "n" denotes the dimensions and p_i and q_i are the data points. After computing the Euclidean distance, the threshold value is set. To set the threshold value, we must perform H_{diff} and then calculate the mean and standard deviation.

$$T_h = \mu + \sigma, \tag{4}$$

$$T_h = \varphi, \tag{5}$$

where "φ" is used to represent the combined value of mean and standard deviation. After setting the T_h value, we compare the measured distance between the consecutive frame and the threshold value, and the choice between the key-frame "K" and the redundant frame "R" is conducted. The elements in key-frames are denoted as "k_N", and the elements of redundant frames are represented as "r_M". Then, the extracted key frames are provided as inputs for pose extraction for gloss prediction. The detailed steps for key frame extraction are provided in Algorithm 1.

Algorithm 1. Key-frame extraction	
	Let I be the input sign video
Input:	$I \in (1\ to\ N)\ I_i \ldots .I_N$
	Let n be the number of frames in I_i
Output:	Set of key-frames f_{key}: f_{key} {1 to m} where m < n
1	for f_{RGB} in n (frames):
2	Convert RGB frames into grayscale frames $f_{RGB} \rightarrow f_{GRAY}$
3	Compute histogram difference H_{diff} between successive frames using Equation (1)
4	Calculate mean μ and standard deviation σ of the H_{diff}
5	Compute threshold value "T_h":
6	Calculate the Euclidean distance E_d using Equation (2)
	f_{GRAY} ={elements of K and elements of R}
	"R" denotes the set of redundant frames
7	Such that,
	$K = \{\mathrm{k}_1, \mathrm{k}_2, \mathrm{k}_3, \ldots\ \mathrm{k}_\mathrm{N}\}$
	$R = \{\mathrm{r}_1, \mathrm{r}_2, \mathrm{r}_4, \ldots\ \mathrm{kr}_\mathrm{M}\}$
8	for I in n:
9	if $E_d > T_h$:
10	$R \backslash K = \{\mathrm{r}_{\mathrm{M}-1}\}$ Element obtained belongs to set of redundant frames but not to set of key-frames Add the frames to the set f_{key}
11	else
12	Discard the frame
13	Repeat steps 1 to 12 for the entire dataset, and once completed, discarding redundant frames stops the process.

3.3. *Pose Estimation from Key-Frame*

Human pose estimation (HPE) refers to representing the orientation of the human body in a graphical format. In other words, locating human body parts and joints using computer vision in images or video. Initially, before the deep learning era, human pose estimation was performed by recording an RGB image using optical sensors and kinect sensors to detect the human pose or an object. The three most common types of human models are skeleton-based, volume-based, and contour-based. Skeleton-based HPE is the most preferred and frequently followed method since it is flexible with stability in the joint locations and orientations. For instance, ankles, wrists, knees, elbows, shoulders, fingertips, etc. There are various standard frameworks for pose estimation. Pischulin et al. proposed a deep-cut algorithm [65] for multi-person pose estimation with joint objectives. This method first locates the person's joints using integer linear programming. The method identifies the joints much more precisely though occlusions appear from person to person, but the process is extremely complex and time-consuming. Further, various other frameworks using deeper cut algorithms [66], PoseNet [67], and OpenPose [68] are used for HPE. In our proposed framework for pose estimation, we use standard pose estimation of vision API for locating the head, body, and hand landmarks from each set of "K" video frames. The landmarks obtained are all 2D and relative to the frames. Its coordinate values for the top right side to the frame are [1, 1], whereas the bottom left corner is denoted as [0, 0]. We use a vision image classifier to spot the presence and absence of individual landmarks or objects. If an object is absent in the relative frame, then the coordinates are represented by 0 and vice versa.

3.4. *Pre-Processing*

After acquiring landmark coordinates in pose estimation, we opt for a pre-processing technique to efficiently enhance the system's generalizing ability. For the system to adapt to different datasets and develop a versatile response, we employ spatial augmentation while training the skeleton data points. Further, the choice of the parameters is random, and they have maintained rationale throughout the signing space for every frame. Additionally, this spatial augmentation technique overrides the overfitting issue. Figure 2 depicts the steps involved in pre-processing and the outcome of augmentation applied on single frames.

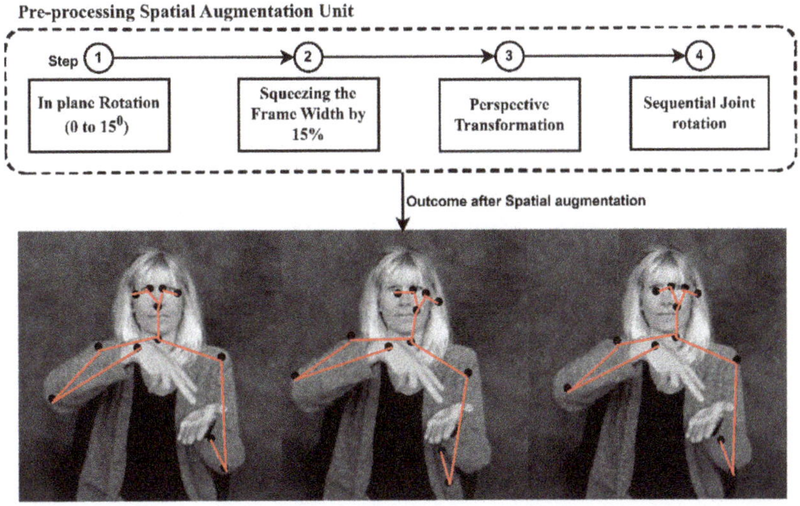

Figure 2. Illustrating the augmentation techniques applied to a single frame while pre-processing.

The initial step in spatial augmentation is applying in-plane rotation to each frame whose angle of rotation, denoted by "θ", lies between 0° to 15°. Therefore, by applying in-plane rotation, the plane is mapped to itself for a particular rotation and does not remain fixed. Perhaps all the vectors in the planes are mapped to other vectors in the same plane by rotation. During rotation, the center of rotation confides on the center of the plane (frames) coordinates (0.5, 0.5). For instance, in a 2D image, the position of point P is represented by the coordinate (P_x, P_y), and the numerical representation of point P in a plane is anonymous until we define a reference coordinate. Once the origin is fixed, the extents of point P from its x and y axis from the origin are its coordinates (P_x, P_y).

Followed by in-plane rotation, the next step we carry out is squeezing the frames on their horizontal sides by setting random proportions ranging up to 15% of the original frame width w_1(right side) and w_2 (left side). Once the squeezing is set, the joint coordinates are recalculated concerning the newly set plane. The third step is perspective augmentation, where the sign video is recorded with a minor shift in angles of inclination applied to the signing video. This method helps the system become accustomed to images with different angles and builds the system's robustness. Like human vision, which can locate and identify an object at any distance and any angle of inclination or distance, perspective augmentation helps the model recognize the same 2D or 3D object on two different projective planes.

By applying perspective transformation, the joint coordinates are made to project on a new plane with spatially defined projection with a slightly inclined angle. The proportion and adjustment to the right and left sides of the single frame are picked randomly by uniform distribution with an interval of [0, 1]. The detailed steps of sequential joint rotation are explained in Algorithm 2.

Algorithm 2. Sequential Joint Rotation	
1	Input image I_{in}, x, and y standard coordinates
2	Initialize center point of the frame as C_{mid}
3	Fix $C_{mid} = 0.5$ Rotate frame f_{rot} according to C_{mid}, *and [x,y]*
4	Standard Rotation Matrix is given as R $R = \begin{pmatrix} \cos\theta & -\sin\theta \\ \sin\theta & \cos\theta \end{pmatrix}$ f_{rot} *with* respect to standard coordinates *[xy]* $\begin{pmatrix} \cos\theta & -\sin\theta \\ \sin\theta & \cos\theta \end{pmatrix} \begin{pmatrix} (x-0.5) \\ (y-0.5) \end{pmatrix}$
5	then the moved state is denoted by x' *and* y' $x' = (x - 0.5)\cos\theta - (y - 0.5)\sin\theta + 0.5$ $y' = (y - 0.5)\cos\theta + (x - 0.5)\sin\theta + 0.5$ $f_{rot}(x'y') = (x - 0.5)\cos\theta - (y - 0.5)\sin\theta + 0.5, (y - 0.5)\cos\theta + (x - 0.5)\sin\theta + 0.5$
6	Angle of rotation $\theta \le 15°$
7	Generate random moving state S_m based on θ and uniform distribution
8	Within the range of C_{mid}, move x based on S_m, then y based on S_m to calculate S_m' to obtain a new range of x and x', y and y' $I_{Augmentation}$ = Augment (I_{in}, x, y) $I_{Augmentation}'$ = Augment (I_{in}, x', y')
9	Calculate recognized image I_{obs} and measure the Euclidean distance E_d **if** $E_d(I_{obs}, C_{mid}) \le E_d(I_{obs}', C_{mid})$ **then**
10	Improve the recognition accuracy **else** stop

3.5. Pose Normalization

Body proportion differs from person to person. Not only this, but positional properties, such as camera distance, capturing angle, angle of rotation, motion transfer, head, face, hand, and palm orientation, etc., vary from signer to signer. Further, input landmark coordinates are associated with values relative to the frame. This leads the model to learn more irrelevant spatial features than the performed sign. In such cases, training and fine-tuning the model will be time-consuming. To overcome this issue, we use the normalization technique, where all such body proportions, distance from the camera, positional properties, motion transfer overheads, and orientation are precluded. Inspired by SL linguistics [69] regarding the use of signing space with body landmarks, we use a 3D space in the signing space in front of the signer and their immediate surroundings. We take the area slightly above the signer's waist, reaching slightly above the signer's head, covering the two loosely bent elbows with projected body landmarks to identify the sign.

In our previous pre-processing step, we applied augmentation techniques to efficiently enhance and bring versatile recognition for different body proportions, orientations, and tilting angles. Though our model is efficient towards generalized input, without normalization, the system picks inappropriate spatial features from the signs performed. So, we use normalization using YOLO version 3 for object detection and pose normalization using anchor boxes. We define a signing space based on a head portion with 7 head units wide and 8 head units high where its horizontal center lies with a nose. Additionally, the vertical side of the anchoring box is fixed, considering the left eye with 0.5 head units upright and 6 units below for the bottom edge. We have two other anchoring boxes for tracking the hand orientations and their shape, which enables the model to target the hand orientations and their corresponding signs, eliminating all other insignificant spatial features relative to the frame. Figure 3 shows the visualization of the normalized pose using YOLO v3 for an independent frame.

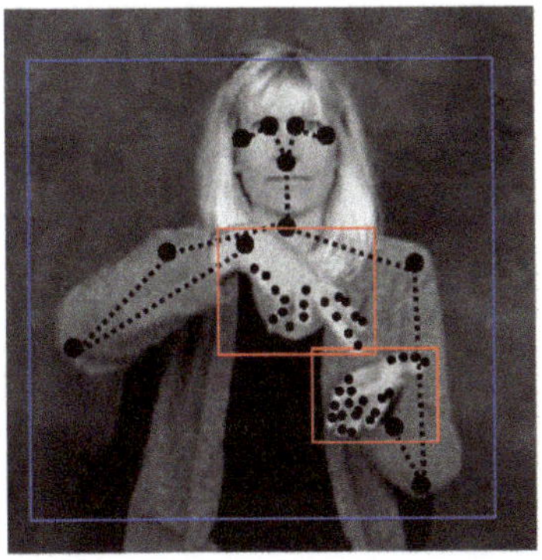

Figure 3. Sample visualization of normalized pose using YOLOv3.

To calculate the anchor box and its normalization, we need to rescale them between 0 and 1 by dividing the image width by its height. The bounding box network format is (x, y, width(w) and height(h), confide). In YOLO v3, the predicted output coordinate of the anchor box is normalized relative to the grid and input image. We do this because we have diverse signers, and among such a diverse dataset, the model should detect the sign performed. The confide value is set to 0.5, and depending on the confide value, the object is detected. As we know, the annotation coordinates are (X_{max}, Y_{max}, X_{min}, Y_{min}), considering (X_1, Y_1) as X and Y coordinates of the top left corner of the bounding box. (X_2, Y_2) are the X and Y coordinates of the bottom right corner of the bounding box and (X_c, Y_c) are the center x and y coordinates of the bounding box.

Where

$$(X_{max}, Y_{max}) = (X_1, Y_1), \tag{6}$$

$$(X_{min}, Y_{min}) = (X_2, Y_2), \tag{7}$$

$$Normalized\ X_{min} = (X_{min} + w/2)/W_{img}, \tag{8}$$

$$Normalized\ Y_{min} = (Y_{min} + h/2)/H_{img}, \tag{9}$$

$$Normalized\ width\ (w) = w/W_{img}, \tag{10}$$

$$Normalized\ height\ (h) = h/H_{img}. \tag{11}$$

The bounding box coordinates, width, and height lie between a particular location of the grid cell, so they balance between 0 and 1. Furthermore, the sum of the square error is calculated only when the object is present.

4. Proposed Architecture

The sequence of movements in body parts provides a lot of information in sign language. Moreover, in our literature study, we analyzed that pose sequences are outstanding records in recognition and detection since the model stays focused on features in the pose images rather than looking into inappropriate components such as background, lighting, and so on. In our proposed architecture, we used the Sign2Pose Gloss prediction transformer, which is a slightly modified version of the transformer with attention [31]. The input to our proposed transformer model is a normalized pose sequence with a 108-dimensional pose vector and 54 joint locations. The Sign2Pose Gloss prediction transformer uses attention skillfully. Figure 4 depicts the entire architecture of our Sign2Pose Gloss prediction transformer. We have an encoder and a decoder layer where the model first translates the sequence of sentences and then applies vectorization, and finally, with the attention layers, transforms them. In our model, we use learned positional encoding rather than spatial positional encoding to define the actual semantics of the sentences and words. Furthermore, we add the positional encodings with 108 dimensions to the individual pose vectors. By adding the learned encodings to the individual pose vectors, we obtain a sequence of input vectors to fetch as input to the encoder layer. There are six layers in both the encoder and decoder, with nine head units in the self-attention module and an input dimension of 108, followed by 108 hidden dimensions and a feed-forward dimension of 2048. As per the standard transformer model, there are two self-attention and feed-forward networks.

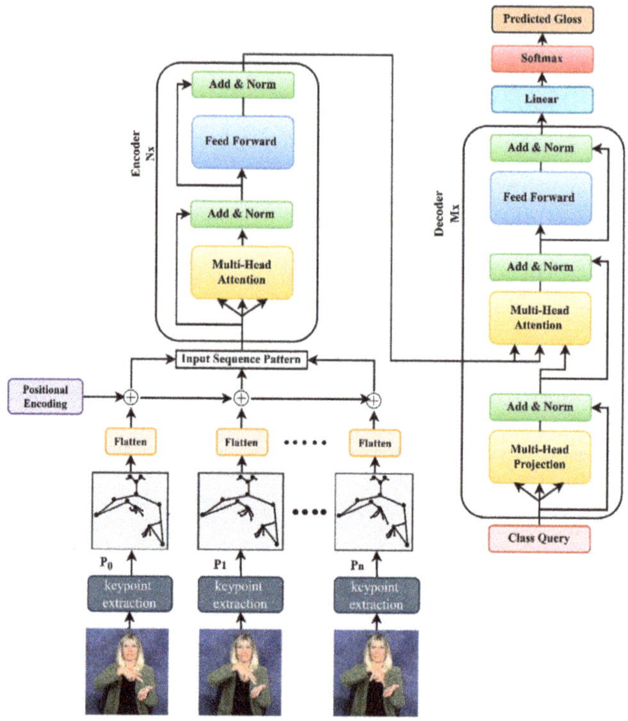

Figure 4. Proposed architecture of the Sign2Pose Gloss prediction transformer.

The standard transformer decoder architecture has query, key, and value vectors as output vectors for each word. For instance, if the input sequence vector to the encoder is "*It's too sunny today*", then the input embeddings present in the sentence are four, and for each input vector/input word, we calculate three output vectors such as query, key, and value. In the above-stated example, we have "n = 4" words. Thus, for "n" words in a sentence, there are nQueries, keys, and vectors to be calculated. In our case, we are proposing a Sign2Pose Gloss prediction transformer for gloss prediction. We use the word-level American sign language dataset, and the sequence vector processed through the entire encoder and decoder process will be a single element. For this purpose, we have one query at the input of the decoder, and that query is called class query since it decodes the class of sign. Since there is only one element to be processed through the entire multi-head projection module present in the decoder, the attention has no influence on key and value vectors, and the SoftMax present in the attention model is always "1". Hence, we calculate the input vector in the value space and there is no requirement for key and query calculation.

After the processing elements pass through the multi-head attention module in the decoder layer, the vectors are concatenated and processed by the linear layer using logit vectors. In this linear layer, we provide the class query input where the confidence of each class is calculated using the SoftMax activation.

5. Experiments

For our experiments, we used a dataset of American sign language at the word level. There are four subsets of datasets, as described in Section 3.1, and we utilized every subset separately for our experimental evaluation. WLASL 100, WLASL 300, WLASL 1000, and WLASL 2000 comprised the evaluation dataset. The suggested transformer model for gloss prediction was trained using the aforementioned datasets. The datasets were split in the ratio of 85:15, out of which 15% of the dataset was used for testing and from 85%, 70% was used for training, 15% for validating, and the remaining 15% for testing. We proposed a novel method to build a robust model which is more flexible in recognizing similar signs, learning different dialects, and coping with different environments with different signers. We have applied a key-frame extraction module to discard the redundant frames and implied augmentation technique. After augmentation, we used YOLO version 3 to normalize the pose vectors in such a way that our system is free from overfitting. With all these pre-processing steps, we input a normalized pose image with all landmarks to the proposed transformer architecture. In addition to the original context of the sign, horizontal flipping was set to 0.5 randomly for all the normalized frames. The details of the parameter tuning of our model are stated in Table 3.

Table 3. Hyperparameter specifications.

Hyperparameter	Tuning Details
Pose vectors	108
Encoder layers	6
Decoder layers	6
Input and hidden dimension	108
Feed Forward dimension	2048
Learning rate	0.001
Weighted decay	0.0001
Optimizer	Stochastic Gradient Descent
Epochs	300

The proposed Sign2Pose Gloss prediction transformer was implemented using TensorFlow in the Anaconda Software tool. The query, key, and value vectors in the standard transformer models were slightly modified to discard the unnecessary computations inside multi-head attention in the decoder module that may occur due to the flow of class query through the module. We run our experiment for 300 epochs, and the learning rate was set to 0.001 with the weighted decay 10^{-4}, and momentum was set to 0. We used a stochastic gradient descent optimizer, and the weights were initialized using uniform distribution ranging (0, 1). This range was randomly fixed, and we used the cross-entropy loss function to score the models' performance in terms of correct gloss prediction.

6. Results and Discussions

We evaluated our pose-based proposed model on all the subdivisions of the publicly available word-level American sign language datasets. As mentioned in Section 3.1, Table 2, WLASL datasets have Top K classes, where subsets/classes WLASL 100, 300, 1000, and 2000 are based on the number of videos. For instance, the first subset of WLASL has 100 classes, and each class represents a particular gloss video with different instances performing the same gloss under the same class category. We compared our results with previous pose-based and appearance-based models to evaluate the models' performance and state-of-the-art outcome achieved by our model in the same dataset. For ease and a prospective comparison of the advancement of the primary data representation streams for SLR, the findings of appearance-based techniques were also considered. Table 4 summarizes the previous pose-based and appearance-based models experimented on subsets of WLASL and other datasets.

Table 4. Summary of different models experimented using WLASL datasets with and without augmentation techniques for prospective comparison of the proposed model.

Model and Dataset	I3D [70]	Pose-GRU [70]	Pose-TGCN [70]	GCN-BERT [38]	ST-GCN [71]	SPOTTER [48]	OURS
Appearance-based	✓	×	×	×	✓	×	×
Pose-based	×	✓	✓	✓	×	✓	✓
Augmentation	✓	✓	✓	×	✓	✓	✓
WLASL 100	✓	✓	✓	✓	✓	✓	✓
WLASL300	✓	✓	✓	✓	✓	✓	✓
WLASL1000	✓	✓	✓	×	✓	×	✓
WLASL 2000	✓	✓	✓	×	✓	×	✓
Other datasets	×	×	×	×	✓	✓	×

As discussed in Section 3.2, sign word videos have multiple frames. In our baseline model, after the extraction of frames from the sign video, we used the key-frame extraction technique to preserve key-frames and discard irrelevant frames. This method reduces processing time complexity and improves the clarity of critical frame predictions for gloss. Further, to make the system more reliable for generalization, we used special augmentation techniques, as mentioned in Section 3.4, and we used YOLOv3 to normalize the pose vectors to fetch as input to the slightly modified standard transformer model proposed by Camgoz et al. [31]. The use of YOLOv3 not only boosts the systems gloss prediction, but our method also overrides overfitting issues. Figure 5a shows an example of the key-frame extraction for the word "Drink". Frames that were pulled out for the gloss "Drink" had transitional frames between repeated and idle frames. We applied a modified histogram difference algorithm and Euclidean distance algorithm to extract the key-frames and discard the redundant frames, as discussed earlier in Section 3.2. Figure 5b shows the sample of the discarded frames eliminating the blurred, idle, and transitional frames and Figure 5c shows the extracted key-frames using Algorithm 1 in Section 3.2.

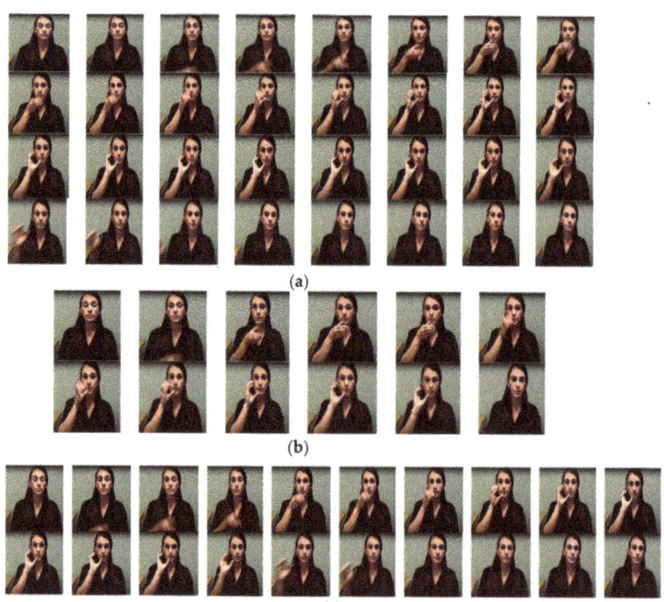

Figure 5. Sample images of key-frame extraction for the Gloss "Drink" from the WLASL 100 dataset (**a**) sample of extracted frames for the mentioned gloss. (**b**) Discarded redundant frames. (**c**) Preserved key-frame sample from extracted frames.

Through this technique, we were able to achieve top 1% accuracy of 80.9 in WLASL 100, 64.21% in WLASL 300, 49.46% in WLASL 1000, and 38.65% in WLASL 2000. However, we contrasted our model with models that are based on both poses and appearances. Our suggested method outperforms the prior state-of-the-art pose-based approach on the WLASL100 by 17 percentage points, attaining 80.9% in top 1% accuracy. On the WLASL300 subset, we also created a state-of-the-art result of 64.21% accuracy, outperforming the prior one by 20 percentage points.

From Figure 6a,b, it is observed that the appearance-based models surpassed pose-based models. Though these appearance-based models (ST-GCN and I3D) outrun our model, we contend that these results come at a substantially higher computational cost owing to the dimensions, which are limited in our system even when coupled with the pose estimation framework. In Figure 7, we observe the model's ability to predict top 1% gloss prediction accuracy during validation with the test samples, and the loss accuracy determines the predicted number of incorrect glosses by our models. The previous pose-based model underperforms in recognizing different words with similar signs, such as "man", "woman" "read", "dance", "wish", "hungry", "cold", "hug", "circle", "turn-around", "runny nose", "head-cold", which slightly vary in their hand orientation. From observing Figure 6c, it is analyzed that our model has seamless improvement top 1% validation accuracy of 80.9% when compared with other pose-based models since the proposed Sign2Pose Gloss prediction transformer uses the hand-crafted input feature representation of body and hand stance that already has sufficient information to decode the notions needed for sign language compared to other appearance-based models. As a result, it needs a much smaller training set to obtain adequate results.

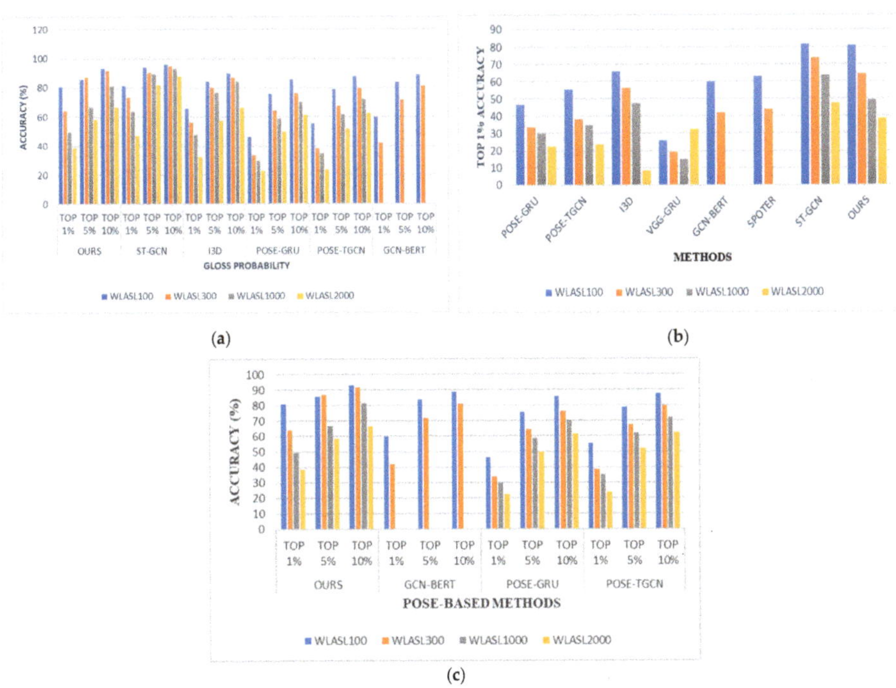

Figure 6. Performance analysis of proposed work with existing appearance and pose-based models. (**a**) Graphical representation comparing our approach with the pose-based as well as appearance-based model. (**b**) Comparing top 1% recognition accuracy on both pose-based and appearance-based models; (**c**) comparing top K macro recognition accuracy on pose-based models.

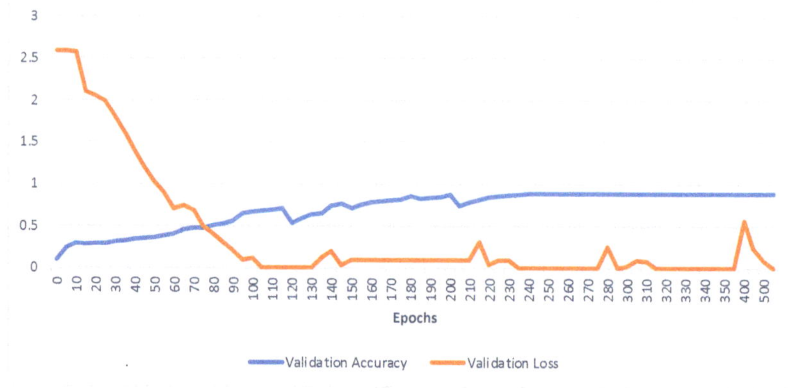

Figure 7. Validation accuracy and validation loss of our model.

From Figure 7, it is observed that the model starts to converge from 150 epochs and attains its maximum top 1% macro recognition accuracy by 220 epochs, and the model performs consistently after 240 epochs attaining 80.9% accuracy as top 1 class accuracy for WLASL 100. We tested the model after training it using a fixed dataset split of 15%. Table 5 shows the compositions of techniques and their top 1% recognition accuracies (%) on four subsets of the WLASL dataset. Our model uses a standard pose estimation algorithm from apple vision API and YOLO V3 for extracting the bounding and anchoring box for the hand. Using this technique, our extraction method is strong and effective, especially in near and different sign viewpoints.

Table 5. Performance analysis on top 1% macro recognition accuracy of proposed Sign2pose Gloss prediction transformer with other pose-based state-of-the-art models.

Pose-Based Models	WLASL100 Top-1% Accuracy	WLASL300 Top-1% Accuracy	WLASL1000 Top-1% Accuracy	WLASL2000 Top-1% Accuracy
POSE-GRU [46]	46.51	33.68	30.1	22.54
POSE-TGCN [46]	55.43	38.32	34.86	23.65
GCN-BERT [38]	60.15	42.18	-	-
SPOTER [48]	63.18	43.78	-	-
Our's	80.9	64.21	49.46	38.65

Figure 8 illustrates how our proposed initiative, which employs hand-crafted feature engineering techniques before the gloss prediction transformer, consistently increased its recognition accuracy in the top k classes WLASL datasets by about 17% compared to prior state-of-the-art models. We have provided a gloss prediction example from our model for ease of comprehension. Although our model is pose-based, we have taken into account key retrieved RGB color mode frames. Table 6 shows that 84.8% of all occurrences presented under this gloss class group were properly predicted, including the tiniest variation, "Baby". Additionally, in contrast to previous pose-based models, the average inference time during validation was 0.03 s. Our approach also performed well on datasets with few instances. In comparison to the previous pose-based architecture, the top 5% and top 10% recognition accuracy for all the WLASL model subsets exhibited a consistent growth of 4 to 10%. In comparison to appearance-based systems such as I3D and ST-GCN, our Sign2Pose Gloss prediction transformer proved to be significantly more suitable for applications in the real world in terms of model size and speed.

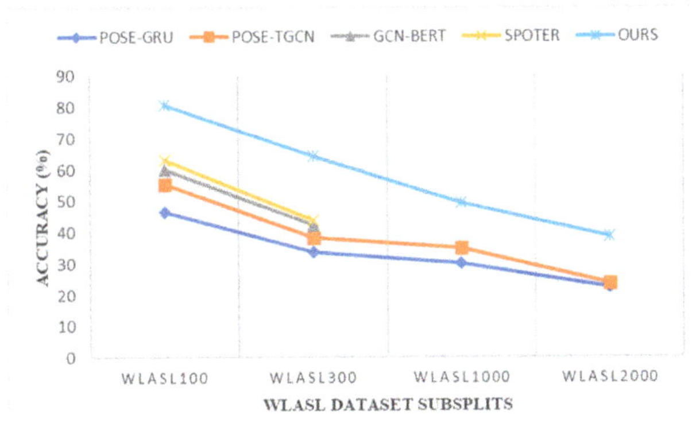

Figure 8. Comparison of the pose-based approaches' top 1 accuracies (%) and scalability on four subsets of the WLASL dataset.

Table 6. Top 1% accuracy of the predicted gloss matching ground truth label.

Extracted Key-Frames	Top 5 Predicted Gloss	Top 1% Accuracy	Ground Truth
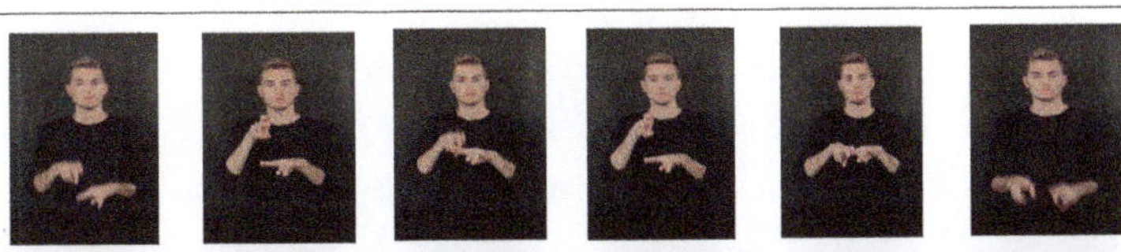	Connect Cut **Chair** Seat Sit	93.6%	Chair
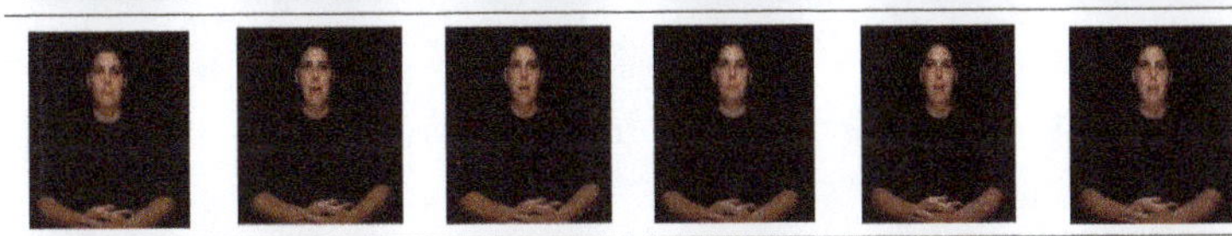	Swing **Baby** Tummy Swaddle Platter	84.8%	Baby
	**Neck** Collar Necklace Lip Smash	88.5%	Neck
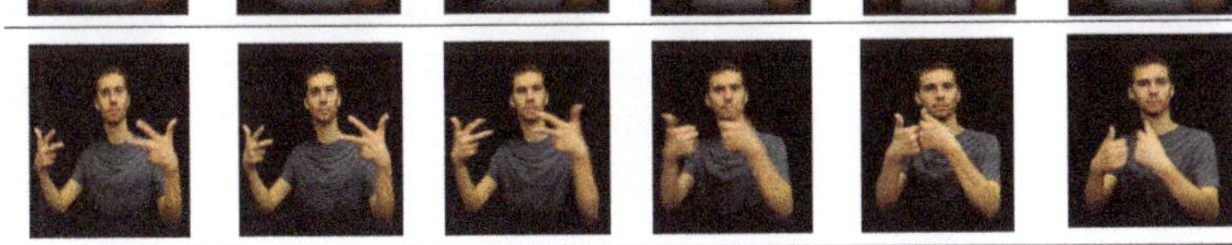	Collide Hit **Match** Unite Relate	90.35%	Match

Figure 8 shows that our suggested strategy consistently beat all other existing models, regardless of the size of the datasets, with an improvement of 17 to 20% over the methods now in use. In the appearance-based model, they face difficulty in predicting gloss words with slight variations in their hand orientation. Additionally, difficulty arises in detecting the bounding box when the sign is performed by the signer in the side view angle. When considering pose-based models, prior approaches could not fully benefit either from normalization or optimizations other than the regular ones carried out to visual data. As our model relies on body pose sequence representations as its foundation, we used insights from sign linguistics to develop a solid normalization methodology using YOLOv3 as well as fresh data augmentation methods tailored to sign language.

7. Conclusions

A real-world challenge for CSLR/SLT systems is determining the clear boundaries of sign words from sign videos. Although alternative techniques have been put out in the past to resolve the end-to-end translation model to ensure efficient mapping of anticipated words with the target sentence, there remain some limitations with the current systems. The performance of the SLT systems' translation is significantly improved by intermediate gloss prediction. In this paper, we proposed a novel approach for gloss prediction using the Sign2Pose Gloss prediction transformer. Instead of relying on pre-trained models to tackle gloss prediction, we used hand-crafted techniques for pose feature extraction that not only predicted gloss more precisely but also decrease processing overheads. With the help of a SignPose2 Gloss prediction transformer, we provided a novel approach for effective gloss prediction that considerably identifies the intermediate gloss for the provided input video sequence. For efficient gloss prediction by our proposed architecture, we used a modified HD algorithm for key-frame extraction to differentiate key frames from redundant frames. We also employed the Euclidean distance algorithm to sort the key-frames and redundant frames based on the threshold value. Further, we equipped our model with augmentation steps, making it more adaptable to any real-time dataset. YOLO v3 was then applied to the pose vectors to detect the precise movements of the hand. The use of YOLO v3 brought a drastic improvement of about 15–20% in our model accuracy which surpassed all the current pose-based methods. In all subsets of the word-level ASL data corpus, our model produced more state-of-the-art results than other pose-based approaches. In the future, we plan to amplify our model with modern skeleton frameworks that allow for further efficient continuous sign translation from intermediate gloss representations. We will also evaluate the proposed work and future frameworks using large-scale annotated data corpora such as AUTSL, MS-ASL, and others.

Author Contributions: Conceptualization, J.E. and D.J.H.; methodology, J.E.; validation, J.E., D.J.H., A.J. and Y.S.; formal analysis, J.E.; investigation, J.E. and D.J.H.; resources, A.J. and Y.S.; data curation, J.E. and D.J.H.; writing—original draft preparation, J.E.; writing—review and editing, D.J.H., A.J. and Y.S.; supervision, D.J.H. All authors have read and agreed to the published version of the manuscript.

Funding: This work was supported by the JSPS KAKENHI grant numbers JP21H03496 and JP22K12157 and supported by JST, PRESTO grant number JPMJPR1934.

Institutional Review Board Statement: Not applicable.

Informed Consent Statement: Not applicable.

Data Availability Statement: The dataset used in this research is publicly available at https://dxli94.github.io/WLASL/ accessed on 1 March 2023.

Acknowledgments: We thank the editors and reviewers for their insightful comments and suggestions to improve the paper. All the authors acknowledge their institution's support.

Conflicts of Interest: The authors declare no conflict of interest.

References

1. Asst professor, B.M.; Dept, C. Automatic Sign Language Finger Spelling Using Convolution Neural Network: Analysis. *Int. J. Pure Appl. Math.* **2017**, *117*, 9–15.
2. Jennifer Eunice R, H.D.J. Deep CNN for Static Indian Sign Language Digits Recognition. In *Frontiers in Artificial Intelligence and Applications*; IOS Press: Amsterdam, The Netherlands, 2022; Volume 347, pp. 437–446.
3. Chajri, Y.; Bouikhalene, B. Handwritten mathematical symbols dataset. *Data Br.* **2016**, *7*, 432–436. [CrossRef] [PubMed]
4. Huang, J.; Zhou, W.; Zhang, Q.; Li, H.; Li, W. Video-based sign language recognition without temporal segmentation. In Proceedings of the 32nd Proceedings of the AAAI Conference on Artificial Intelligence, New Orleans, LA, USA, 2–7 February 2018; pp. 2257–2264.
5. Tolentino, L.K.S.; Serfa Juan, R.O.; Thio-ac, A.C.; Pamahoy, M.A.B.; Forteza, J.R.R.; Garcia, X.J.O. Static sign language recognition using deep learning. *Int. J. Mach. Learn. Comput.* **2019**, *9*, 821–827. [CrossRef]
6. Liao, Y.; Xiong, P.; Min, W.; Min, W.; Lu, J. Dynamic Sign Language Recognition Based on Video Sequence with BLSTM-3D Residual Networks. *IEEE Access* **2019**, *7*, 38044–38054. [CrossRef]

7. Kumar, P.; Gauba, H.; Roy, P.P.; Dogra, D.P. Coupled HMM-based Multi-Sensor Data Fusion for Sign Language Recognition. *Pattern Recognit. Lett.* **2016**, *86*, 1–8. [CrossRef]
8. Chabchoub, A.; Hamouda, A.; Al-Ahmadi, S.; Barkouti, W.; Cherif, A. Hand Sign Language Feature Extraction Using Image Processing. *Adv. Intell. Syst. Comput.* **2020**, *1070*, 122–131. [CrossRef]
9. Ong, E.J.; Bowden, R. A boosted classifier tree for hand shape detection. In Proceedings of the Sixth IEEE International Conference on Automatic Face and Gesture Recognition, Seoul, Republic of Korea, 19 May 2004; pp. 889–894. [CrossRef]
10. Charles, J.; Pfister, T.; Everingham, M.; Zisserman, A. Automatic and efficient human pose estimation for sign language videos. *Int. J. Comput. Vis.* **2014**, *110*, 70–90. [CrossRef]
11. Liu, J.; Liu, B.; Zhang, S.; Yang, F.; Yang, P.; Metaxas, D.N.; Neidle, C. Non-manual grammatical marker recognition based on multi-scale, spatio-temporal analysis of head pose and facial expressions. *Image Vis. Comput.* **2014**, *32*, 671–681. [CrossRef]
12. Cheng, K.L.; Yang, Z.; Chen, Q.; Tai, Y.W. Fully Convolutional Networks for Continuous Sign Language Recognition. In *Lecture Notes in Computer Science*; Including subseries Lecture Notes in Artificial Intelligence and Lecture Notes in Bioinformatics; Springer: Berlin/Heidelberg, Germany, 2020; Volume 12369 LNCS, pp. 697–714. [CrossRef]
13. Koller, O.; Ney, H.; Bowden, R. Deep Hand: How to Train a CNN on 1 Million Hand Images When Your Data Is Continuous and Weakly Labelled. In Proceedings of the IEEE Conference on Computer Vision and Pattern Recognition, Las Vegas, NV, USA, 27–30 June 2016.
14. Koller, O.; Zargaran, S.; Ney, H. Resign: Re-aligned end-to-end sequence modelling with deep recurrent CNN-HMMs. In Proceedings of the 30th IEEE Conference on Computer Vision and Pattern Recognition (CVPR), Honolulu, HI, USA, 21 July 2017–26 July 2017; pp. 3416–3424. [CrossRef]
15. Zhang, F.; Sheng, J. Gesture Recognition Based on CNN and DCGAN for Calculation and Text Output. *IEEE Access* **2019**, *7*, 28230–28237. [CrossRef]
16. Rastgoo, R.; Kiani, K.; Escalera, S. Word separation in continuous sign language using isolated signs and post-processing. *arXiv* **2022**, arXiv:2204.00923.
17. Guo, D.; Zhou, W.; Li, H.; Wang, M. Hierarchical LSTM for sign language translation. In Proceedings of the 32nd AAAI Conference on Artificial Intelligence, New Orleans, LA, USA, 2–7 February 2018; pp. 6845–6852.
18. Agha, R.A.A.R.; Sefer, M.N.; Fattah, P. A comprehensive study on sign languages recognition systems using (SVM, KNN, CNN and ANN). In Proceedings of the Proceedings of the First International Conference on Data Science, E-learning and Information Systems-DATA'18, New York, NY, USA, 1–2 October 2018; ACM Press: New York, NY, USA, 2018; pp. 1–6.
19. Rahim, M.A.; Islam, M.R.; Shin, J. Non-touch sign word recognition based on dynamic hand gesture using hybrid segmentation and CNN feature fusion. *Appl. Sci.* **2019**, *9*, 3790. [CrossRef]
20. Wu, Y.; Zhou, Y.; Zeng, W.; Qian, Q.; Song, M. An Attention-based 3D CNN with Multi-scale Integration Block for Alzheimer' s Disease Classification. *IEEE J. Biomed. Health Inform.* **2022**, *26*, 5665–5673. [CrossRef] [PubMed]
21. Neto, G.M.R.; Junior, G.B.; de Almeida, J.D.S.; de Paiva, A.C. Sign Language Recognition Based on 3D Convolutional Neural Networks. In *Lecture Notes in Computer Science*; Including Subseries Lecture Notes in Artificial Intelligence and Lecture Notes in Bioinformatics; Springer: Berlin/Heidelberg, Germany, 2018; Volume 10882 LNCS, pp. 399–407.
22. Sutskever, I.; Vinyals, O.; Le, Q.V. Sequence to sequence learning with neural networks. *Adv. Neural Inf. Process. Syst.* **2014**, *4*, 3104–3112.
23. Chen, Y.; Wei, F.; Sun, X.; Wu, Z.; Lin, S. A Simple Multi-Modality Transfer Learning Baseline for Sign Language Translation. In Proceedings of the IEEE/CVF Conference on Computer Vision and Pattern Recognition, New Orleans, LA, USA, 19–20 June 2022; pp. 5110–5120. [CrossRef]
24. Camgoz, N.C.; Hadfield, S.; Koller, O.; Ney, H.; Bowden, R. Neural Sign Language Translation. In Proceedings of the IEEE Conference on Computer Vision and Pattern Recognition, Salt Lake City, UT, USA, 18–23 June 2018; pp. 7784–7793. [CrossRef]
25. Jin, T.; Zhao, Z.; Zhang, M.; Zeng, X. Findings of the Association for Computational Linguistics Prior Knowledge and Memory Enriched Transformer for Sign Language Translation. *Assoc. Comput. Linguist.* **2022**, *2022*, 3766–3775.
26. Camgoz, N.C.; Koller, O.; Hadfield, S.; Bowden, R. Sign language transformers: Joint end-to-end sign language recognition and translation. In Proceedings of the Proceedings of the IEEE/CVF Conference on Computer Vision and Pattern Recognition, Seattle, WA, USA, 13–19 June 2020; pp. 10023–10033.
27. Xu, Y.; Seneff, S. Two-Stage Translation: A Combined Linguistic and Statistical Machine Translation Framework. In Proceedings of the Conference of the Association for Machine Translation in the Americas, Waikiki, HI, USA, 21–25 October 2008.
28. Jang, J.Y.; Park, H.; Shin, S.; Shin, S.; Yoon, B.; Gweon, G. Automatic Gloss-level Data Augmentation for Sign Language Translation. In Proceedings of the 2022 Language Resources and Evaluation Conference, LREC 2022, Marseille, France, 20–25 June 2022; pp. 6808–6813.
29. Sehyr, Z.S.; Caselli, N.; Cohen-Goldberg, A.M.; Emmorey, K. The ASL-LEX 2.0 Project: A Database of Lexical and Phonological Properties for 2,723 Signs in American Sign Language. *J. Deaf Stud. Deaf Educ.* **2021**, *26*, 263–277. [CrossRef]
30. Caselli, N.K.; Sehyr, Z.S.; Cohen-Goldberg, A.M.; Emmorey, K. ASL-LEX: A lexical database of American Sign Language. *Behav. Res. Methods* **2017**, *49*, 784–801. [CrossRef]
31. Vaswani, A.; Shazeer, N.M.; Parmar, N.; Uszkoreit, J.; Jones, L.; Gomez, A.N.; Kaiser, L.; Polosukhin, I. Attention is All you Need. *arXiv* **2017**, arXiv:1706.03762.

32. Koller, O.; Zargaran, S.; Ney, H.; Bowden, R. Deep sign: Hybrid CNN-HMM for continuous sign language recognition. In Proceedings of the British Machine Vision Conference 2016, York, UK, 19–22 September 2016; pp. 136.1–136.12. [CrossRef]
33. Wu, D.; Pigou, L.; Kindermans, P.J.; Le, N.D.H.; Shao, L.; Dambre, J.; Odobez, J.M. Deep Dynamic Neural Networks for Multimodal Gesture Segmentation and Recognition. *IEEE Trans. Pattern Anal. Mach. Intell.* **2016**, *38*, 1583–1597. [CrossRef]
34. Koller, O.; Forster, J.; Ney, H. Continuous sign language recognition: Towards large vocabulary statistical recognition systems handling multiple signers. *Comput. Vis. Image Underst.* **2015**, *141*, 108–125. [CrossRef]
35. Cui, R.; Liu, H.; Zhang, C. A Deep Neural Framework for Continuous Sign Language Recognition by Iterative Training. *IEEE Trans. Multimed.* **2019**, *21*, 1880–1891. [CrossRef]
36. Sharma, S.; Gupta, R.; Kumar, A. Continuous sign language recognition using isolated signs data and deep transfer learning. *J. Ambient Intell. Humaniz. Comput.* **2021**, *1*, 1531–1542. [CrossRef]
37. Niu, Z.; Mak, B. Stochastic Fine-Grained Labeling of Multi-state Sign Glosses for Continuous Sign Language Recognition. In Proceedings of the Computer Vision—ECCV 2020, Glasgow, UK, 23–28 August 2020; Vedaldi, A., Bischof, H., Brox, T., Frahm, J.-M., Eds.; Springer International Publishing: Cham, Switzerland, 2020; pp. 172–186.
38. Tunga, A.; Nuthalapati, S.V.; Wachs, J. Pose-based Sign Language Recognition using GCN and BERT. In Proceedings of the IEEE/CVF Winter Conference on Applications of Computer Vision, Waikoloa, HI, USA, 3–8 January 2021; pp. 31–40. [CrossRef]
39. Cui, R.; Liu, H.; Zhang, C. Recurrent convolutional neural networks for continuous sign language recognition by staged optimization. In Proceedings of the 30th IEEE Conference on Computer Vision and Pattern Recognition (CVPR), Honolulu, HI, USA, 21–26 July 2016; pp. 1610–1618. [CrossRef]
40. Zhao, J.; Qi, W.; Zhou, W.; Duan, N.; Zhou, M.; Li, H. Conditional Sentence Generation and Cross-Modal Reranking for Sign Language Translation. *IEEE Trans. Multimed.* **2022**, *24*, 2662–2672. [CrossRef]
41. Kim, Y.; Kwak, M.; Lee, D.; Kim, Y.; Baek, H. Keypoint based Sign Language Translation without Glosses. *arXiv* **2022**, arXiv:2204.10511.
42. Du, Y.; Xie, P.; Wang, M.; Hu, X.; Zhao, Z.; Liu, J. Full transformer network with masking future for word-level sign language recognition. *Neurocomputing* **2022**, *500*, 115–123. [CrossRef]
43. Camgöz, N.C.; Koller, O.; Hadfield, S.; Bowden, R. Sign Language Transformers: Joint End-to-end Sign Language Recognition and Translation. *arXiv* **2020**, arXiv:2003.13830v1.
44. Ko, S.K.; Kim, C.J.; Jung, H.; Cho, C. Neural sign language translation based on human keypoint estimation. *Appl. Sci.* **2019**, *9*, 2683. [CrossRef]
45. Read, J.; Polytechnique, E. Better Sign Language Translation with STMC-Transformer. *arXiv* **2017**, arXiv:2004.00588.
46. Walczynska, J. HandTalk: American Sign Language Recognition by 3D-CNNs. Ph.D. Thesis, University of Groningen, Groningen, The Netherlands, 2022.
47. Papastratis, I.; Dimitropoulos, K.; Daras, P. Continuous Sign Language Recognition through a Context-Aware Generative Adversarial Network. *Sensors* **2021**, *21*, 2437. [CrossRef]
48. Bohacek, M.; Hruz, M. Sign Pose-based Transformer for Word-level Sign Language Recognition. In Proceedings of the IEEE/CVF Winter Conference on Applications of Computer Vision (WACV) Workshops, Waikoloa, HI, USA, 4–8 January 2022; pp. 182–191. [CrossRef]
49. Inan, M.; Zhong, Y.; Hassan, S.; Quandt, L.; Alikhani, M. Modeling Intensification for Sign Language Generation: A Computational Approach. *arXiv* **2022**, arXiv:2203.09679.
50. Jiang, S.; Sun, B.; Wang, L.; Bai, Y.; Li, K.; Fu, Y. Sign Language Recognition via Skeleton-Aware Multi-Model Ensemble. *arXiv* **2021**, arXiv:2110.06161v1.
51. Yao, P. Key Frame Extraction Method of Music and Dance Video Based on Multicore Learning Feature Fusion. *Sci. Program.* **2022**, *2022*, 9735392. [CrossRef]
52. Wang, J.; Zeng, C.; Wang, Z.; Jiang, K. An improved smart key frame extraction algorithm for vehicle target recognition. *Comput. Electr. Eng.* **2022**, *97*, 107540. [CrossRef]
53. Li, Z.; Li, Y.; Tan, B.; Ding, S.; Xie, S. Structured Sparse Coding With the Group Log-regularizer for Key Frame Extraction. *IEEE/CAA J. Autom. Sin.* **2022**, *9*, 1818–1830. [CrossRef]
54. Nie, B.X.; Xiong, C.; Zhu, S.C. Joint action recognition and pose estimation from video. In Proceedings of the IEEE Conference on Computer Vision and Pattern Recognition, Boston, MA, USA, 7–12 June 2015; pp. 1293–1301. [CrossRef]
55. Gan, S.; Yin, Y.; Jiang, Z.; Xie, L.; Lu, S. Skeleton-Aware Neural Sign Language Translation. In Proceedings of the 29th ACM International Conference on Multimedia, Virtual, 20–24 October 2021; pp. 4353–4361. [CrossRef]
56. Novopoltsev, M.; Verkhovtsev, L.; Murtazin, R.; Milevich, D.; Zemtsova, I. Fine-tuning of sign language recognition models: A technical report. *arXiv* **2023**, arXiv:2302.07693. [CrossRef]
57. Shalev-Arkushin, R.; Moryossef, A.; Fried, O. Ham2Pose: Animating Sign Language Notation into Pose Sequences. *arXiv* **2022**, arXiv:2211.13613. [CrossRef]
58. Liu, F.; Dai, Q.; Wang, S.; Zhao, L.; Shi, X.; Qiao, J. Multi-relational graph convolutional networks for skeleton-based action recognition. In Proceedings of the 2020 IEEE Intl Conf on Parallel & Distributed Processing with Applications, Big Data & Cloud Computing, Sustainable Computing & Communications, Social Computing & Networking (ISPA/BDCloud/SocialCom/SustainCom), Exeter, UK, 17–19 December 2020; pp. 474–480. [CrossRef]

59. De Coster, M.; Van Herreweghe, M.; Dambre, J. Isolated sign recognition from RGB video using pose flow and self-attention. In Proceedings of the IEEE/CVF Conference on Computer Vision and Pattern Recognition, Nashville, TN, USA, 20–25 June 2021; pp. 3436–3445. [CrossRef]
60. Li, D.; Opazo, C.R.; Yu, X.; Li, H. Word-level deep sign language recognition from video: A new large-scale dataset and methods comparison. In Proceedings of the IEEE/CVF Winter Conference on Applications of Computer Vision, Snowmass Village, CO, USA, 1–5 March 2020; pp. 1448–1458. [CrossRef]
61. Madadi, M.; Escalera, S.; Carruesco, A.; Andujar, C.; Baró, X.; Gonzàlez, J. Occlusion Aware Hand Pose Recovery from Sequences of Depth Images. In Proceedings of the 2017 12th IEEE International Conference on Automatic Face & Gesture Recognition (FG 2017), Washington, DC, USA, 30 May 2017–3 June 2017; pp. 230–237.
62. Joze, H.R.V.; Koller, O. MS-ASL: A large-scale data set and benchmark for understanding American sign language. In Proceedings of the 30th British Machine Vision Conference 2019, BMVC 2019, Cardiff, UK, 9–12 September 2019.
63. Kagirov, I.; Ivanko, D.; Ryumin, D.; Axyonov, A.; Karpov, A. TheRuSLan: Database of Russian sign language. In Proceedings of the 12th Language Resources and Evaluation Conference, Marseille, France, 11–16 May 2020; pp. 6079–6085.
64. Sincan, O.M.; Keles, H.Y. AUTSL: A large scale multi-modal Turkish sign language dataset and baseline methods. *IEEE Access* **2020**, *8*, 181340–181355. [CrossRef]
65. Pishchulin, L.; Insafutdinov, E.; Tang, S.; Andres, B.; Andriluka, M.; Gehler, P.; Schiele, B. DeepCut: Joint subset partition and labeling for multi person pose estimation. In Proceedings of the IEEE Conference on Computer Vision and Pattern Recognition, Las Vegas, NV, USA, 27–30 June 2016; pp. 4929–4937. [CrossRef]
66. Feng, J.; Wang, X.; Liu, W. Deep graph cut network for weakly-supervised semantic segmentation. *Sci. China Inf. Sci.* **2021**, *64*, 130105. [CrossRef]
67. Li, M.; Qin, J.; Li, D.; Chen, R.; Liao, X.; Guo, B. VNLSTM-PoseNet: A novel deep ConvNet for real-time 6-DOF camera relocalization in urban streets. *Geo-Spatial Inf. Sci.* **2021**, *24*, 422–437. [CrossRef]
68. Kitamura, T.; Teshima, H.; Thomas, D.; Kawasaki, H. Refining OpenPose with a new sports dataset for robust 2D pose estimation. In Proceedings of the IEEE/CVF Winter Conference on Applications of Computer Vision, Waikoloa, HI, USA, 3–8 January 2022; pp. 672–681. [CrossRef]
69. Bauer, A. *The Use of Signing Space in a Shared Sign Language of Australia*; De Gruyter Mouton: Berlin, Germany, 2013; ISBN 9781614515470.
70. Senanayaka, S.A.M.A.S.; Perera, R.A.D.B.S.; Rankothge, W.; Usgalhewa, S.S.; Hettihewa, H.D.; Abeygunawardhana, P.K.W. Continuous American Sign Language Recognition Using Computer Vision And Deep Learning Technologies. In Proceedings of the 2022 IEEE Region 10 Symposium (TENSYMP), Mumbai, India, 1-03 July 2022; IEEE: Piscataway, NJ, USA, 2022; pp. 1–6.
71. Maruyama, M.; Singh, S.; Inoue, K.; Roy, P.P.; Iwamura, M.; Yoshioka, M. Word-Level Sign Language Recognition with Multi-Stream Neural Networks Focusing on Local Regions and Skeletal Information. *arXiv* **2021**, arXiv:2106.15989. [CrossRef]

sensors

MDPI

Article

Sensor-Based Activity Recognition Using Frequency Band Enhancement Filters and Model Ensembles

Hyuga Tsutsumi, Kei Kondo, Koki Takenaka and Tatsuhito Hasegawa *

Graduate School of Engineering, University of Fukui, Fukui 910-8507, Japan
* Correspondence: t-hase@u-fukui.ac.jp

Abstract: Deep learning methods are widely used in sensor-based activity recognition, contributing to improved recognition accuracy. Accelerometer and gyroscope data are mainly used as input to the models. Accelerometer data are sometimes converted to a frequency spectrum. However, data augmentation based on frequency characteristics has not been thoroughly investigated. This study proposes an activity recognition method that uses ensemble learning and filters that emphasize the frequency that is important for recognizing a certain activity. To realize the proposed method, we experimentally identified the important frequency of various activities by masking some frequency bands in the accelerometer data and comparing the accuracy using the masked data. To demonstrate the effectiveness of the proposed method, we compared its accuracy with and without enhancement filters during training and testing and with and without ensemble learning. The results showed that applying a frequency band enhancement filter during training and testing and ensemble learning achieved the highest recognition accuracy. In order to demonstrate the robustness of the proposed method, we used four different datasets and compared the recognition accuracy between a single model and a model using ensemble learning. As a result, in three of the four datasets, the proposed method showed the highest recognition accuracy, indicating the robustness of the proposed method.

Keywords: frequency emphasis; ensemble learning; deep learning

Citation: Tsutsumi, H.; Kondo, K.; Takenaka, K.; Hasegawa, T. Sensor-Based Activity Recognition Using Frequency Band Enhancement Filters and Model Ensembles. *Sensors* **2023**, *23*, 1465. https://doi.org/10.3390/s23031465

Academic Editor: Giovanni Saggio

Received: 17 November 2022
Revised: 19 January 2023
Accepted: 23 January 2023
Published: 28 January 2023

1. Introduction

In recent years, the widespread use of smartphones and wearable devices has facilitated user activity sensing. These devices can perform activity recognition using accelerometer and gyroscope data as time-series data [1,2]. Activity recognition can be used, for example, to determine a user's health status [3,4]. Activity recognition technology can also be applied to sports such as volleyball and badminton [5,6]. For enhanced service applications, it is desirable to recognize activities accurately and in detail. For this purpose, Sikder et al. [7] transformed accelerometer and gyroscope data into frequency and power spectrum and used them as input to a convolutional neural network (CNN) to classify six types of activities. That study used the frequency spectrum as input for the model and evaluate recognition accuracy but did not consider the difference in frequency characteristics between activity. Other studies have focused on the frequency characteristics of activities. Ooue et al. [8] converted accelerometer data into a power spectrum to determine the frequency characteristics of different walking patterns and found that they differed between normal walking and walking with a limp. Therefore, it is likely that the frequency characteristics of each activity will differ in activity recognition, and there may be important frequencies for the prediction of each activity. Liu et al. [9] analyzed the power spectrum of input data to obtain the major frequency bands and proposed a tree-structured wavelet neural network (T-WaveNet) for time-series signal analysis but did not perform frequency enhancement of the input data. In this study, we propose an activity recognition method that identifies the important frequency for recognizing a certain activity, applies a filter that emphasizes each frequency in the input, and performs ensemble learning during training and testing. The aim is to improve the accuracy of activity recognition and to facilitate the development of various activity recognition services using a general-purpose method based on frequency enhancement and ensemble learning. As discussed below, parts of studies on activity recognition proposed converting accelerometer data into a frequency spectrum as the input of CNN. In contrast, this study makes the following contributions:

- We experimentally identified the important frequency of various activities using the Human Activity Sensing Consortium (HASC) activity recognition dataset [10].
- We developed a new method to improve the accuracy of activity recognition by creating a filter that emphasizes the important frequency of each activity and applying it to training and testing data, training the model with the data, and using ensemble learning.

2. Related Research

2.1. Sensor-Based Activity Recognition

Various methods for sensor-based activity recognition, including CNN and ensemble learning, have been developed. Shaohua et al. [11] used three-axis smartphone accelerometer data to perform activity recognition using CNN, Long Short-Term Memory (LSTM), Bidirectional LSTM, Multilayer Perceptron, and support vector machine (SVM) models, and compared their accuracies using two large datasets. According to their experimental results, the CNN model had the highest accuracy. Ito et al. [12] performed Fourier transform processing of accelerometer and gyroscope data to create a spectrum image, which was used as input to a CNN model for activity recognition. This model had three convolutional layers and three pooling layers. After integrating the features of the spectrogram images from the accelerometer and gyroscope, classification was performed on all three fully-connected layers. The best convolution size was obtained by comparing the accuracy of different convolution sizes in the time and frequency directions. Subasi et al. [13] used ensemble learning to classify seven types of activities using random forest and SVM methods and compared their recognition accuracies with that of activity recognition using Adaptive Boosting combined with these methods. Sakorn et al. [14] used acceleration and gyro data collected by smartphones for activity recognition. They proposed a method that combines a 4-layer CNN and an LSTM network, and showed that it improves the average accuracy by up to 2.24% compared to state-of-the-art methods. Others have proposed models that combine CNNs and BiGRUs, and have shown to significantly outperform the recognition accuracy of other RNN models [15]. Nadeem et al. [16] proposed a method for extracting optimal features using sequential floating forward search (SFFS), and showed that the recognition accuracy is about 6% higher than when no features are selected. Muhammad et al. [17] proposed a two-level model and performed data recognition when multiple activities are combined. All these studies used data obtained from accelerometers and gyroscopes as input or spectrogram images to recognize activity. However, none of them used data that utilized the characteristics of each activity.

2.2. Frequency Characteristics in Activity Recognition

Some studies have used the frequency characteristics of activities. Yoshizawa et al. [18] used an Infinite Impulse Response (IIR) bandpass filter to detect change points from one moving activity to another. A change point was detected when the sum of the fluctuations of each component of the accelerometer data exceeded a certain value. The authors also identified the important frequency by changing the spectrum coefficients used in the change point detection method to determine the number of filters and pass frequencies of the IIR bandpass filter. Fujiwara et al. [19] applied short-term Fourier transform to Doppler sensor data to calculate the frequency components as features used to construct a lifestyle activity recognition model. To reduce the dimensionality of the feature values, they used only a portion of the frequency components. They determined the frequency components reduced by examining changes in recognition accuracy while reducing high- and low-frequency components. They found that recognition accuracy was highest when the bandwidth of the frequency components used as feature values ranged from 0 to 5 Hz. These studies have demonstrated that there is an important frequency for activity recognition. However, such frequency has been used mainly for model analysis or feature reduction and rarely for improving the accuracy of activity recognition.

2.3. Activity Recognition Using Ensemble Learning

Irvine et al. [20] proposed a neural network ensemble learning method for the recognition of daily activities in a smart home. Zhu et al. [21] used an ensemble learning of two CNN models to classify seven types of activities. First, they made predictions using a model that classified the seven types. Subsequently, if the results were of two specific classes, they made predictions using another model that classified these two types. They then obtained the final output by performing weighted voting on the outputs of the two models. Yiming et al. [22] proposed a method that combines extreme learning machines (ELMs) with pairwise diversity measure and glowworm swarm optimization-based selective ensemble learning (DMGSOSEN), which achieves higher recognition accuracy with fewer models than the comparison method. Other methods include a CELearning model using multiple layers of four different classifiers [23], an ensemble learning model using Adaboost and SVM [24], a model combining gated recurrent units (GRU), CNN, and deep neural networks (DNN) [25], and ensemble learning with multiple deep learning models [26]. Another study [27] applied multiple data augmentation to input data to perform activity recognition using ensemble learning but did not focus on frequency characteristics.

3. Proposed Method

Figure 1 shows an overview of the proposed method. The proposed method improves the accuracy of activity recognition by identifying important frequency bands for each activity, creating a filter to enhance them, and applying each technique (DA: frequency emphasis in training, TTA: frequency emphasis in testing, and EL: ensemble learning). The method consists of three phases described in Sections 3.1–3.3.

3.1. Phase 1: Finding the Important Frequency for Each Activity

In this phase, the important frequency for each activity is obtained as follows:

1. The CNN model M is trained using the original accelerometer data x_{train} as in general activity recognition.
2. For acceleration data x_{valid} the subjects of which differ from that of x_{train}, some frequencies are masked by changing f in Equation (1) between (0, $f_s/2$]:

$$x' = F_m(x, f) = ifft(P(fft(x), f)). \tag{1}$$

3. Using the model M trained in step 1, the change in the recognition accuracy of the data masked in step 2 is examined.
4. Step 3 is performed for each activity c to obtain the set of frequency bands to be emphasized: $\mathcal{F} = \{f^c|_c \in \mathcal{C}\}$ (Figure 1a).

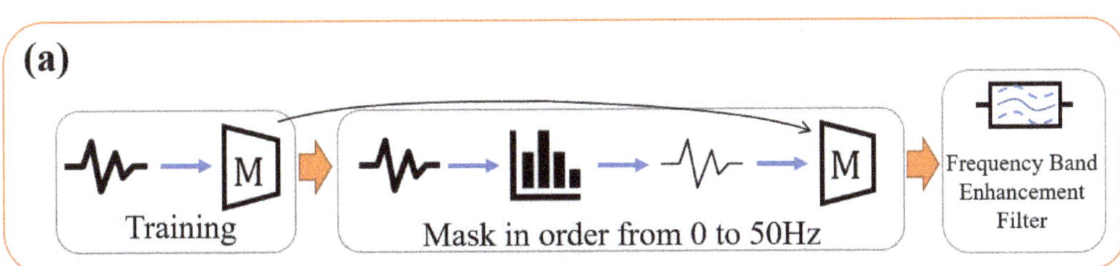

Figure 1. *Cont.*

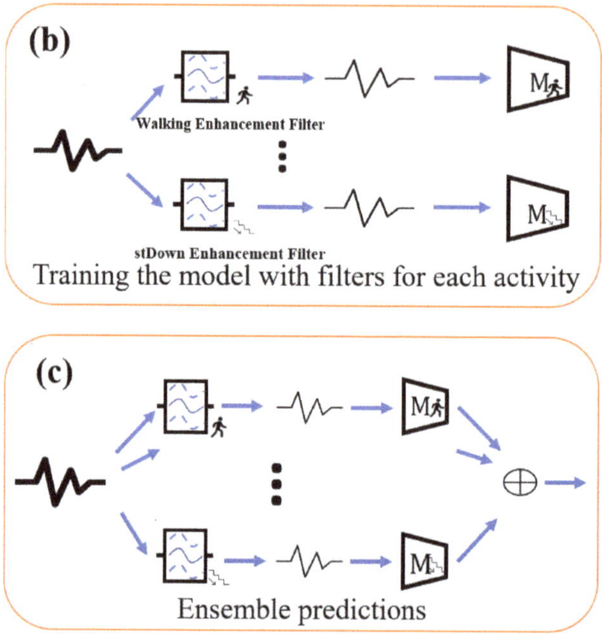

Figure 1. Overview of the proposed method. (**a**) Phase 1: Finding the important frequencies of each activity; (**b**) Phase 2: Emphasis during training; (**c**) Phase 3: Emphasis during testing.

Note that $x \in \mathbb{R}^{3\times w}$ is the triaxial accelerometer data (w is the window size), $fft(\cdot)$ is the Fourier transform, $ifft(\cdot)$ is the inverse Fourier transform, $P(\cdot)$ is the process of masking frequency bands, $F_m(x, f)$ is the data after mask processing, f^c is the important frequency at a given activity c, and f_s is the sampling frequency of the accelerometer data. The frequency of 0 Hz is not masked because it is a DC component. The maximum frequency to be masked is $f_s/2$ because the frequency of the Fourier-transformed data has a maximum value of 1/2 of the sampling frequency. The frequency at which the recognition accuracy decreases is considered the important frequency.

3.2. Phase 2: Emphasis during Training

In this phase, the CNN model M_c is trained on the training data x_{train} using $\mathcal{F}$ calculated as described in Section 3.1, with the frequency band enhancement filter of Equation (1) applied to the data (Figure 1b). The number of models is $|C|$ because the models are trained using data enhancing the important frequency of each activity. The frequency band weighting filter is implemented as Equation (1) where $P(\cdot)$ is the process of frequency band enhancement. The f^c obtained in Phase 1 is input to f in Equation (1).

In this study, four types of window functions were used as filters to enhance the frequency bands. Examples of the filters used are shown in Figure 2. The peak window does not change the amplitude spectrum of the important frequency of each activity as determined experimentally but multiplies the amplitude spectrum of the other frequencies by a factor of 1/2. The Gaussian window is a normally distributed window, with the important frequency of each activity as the mean and a standard deviation of 10 adjusted so that the maximum value is 1 and the minimum value is 0.5. The triangular window is a window with the amplitude spectrum of the important frequency of each activity as the vertex. The minimum value is set to 0.5. Random window is a random value of 0.5–1 applied to the (0,7.8] Hz portion of the amplitude spectrum. Using the random window, we determined whether the emphasis on the important frequency of each activity contributes to improving the accuracy of activity recognition. Figure 2 shows the filter for $f^c = 3$ Hz.

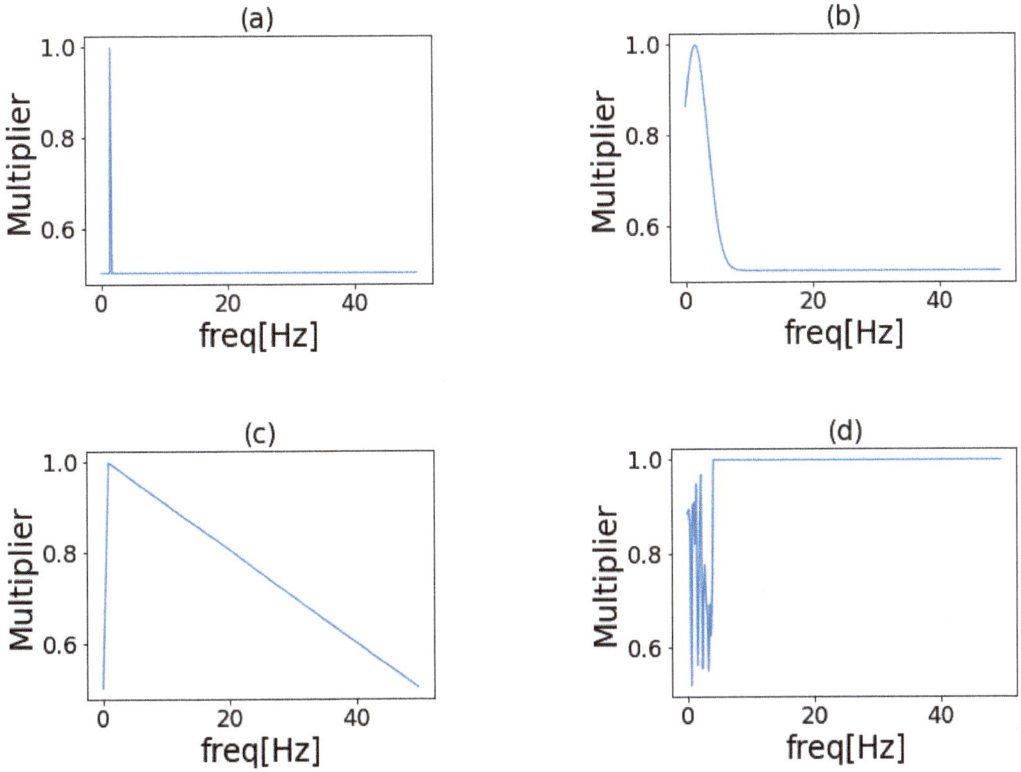

Figure 2. Examples of frequency band enhancement filters. (**a**) Peak window; (**b**) Gaussian window; (**c**) Triangular window; (**d**) Random window.

3.3. Phase 3: Emphasis during Testing

In this phase, Equation (1) was applied to the testing data x_{test} and inputted to the model M_c trained in the previous phase. The final output is the result of the majority voting on the output of each model (Figure 1c). This method can be regarded as a kind of Test Time Augmentation (TTA) [28], a method in which the testing data are processed to create several types of data, where the input data are augmented by frequency band enhancement filters.

Note that in this study, in order to eliminate differences in recognition accuracy due to differences in model structure, we used VGG16 [29] as the unified model used in Phases 1, 2, and 3.

4. Evaluation Experiment

4.1. Experiment Summary

We first conducted an experiment to determine the important frequency band for each activity. We masked some frequencies in the accelerometer data and used these data as input to the model to examine changes in accuracy and identify the important frequency (i.e., the frequency at which accuracy decreased). Next, using the obtained frequency, we created a frequency band enhancement filter for each activity and applied it to the accelerometer data. We then conducted an ablation study to evaluate the contribution of the three components of the proposed method (frequency emphasis during training, frequency emphasis during testing, and ensemble learning) to recognition accuracy.

4.2. *Experimental Setup*

4.2.1. Model Structure and Training Procedures

In the experiments, we used VGG16 as an activity recognition model modified for 1D data. To reduce the influence of the model's classifier, we applied a shallow classifier using global average pooling, and the classifier was a single fully-connected layer. In training, the batch size was set to 256, the learning rate was set to 0.001, and the number of epochs was set to 200. The kernel size was set to three, the stride width was set to one, no padding, Rectified Linear Unit was the activation function, and the pooling size was set to two.

4.2.2. Dataset

The HASC dataset was used for activity recognition. The sampling frequency was 100 Hz. We randomly sampled the acceleration data from 80 persons for training, 20 for validation, and another 30 for testing. The window size was 256 samples, divided into time series. Six activity labels were used: stay (standing still), walk (walking), jog (jogging), skip (skipping), stUp (climbing up a staircase), and stDown (climbing down a staircase). Accelerometer data contain noise; however, in this study, we assumed that the deep learning model could solve the classification problem even if the raw acceleration data have noise. As a preprocessing step, we divided the data into time series using a sliding window method, and we did not conduct further preprocessing.

4.3. *Experiment Conducted to Identify Important Frequency*

Figure 3 shows the results of the experiment conducted to determine the important frequency band for each activity. Figure 3a shows that the accuracy of stay did not change after the experiment, suggesting that the DC component at 0 Hz was important. Figure 3c shows that the accuracy increased when the frequency around 1 Hz was masked. Figure 3e shows that the recognition accuracy decreased when the frequency around 1 Hz was masked. This frequency was important for stUp. Masking presumably improved accuracy because it enabled the correct classification of jog data that had been misclassified as stUp. Table 1 shows the important frequency for each activity. Relatively slow-moving activities, such as walk and stUp, had low important frequency, while relatively fast-moving activities, such as jog and stDown, had high important frequency. Based on these results, we created a filter that emphasized the frequency around the selected frequency.

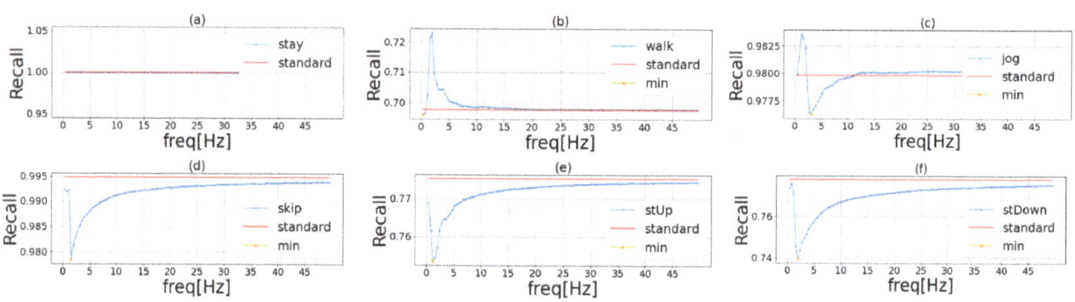

Figure 3. Important frequency of each activity in HASC. The blue line in the graph shows the recognition accuracy when we masked the frequencies of the original sensor data in order. The red line shows the recognition accuracy when we used the original data. The yellow point is the lowest recognition accuracy when we use the frequency masked data. Each activity is (**a**) stay, (**b**) walk, (**c**) jog, (**d**) skip, (**e**) stUp, and (**f**) stDown.

4.4. *Ablation Study*

4.4.1. Experimental Procedure

In the training emphasis phase, six models were trained since HASC has six different activities. In ablation study, we compared eight models listed in Table 2 to evaluate the effectiveness of DA, TTA, and EL proposed in this paper. (a) is our proposed method. (b) uses DA and TTA with a single model. (c) uses DA with EL in which each branch uses the same original sensor value in testing phase. (d) uses only DA with a single model. (e) uses TTA with EL in which each branch is trained using the same original sensor value. (f) only uses TTA. (g) is a simple ensemble learning, and (h) is a simple single model. In (b) and (f), a single model makes six predictions, applying a different enhancement filter to the test data x_{test} each time a prediction is made.

Table 1. Important frequency of each activity in HASC.

Activity	Frequency (Hz)
stay	0.00
walk	0.78
jog	3.13
skip	1.56
stUp	1.17
stDown	1.95

Table 2. Accuracy of the eight methods used in the experiment. DA, TTA, and EL are denoted by ✓for those applied and x for those not applied. The highest accuracy is underlined and bolded, and the second highest is underlined.

Method	DA	TTA	EL	Accuracy
(a)	✓	✓	✓	**<u>0.890</u>**
(b)	✓	✓	x	0.877
(c)	✓	x	✓	<u>0.881</u>
(d)	✓	x	x	0.876
(e)	x	✓	✓	0.880
(f)	x	✓	x	0.845
(g)	x	x	✓	0.880
(h)	x	x	x	0.873

4.4.2. Results

Table 2 shows the validation results: the highest accuracy in bold and underlined and the second-highest accuracy in underlined. Our proposed method (a) comprised of the ensemble learning method with the frequency band enhancement filter applied during training and testing had the highest accuracy, demonstrating the effectiveness of the proposed method. Comparing (a) with (b), (c), and (e), the difference between (b) and (a) was the largest. This suggests that ensemble learning contributed the most to the improvement in accuracy. Comparing (a) and (g), the accuracy of (a) was about 1% higher than that of (g), suggesting that applying a frequency band enhancement filter to the dataset was effective. Between (f) and (h), (f) had lower accuracy. This may be because an enhancement filter was used only during testing, and data that could not be classified by the features learned during training were inputted, resulting in low accuracy. Between (b) and (h), (b) had higher accuracy, suggesting that the use of the enhancement filter during testing was more effective when combined with its use during training.

4.5. *Effects of Window Functions*

Table 3 shows the validation results: the highest accuracy in bold and underlined and the second-highest accuracy underlined. Accuracy was highest when using the Gaussian window, suggesting that a Gaussian window is appropriate for creating a frequency band enhancement filter. The lowest accuracy was obtained when a random filter was applied, suggesting that an emphasis on the important frequency band of each activity when creating a filter contributes to higher activity recognition accuracy. Furthermore, recognition accuracy was lower when the peak window was applied than when the Gaussian window was applied, suggesting that it is more effective to emphasize the frequency around the important frequency than a single frequency.

Table 3. Accuracy when each filter was applied. The highest accuracy is underlined and bolded, the second highest is underlined.

Filter Type	Accuracy
Peak window	<u>0.890</u>
Gaussian window	<u>**0.896**</u>
Triangular window	0.888
Random	0.872

4.6. Validation Using Multiple Datasets

4.6.1. Datasets

To evaluate the robustness of the proposed method, we conducted experiments comparing some public datasets: HASC, UniMiB [30], PAMAP2 [31], and HHAR [32]. In this experiment, we adopted "VGG16" as a single baseline model, "Ensemble learning" as a simple ensemble model, and "Proposed method" combining DA using Gaussian window, TTA, and EL. In UniMiB, we randomly sampled the acceleration data from 20 persons for training, five for validation, and another five for testing. The window size was 151 samples, divided into time series. There were 17 activities in total. In PAMAP2, we randomly sampled the acceleration data from five persons for training, two for validation, and another two for testing. The window size was 256 samples, and the stride size was 128 samples for time series segmentation. There were 12 activities in total. In HHAR, we randomly sampled the acceleration data from five persons for training, two for validation, and another two for testing. The window size was 256 samples, and the stride size was 256 samples for time series partitioning. There were six different activities.

4.6.2. Results

Figures 4–6 show the results of phase 1 of the proposed method to investigate the important frequency of different activities in UniMiB, PAMAP2, and HHAR, respectively. Table 4 shows the accuracy of the three models using each dataset. The highest accuracy for each dataset is shown in bold. The proposed method had higher accuracy than the ensemble learning when using HASC, PAMAP2, and HHAR and lower accuracy than the ensemble learning when using UniMiB. Thus, the effectiveness of the proposed method was demonstrated in three of the four datasets. This indicates that the proposed method is robust in different domains. Table 4 shows that the difference in accuracy between ensemble learning and the proposed method is smaller than the difference in accuracy between VGG16 and ensemble learning. This suggests that the effect of the improvement in accuracy by ensemble learning is greater than the application of the frequency-enhancement filter. In PAMAP2 the accuracy of the proposed method is a little less than the conventional ensemble method but reaches almost the same estimation accuracy. The proposed method employs a frequency-enhanced method for each activity label compared to the conventional ensemble. Therefore, the proposed method may not be more effective than the conventional ensemble method when there are a very large number of behaviors and when similar behaviors are included. It can also be seen that the accuracy of ensemble learning is higher than the accuracy of the proposed method when the UniMiB dataset is used. This may be due to the fact that, as shown in (b), (c), and (p) in Figure 6, there are more activities with a smaller decrease in recognition accuracy when mask processing is performed than in the other datasets.

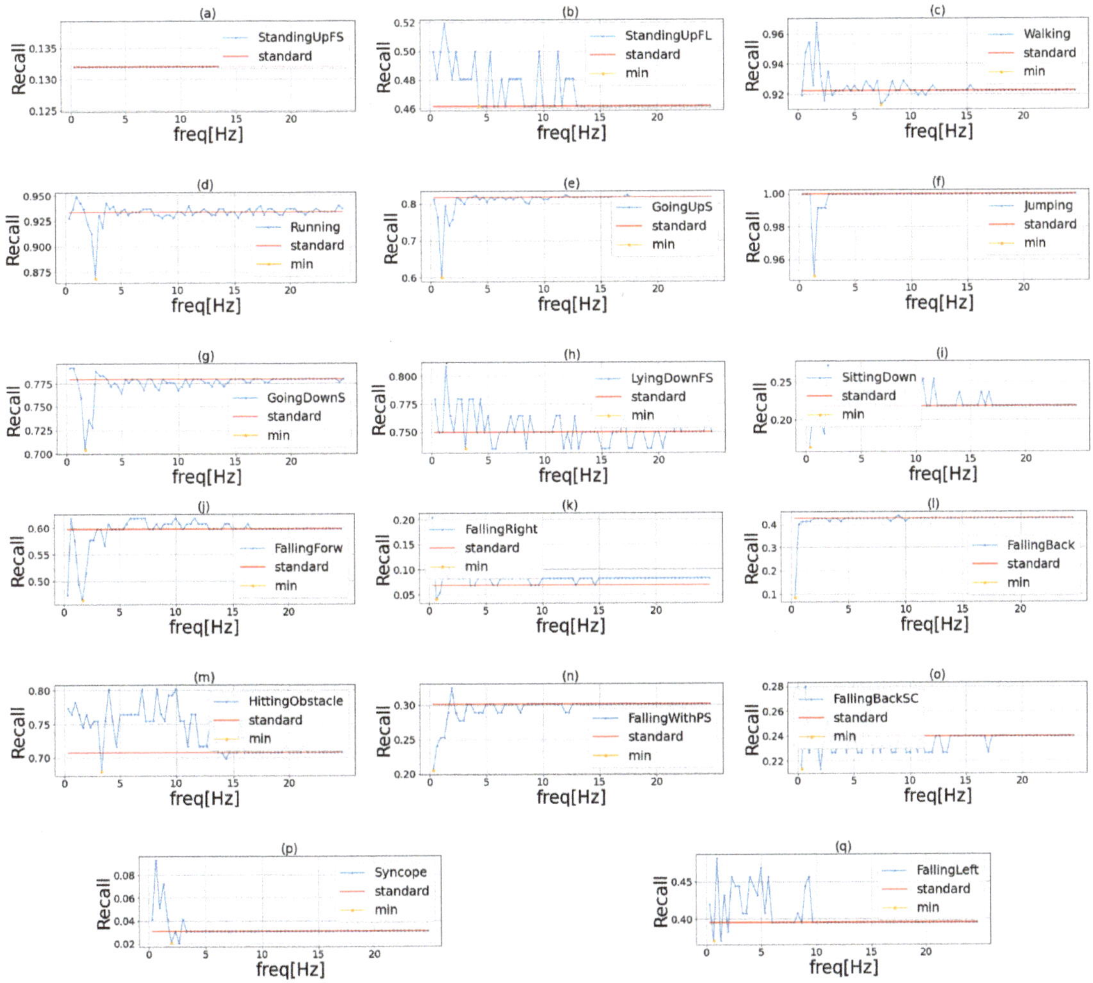

Figure 4. Important frequency of each activity in UniMiB. The blue line in the graph shows the recognition accuracy when we masked the frequencies of the original sensor data in order. The red line shows the recognition accuracy when we used the original data. The yellow point is the lowest recognition accuracy when we use the frequency masked data. Each activity is (**a**) StandingUpFS, (**b**) StandingUpFL, (**c**) Walking, (**d**) Running, (**e**) GoingUpS, (**f**) Jumping, (**g**) GoingDownS, (**h**) LyingDownFS, (**i**) SittingDown, (**j**) FallingForw, (**k**) FallingRight, (**l**) FallingBack, (**m**) HittingObstacle, (**n**) FallingWithPS, (**o**) FallingBackSC, (**p**) Syncope, (**q**) FallingLeft.

Table 4. Accuracy of the three models using each dataset. Bold type indicates the highest accuracy using the respective dataset.

Method	HASC	UniMiB	PAMAP2	HHAR
VGG16	0.873	0.663	0.787	0.715
Ensemble learning	0.880	**0.707**	0.812	0.753
Proposed method	**0.896**	0.703	**0.818**	**0.760**

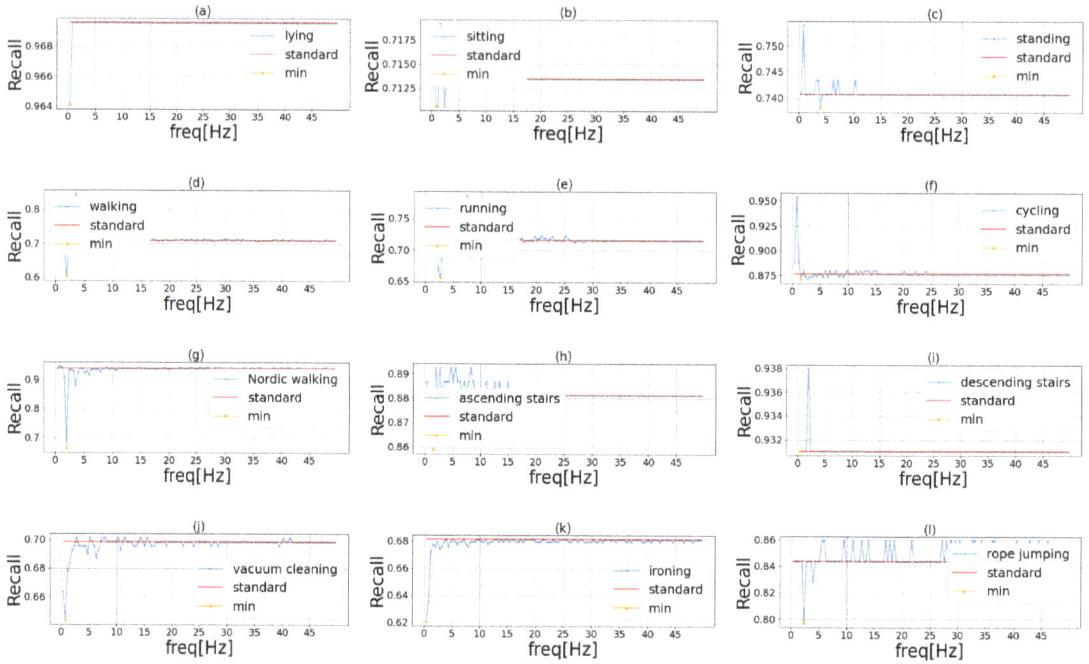

Figure 5. Important frequency of each activity in PAMAP2. The blue line in the graph shows the recognition accuracy when we masked the frequencies of the original sensor data in order. The red line shows the recognition accuracy when we used the original data. The yellow point is the lowest recognition accuracy when we used the frequency masked data. Each of these activities is (**a**) lying, (**b**) sitting, (**c**) standing, (**d**) walking, (**e**) running, (**f**) cycling, (**g**) Nordic walking, (**h**) ascending stairs, (**i**) descending stairs, (**j**) vacuum cleaning, (**k**) ironing, and (**l**) rope jumping.

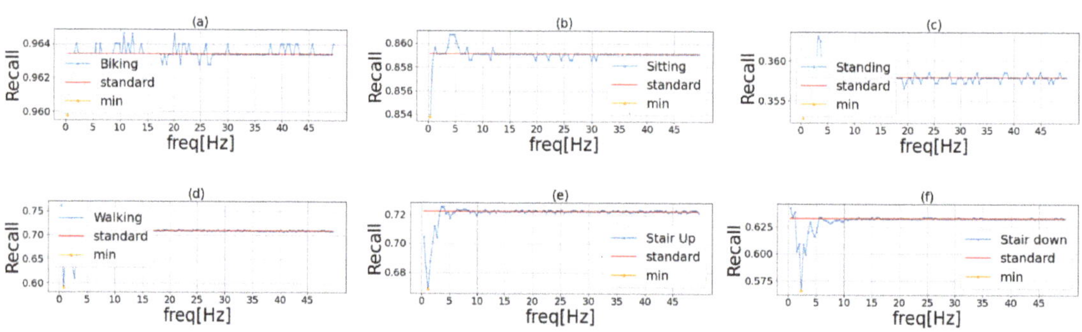

Figure 6. Important frequency of each activity in HHAR. The blue line in the graph shows the recognition accuracy when we masked the frequencies of the original sensor data in order. The red line shows the recognition accuracy when we used the original data. The yellow point is the lowest recognition accuracy when we used the frequency masked data. Each activity is (**a**) Biking, (**b**) Sitting, (**c**) Standing, (**d**) Walking, (**e**) Stair Up, and (**f**) Stair down.

Table 5 shows that the important frequencies for Falling Right and Falling Left are identical. This is thought to be because they are almost identical activities, differing only in the direction of falling. Table 6 shows that, similar to HASC, PAMAP2 was less important for relatively slow-moving activities such as lying and sitting and more important for relatively fast-moving activities such as running and rope jumping. Table 7 shows that the

HHAR includes only relatively slow-moving activities, which may account for the lower important frequencies.

Table 5. Important frequency of each activity in UniMiB.

Activity	Frequency (Hz)
StandingUpFS	0.00
StandingUpFL	4.30
Walking	7.28
Running	2.65
GoingUpS	0.99
Jumping	1.32
GoingDownS	1.65
LyingDownFS	2.98
SittingDown	0.33
FallingForw	1.65
FallingRight	0.66
FallingBack	0.33
HittingObstacle	3.31
FallingWithPS	0.33
FallingBackSC	0.33
Syncope	1.98
FallingLeft	0.66

Table 6. Important frequency of each activity in PAMAP2.

Activity	Frequency (Hz)
lying	0.39
sitting	0.78
standing	3.91
walking	1.95
running	2.73
cycling	1.56
Nordic walking	1.95
ascending stairs	1.56
descending stairs	0.39
vacuum cleaning	0.78
ironing	0.39
rope jumping	2.34

Table 7. Important frequency of each activity in HHAR.

Activity	Frequency (Hz)
Biking	0.39
Sitting	0.39
Standing	0.39
Walking	0.78
Stair Up	1.17
Stair Down	2.34

5. Conclusions

In this study, in order to improve the accuracy of activity recognition prediction and to develop a variety of activity recognition services, we proposed a general-purpose method based on frequency enhancement and ensembles. The proposed method (1) finds important frequency in predicting each activity and creates a filter that emphasizes the found frequency, (2) trains the model by applying the filter to training data, and (3) performs ensemble learning by applying the filter to testing data.

The experiments conducted to identify the important frequency of each activity revealed that the DC component of stay (0 Hz) was important. Relatively slow-moving activities are expected to have a lower important frequency, while relatively fast-moving activities are expected to have a higher important frequency.

Ablation study results showed that the proposed method combining emphasis during training and testing and ensemble learning resulted in the highest recognition accuracy. Ensemble learning was the element that contributed the most to the accuracy of the proposed method. The frequency band enhancement filter was effective when applied to both the training and testing data but not when applied only to the testing data.

In an experiment conducted to examine the effect of the window function, four different filtering patterns were tested and compared in terms of recognition accuracy. Accuracy was highest when the filter was created with a Gaussian window and lowest when a random filter was applied, suggesting that emphasizing important frequency when creating filter results in higher accuracy. In addition, although this study proposes a method of emphasizing important frequencies for each activity, it is thought that the accuracy of recognition may be further improved by emphasizing or weakening the frequencies according to their importance.

An experiment was conducted to verify the robustness of the proposed method in different domains. The results showed that the proposed method performed better than an ensemble learning method in three out of four datasets (HASC, PAMAP2, and HHAR), demonstrating its robustness in different domains.

In this study, we used VGG16 in the phase of finding important frequencies and would like to experiment to see if the important frequencies change depending on the structure of the model. Additionally, the most important frequency of each activity is emphasized to improve the estimation accuracy of activity recognition. In addition to emphasizing the most important frequencies, we believe that the recognition accuracy can be further improved by emphasizing or de-emphasizing the frequencies according to their importance. We would like to create a frequency band enhancement filter other than the one used in this study and verify the change in accuracy. In addition, since the range of values applied to the amplitude spectrum in this study was between 0.5 and 1, we would like to investigate how the accuracy changes when the values are varied. We would also like to further improve the accuracy by using deep learning to create the frequency filter itself. As described above, we believe that the recognition accuracy can be improved over the current accuracy by changing the method of creating the frequency band enhancement filter.

Author Contributions: Conceptualization, H.T., T.H.; methodology, H.T., K.K., K.T. and T.H.; software, H.T., K.K.; validation, H.T.; formal analysis, H.T.; investigation, H.T.; resources, T.H.; data curation, H.T.; writing—original draft preparation, H.T.; writing—review and editing, K.K., K.T. and T.H.; visualization, H.T.; supervision, T.H.; project administration, T.H.; funding acquisition, T.H. All authors have read and agreed to the published version of the manuscript.

Funding: This work was supported in part by the Japan Society for the Promotion of Science (JSPS) KAKENHI Grant-in-Aid for Early-Career Scientists under Grant 19K20420.

Institutional Review Board Statement: Not applicable.

Informed Consent Statement: Not applicable.

Data Availability Statement: Data sharing not applicable.

Conflicts of Interest: The authors declare no conflict of interest.

References

1. Prasad, A.; Tyagi, A.K.; Althobaiti, M.M.; Almulihi, A.; Mansour, R.F.; Mahmoud, A.M. Human Activity Recognition Using Cell Phone-Based Accelerometer and Convolutional Neural Network. *Appl. Sci.* **2021**, *11*, 12099. [CrossRef]
2. Zhou, B.; Yang, J.; Li, Q. Smartphone-Based Activity Recognition for Indoor Localization Using a Convolutional Neural Network. *Sensors* **2019**, *19*, 621. [CrossRef] [PubMed]

3. Robben, S.; Pol, M.; Kröse, B. Longitudinal ambient sensor monitoring for functional health assessments: A case study. In Proceedings of the 2014 ACM International Joint Conference on Pervasive and Ubiquitous Computing: Adjunct Publication (UbiComp '14 Adjunct), Seattle, WA, USA, 13–17 September 2014; Association for Computing Machinery: New York, NY, USA, 2014; pp. 1209–1216.
4. Fridriksdottir, E.; Bonomi, A.G. Accelerometer-Based Human Activity Recognition for Patient Monitoring Using a Deep Neural Network. *Sensors* **2020**, *20*, 6424. [CrossRef]
5. Haider, F.; Salim, F.A.; Postma, D.B.W.; van Delden, R.; Reidsma, D.; van Beijnum, B.-J.; Luz, S. A Super-Bagging Method for Volleyball Action Recognition Using Wearable Sensors. *Multimodal Technol. Interact.* **2020**, *4*, 33. [CrossRef]
6. Steels, T.; Van Herbruggen, B.; Fontaine, J.; De Pessemier, T.; Plets, D.; De Poorter, E. Badminton Activity Recognition Using Accelerometer Data. *Sensors* **2020**, *20*, 4685. [CrossRef] [PubMed]
7. Sikder, M.S.N.; Chowdhury, A.S.M.A.; Nahid, A.-A. Human Activity Recognition Using Multichannel Convolutional Neural Network. In Proceedings of the 2019 5th International Conference on Advances in Electrical Engineering (ICAEE), Dhaka, Bangladesh, 26–28 September 2019; pp. 560–565.
8. Ooue, H.; Hashiyama, T.; Iwata, M.; Tano, S. Classification of walking pattern using 3-axis acceleration sensors. In Proceedings of the 22nd Fuzzy System Symposium, Sapporo, Japan, 6–8 September 2006; pp. 507–511.
9. Liu, M.; Zeng, A.; LAI, Q.; Gao, R.; Li, M.; Qin, J.; Xu, Q. T-WaveNet: A Tree-Structured Wavelet Neural Network for Time Series Signal Analysis. In Proceedings of the International Conference on Learning Representations, Virtual, 25 April 2022.
10. Nobuo, K.; Ogawa, N.; Iwasaki, Y.; Kaji, K.; Terada, T.; Maruo, K.; Inoue, S.; Kawahara, Y.; Sumi, Y.; Nishio, N. HASC Challenge: Gathering Large Scale Human Activity Corpus for the Real-World Activity Understandings. In *AH 2011, Proceedings of the 2nd Augmented Human International Conference, Tokyo, Japan, 13 March 2011*; Association for Computing Machinery: New York, NY, USA, 2011; pp. 1–5.
11. Wan, Shaohua, Lianyong Qi, Xiaolong Xu, Chao Tong and Zonghua Gu, Deep Learning Models for Real-Time Human Activity Recognition with Smartphones. *Mob. Netw. Appl.* **2020**, *25*, 743–755. [CrossRef]
12. Ito, C.; Cao, X.; Shuzo, M.; Maeda, E. Application of CNN for Human Activity Recognition with FFT Spectrogram of Acceleration and Gyro Sensors. In *UbiComp '18, Proceedings of the Ubiquitous Computing and Wearable Computers, Singapore, Singapore*; Association for Computing Machinery: New York, NY, USA, 2018; pp. 1503–1510.
13. Subasi, A.; Dammas, D.H.; Alghamdi, R.D.; Makawi, R.A.; Albiety, E.A.; Brahimi, T.; Sarirete, A. Sensor Based Human Activity Recognition Using Adaboost Ensemble Classifier. *Procedia Comput. Sci.* **2018**, *140*, 104–111. [CrossRef]
14. Mekruksavanich, S.; Jitpattanakul, A. LSTM Networks Using Smartphone Data for Sensor-Based Human Activity Recognition in Smart Homes. *Sensors* **2021**, *21*, 1636. [CrossRef] [PubMed]
15. Mekruksavanich, S.; Jitpattanakul, A. Deep Convolutional Neural Network with RNNs for Complex Activity Recognition Using Wrist-Worn Wearable Sensor Data. *Electronics* **2021**, *10*, 1685. [CrossRef]
16. Ahmed, N.; Rafiq, J.I.; Islam, M.R. Enhanced Human Activity Recognition Based on Smartphone Sensor Data Using Hybrid Feature Selection Model. *Sensors* **2020**, *20*, 317. [CrossRef] [PubMed]
17. Nisar, M.A.; Shirahama, K.; Li, F.; Huang, X.; Grzegorzek, M. Rank Pooling Approach for Wearable Sensor-Based ADLs Recognition. *Sensors* **2020**, *20*, 3463. [CrossRef] [PubMed]
18. Yoshizawa, M.; Takasaki, W.; Ohmura, R. Parameter exploration for response time reduction in accelerometer-based activity recognition. In Proceedings of the 2013 ACM Conference on Pervasive and Ubiquitous Computing Adjunct Publication (UbiComp '13 Adjunct), Zurich, Switzerland, 8–12 September 2013; Association for Computing Machinery: New York, NY, USA, 2013; pp. 653–664.
19. Fujiwara, M.; Fujimoto, M.; Arakawa, Y.; Yasumoto, K. *Development and Evaluation of In-Home Activity Recognition Utilizing Doppler Sensor*; 2018 Information Processing Society of Japan (IPSJ SIG) Technical Report; IPSJ: Tokyo, Japan, 2018; pp. 1–8.
20. Naomi, I.; Nugent, C.; Zhang, S.; Wang, H.; Ng, W.W.Y. Neural Network Ensembles for Sensor-Based Human Activity Recognition within Smart Environments. *Sensors* **2020**, *20*, 1–26.
21. Zhu, R.; Xiao, Z.; Li, Y.; Yang, M.; Tan, Y.; Zhou, L.; Lin, S.; Wen, H. Efficient Human Activity Recognition Solving the Confusing Activities via Deep Ensemble Learning. *IEEE Access* **2019**, *7*, 75490–75499. [CrossRef]
22. Tian, Y.; Zhang, J.; Chen, L.; Geng, Y.; Wang, X. Selective Ensemble Based on Extreme Learning Machine for Sensor-Based Human Activity Recognition. *Sensors* **2019**, *19*, 3468. [CrossRef] [PubMed]
23. Xu, S.; Tang, Q.; Jin, L.; Pan, Z. A Cascade Ensemble Learning Model for Human Activity Recognition with Smartphones. *Sensors* **2019**, *19*, 2307. [CrossRef] [PubMed]
24. Subasi, A.; Fllatah, A.; Alzobidi, K.; Brahimi, T.; Sarirete, A. Smartphone-Based Human Activity Recognition Using Bagging and Boosting. *Procedia Comput. Sci.* **2019**, *163*, 54–61. [CrossRef]
25. Tan, T.-H.; Wu, J.-Y.; Liu, S.-H.; Gochoo, M. Human Activity Recognition Using an Ensemble Learning Algorithm with Smartphone Sensor Data. *Electronics* **2022**, *11*, 322. [CrossRef]
26. Bhattacharya, D.; Sharma, D.; Kim, W.; Ijaz, M.F.; Singh, P.K. Ensem-HAR: An Ensemble Deep Learning Model for Smartphone Sensor-Based Human Activity Recognition for Measurement of Elderly Health Monitoring. *Biosensors* **2022**, *12*, 393. [CrossRef] [PubMed]
27. Kondo, K.; Hasegawa, T. Sensor-Based Human Activity Recognition Using Adaptive Class Hierarchy. *Sensors* **2021**, *21*, 7743. [CrossRef] [PubMed]

28. Shanmugam, D.; Blalock, D.; Balakrishnan, G.; Guttag, J. Better aggregation in test-time augmentation. In Proceedings of the IEEE/CVF International Conference on Computer Vision, Montreal, QC, Canada, 10–17 October 2021; Computer Vision Foundation: New York, NY, USA, 2021; pp. 1214–1223.
29. Simonyan, K.; Zisserman, A. Very Deep Convolutional Networks for Large-Scale Image Recognition. In Proceedings of the 3rd International Conference on Learning Representations (ICLR 2015), San Diego, CA, USA, 7–9 May 2015; Computational and Biological Learning Society: Lincroft, NJ, USA, 2015; pp. 1–14.
30. Micucci, D.; Mobilio, M.; Napoletano, P. UniMiB SHAR: A Dataset for Human Activity Recognition Using Acceleration Data from Smartphones. *Appl. Sci.* **2017**, *7*, 1101. [CrossRef]
31. Reiss, A.; Stricker, D. Introducing a New Benchmarked Dataset for Activity Monitoring. In Proceedings of the 2012 16th IEEE International Symposium on Wearable Computers (ISWC), Newcastle, UK, 18–22 June 2012; pp. 108–109.
32. Stisen, A.; Blunck, H.; Bhattacharya, S.; Prentow, T.S.; Kjærgaard, M.B.; Dey, A.; Sonne, T.; Møller Jensen, M.M. Smart Devices are Different: Assessing and Mitigating Mobile Sensing Heterogeneities for Activity Recognition. In *SenSys 2015, Proceedings of the 13th ACM Conference on Embedded Networked Sensor Systems, Seoul, Korea*; Association for Computing Machinery: New York, NY, USA, 2015; pp. 127–140.

sensors

Article

k-Tournament Grasshopper Extreme Learner for FMG-Based Gesture Recognition

Rim Barioul * and **Olfa Kanoun**

Chair of Measurement and Sensor Technology, Technische Universitat Chemnitz, 09126 Chemnitz, Germany
* Correspondence: rim.barioul@ieee.org

Abstract: The recognition of hand signs is essential for several applications. Due to the variation of possible signals and the complexity of sensor-based systems for hand gesture recognition, a new artificial neural network algorithm providing high accuracy with a reduced architecture and automatic feature selection is needed. In this paper, a novel classification method based on an extreme learning machine (ELM), supported by an improved grasshopper optimization algorithm (GOA) as a core for a weight-pruning process, is proposed. The k-tournament grasshopper optimization algorithm was implemented to select and prune the ELM weights resulting in the proposed k-tournament grasshopper extreme learner (KTGEL) classifier. Myographic methods, such as force myography (FMG), deliver interesting signals that can build the basis for hand sign recognition. FMG was investigated to limit the number of sensors at suitable positions and provide adequate signal processing algorithms for perspective implementation in wearable embedded systems. Based on the proposed KTGEL, the number of sensors and the effect of the number of subjects was investigated in the first stage. It was shown that by increasing the number of subjects participating in the data collection, eight was the minimal number of sensors needed to result in acceptable sign recognition performance. Moreover, implemented with 3000 hidden nodes, after the feature selection wrapper, the ELM had both a microaverage precision and a microaverage sensitivity of 97% for the recognition of a set of gestures, including a middle ambiguity level. The KTGEL reduced the hidden nodes to only 1000, reaching the same total sensitivity with a reduced total precision of only 1% without needing an additional feature selection method.

Keywords: extreme learning machine; force myography; grasshopper optimization algorithm; k-tournament selection

Citation: Barioul, R.; Kanoun, O. k-Tournament Grasshopper Extreme Learner for FMG-Based Gesture Recognition. *Sensors* **2023**, *23*, 1096. https://doi.org/10.3390/s23031096

Academic Editor: Giovanni Saggio

Received: 14 November 2022
Revised: 30 December 2022
Accepted: 10 January 2023
Published: 18 January 2023

1. Introduction

Hand gestures are part of behavioral attributes that are authentic (emphasize or help to express a thought or feeling), distinguishable (present a known meaning that depends on culture, language, and use case), and have unique physiological patterns (physiological signals and phenomena resulted from various hand gestures present varying unique properties). Hand gesture recognition is essential in several applications, such as sign language, mobile security systems, smart homes, and other IoT-based applications. In addition, hand gesture recognition involves several challenges concerning the sensors and machine learning algorithms, including the system design, which needs to fit different persons, and the influence of the physiological state of the subject on the collected signal quality. Hand grasp recognition and hand sign recognition are the main subtopics of hand gesture recognition. The first is dedicated to the identification of the grasping nature, and the object-handling tasks while sign recognition is dedicated mainly to communication between persons or between persons and intelligent agents. Hand sign recognition is valuable, e.g., for communication over long distances, in noisy environments, and with people with disabilities. Identifying hand signs with camera-based systems is challenging in such environments and suffers from limited resolution, significant distances, and varying light conditions. Myographic measurement methods and sensors, which allow the direct collection of information on the muscle state during the gesture performance, can be of great importance in overcoming these limitations. Techniques such as surface electromyography, force myography, and surface electrical impedance myography show promising performance for gesture detection, even if only a few current investigations exist for sign language recognition based on myographic signals. Another challenge for hand sign classification is in the level of algorithms and features. The classification algorithm must get a suitable feature subset to be able to realize a high classification accuracy. In addition, the quality of the classification is variable, along with the number of features. Hence, the control of the feature number is essential since a limited feature number may cause data overlapping, which means that the classification becomes not sufficiently grounded. Too many features increase the dimension of the problem, and more complex classification algorithms will be needed. Thus, the goal of feature selection is to define the best subset of features by directly removing the irrelevant and redundant features from the data and improving the classification performance and stability. Moreover, reduced resource consumption is required to ensure the suitability of the classification algorithm with wearable hand gesture recognition systems. Most investigations adopt a feature selection based on metaheuristic optimization methods in binary format. The classification accuracy depends on many factors, including the gesture types and numbers, the measurement accuracy of the myographic signals, and the choice of the classifier itself. Furthermore, the classification method should be suitable for solving multiclass problems with minimal calculation. Such property is reported to be insured by an extreme learning machine (ELM). It is a single-layer feed-forward network (SLFN) with randomly generated input weights and biases and output layer weights calculated via linear algebra methods allowing fast training in only one iteration, even in multiclassification problems. However, ELM suffers from the incertitude caused by this random weight generation. Many optimization methods have been suggested in the literature to solve this problem, including controlling the randomization and pruning the hidden nodes. However, weight pruning is not sufficiently investigated for the ELM architecture's optimization. This work proposes a new approach for ELM network optimization based on a coupled weight and feature selection that allows not only the elimination of irrelevant weights in the network but also an integrated feature selection and hidden node number reduction.

The paper is structured as follows: In Section 2, related works are described, which provide information on the state of the art of ELM pruning and FMG-based gesture recognition. In Section 3, the methodology of implementation of a k-tournament grasshopper extreme learner, the ELM weight selection concept, and the proposed KTGEL is detailed. Section 4 shows the study of the number of FMG sensors for an efficient hand sign recognition system and the influence of the number of subjects on the KTGEL performance. This section also provides the experimental investigation on the performance of the KTGEL compared with the state of the art and with a variation of the ambiguity level in the data set followed by the conclusion.

2. Related Work

In the first part of this section, we present an overview of applied methods for pruning an extreme learning machine to reduce its model architecture while keeping its good performance and exposing the gap in approaches exploited to fulfill this aim. In the second part, an overview of hand gesture recognition based on FMG sensors is presented, focusing on the number of sensors, the features, the number of subjects, and the American Sign Language recognition as an application.

2.1. Pruning of Extreme Learning Machine

An extreme learning machine (ELM) is a single-layer feed-forward network (SLFN) where the fundamental concept is that the weights and biases of the hidden layer are randomly generated. Moreover, the output layer weights are calculated using a least-squares solution defined by the outputs of the hidden layer and the target [1]. Thus, the weights that connect the hidden nodes to the outputs can be trained very fast in one iteration according to the pseudocode presented in Algorithm 1.

Algorithm 1: Pseudocode of an extreme learning machine [2].

1 Given a training set $\mathbf{N} = \{(x_i, t_i) | x_i \in R^n, t_i \in R, i = 1 \cdots N\}$, activation function $G(w, b, x)$, and number of hidden nodes $\widetilde{N}$;
2 Assign random input weights w_i, and biases b_i, for $i = 1 \cdots \widetilde{N}$;
3 Calculate the hidden layer output matrix H;
4 Calculate the output weight matrix

$$\beta = \mathbf{H}^{\dagger}\mathbf{T} \tag{1}$$

where $\mathbf{H}^{\dagger}$ is the Moore–Penrose generalized inverse of matrix H and $T = [t_1 \cdots t_N]^T$; The output weight matrix β;

Since its first introduction, the ELM has been a subject for optimization as it represents a promising possibility for embedded systems and online real-time classification tasks. However, it also presents some limitations, especially in its hidden node number and weights' randomization method. An ELM also randomly generates the input weights and the bias of hidden nodes, which has the following consequences: first, a slow learning speed caused by the minor roles played by some hidden nodes with too small output weights on the network's output; second, a slow error reduction during the training process is caused by these invalid hidden layer neurons, which increase the network complexity [3]. To solve this, most of the proposed algorithms focus on simplifying the computation process, finding the optimized depth of the SLFNs, or expanding the range of the generalized methods via multilayers or a complex domain. However, for random weight optimization, the proposed solutions tend to replace the completely randomly generated input weight and bias with fully controllable metrics, which turns the ELM into a controlled method and reduces the benefits of the weights' randomness in the ELM results. The optimally pruned ELM (OP-ELM) was proposed by Miche et al. [4] based on the ELM algorithm in terms of kernel selection and using the methodology of pruning the neurons, leading to more efficient algorithms and improving the ELM problems experienced when using irrelevant or correlated data [4,5]. Compared to the ELM, the OP-ELM enhanced the robustness and accuracy of the network. However, it had a higher computational time, affecting the accuracy and training time [6,7]. Genetic algorithms for pruned ELM (GPA-ELM) were proposed by Alencar et al. [8] to prune the hidden layer neurons based on multiobjective GAs. It combined the advantages of ELMs and GAs to optimize the performance of the ELM classifiers and prune the maximum possible number of hidden neurons. In [9], the authors proposed the PSO-ELM for optimizing the input feature subset selection and the number of hidden nodes to enhance the classification performance of ELM in the application of power system disturbances classification. The experimental results showed that the proposed PSO-ELM was faster and more accurate than the original ELM algorithm. However, the PSO which was used to perform those optimizations was reported to be outperformed by other newly introduced swarm intelligence optimization methods, including the GOA [10–12]. In the literature, the main difference between the various pruned ELM versions is the different optimization methods implemented to modify the ELM architecture to realize the hidden nodes' pruning. However, there is no specific idea proposed so far about weight selection without controlling the random initialization or connection pruning optimization, which is an integral part of extreme machine learning in data classification. Hence, in this work, optimizing the ELM by proposing a weight selection by an improved version of the GOA after the initial random initialization is presented as a methodology for connection pruning in ELMs.

2.2. Sensors for FMG

Since it is possible to perform FMG with either pressure or strain sensors, unlimited choices of sensors are available. However, in 2006, Amft et al. [13] compared the force-sensitive resistor (FSR) as a pressure sensor with a fabric stretch sensor (FSS) as a strain sensor and surface electromyography (sEMG) for monitoring muscles' contraction for grasping, upper-hand activities, and object lifting. The feasibility of muscle activity detection by the strain and pressure sensors as alternatives to sEMG was confirmed in that study. Moreover, the experimental results showed that the pressure sensors were more suitable as a future alternative to sEMG for gesture recognition applications as they were able to monitor the contraction of more muscle groups than the strain sensor. Hence, the FSR pressure sensor for FMG measurement was chosen for this study. Moreover, as commercial sensors were more suitable for this work's aims, a study of the FSR sensors market and publications was conducted. As the cost for various FSR sensors were almost similar, and the FSR sensor by Interlink Electronics and the FlexiForce™ by Tekscan Ink were the most popular commercial sensors, which were used in 55% of publications about FMG applications until 2019 [14], the sensor choice range was limited between both these sensors. Their characteristics extracted from their data sheets are shown in Table 1.

Table 1. FSR Interlink and Flexiforce properties from data sheets [14].

Title 1	Interlink FSR (FSR402)	Flexiforce (FLX-A201-F)
Minimum actuation force (N)	0.1	N/A
Force sensitivity range (N)	0.1–10	0 to 4.4, 0 to 445
Single-part force repeatability	±2%	±2.5%
Part-to-part force repeatability	±6%	±40%
Drift	<5% per log10 (time)	<5% per log10 (time)
Hysteresis	+10%	<4.5%
Response time (μs)	<3	<5
Linearity error	N/A	<±3%

Vecchi et al. compared the previous sensors on several points, such as repeatability, time drift, or dynamic force measurement via an experimental process. The results showed that the FlexiForce sensors had better performance in terms of linearity, repeatability, time drift, and dynamic accuracy. However, Interlink's FSR was more robust [15]. Another study that compared the same sensors with the LuSense PS3 (Standard 151) sensor was conducted in 2006 and concluded that the FlexiForce had not only the highest precision but also the highest noise with the slowest response time and the highest resistance dropping from the nominal value during subsequent tests [16]. Hence, each sensor has its pros and cons. The choice was based on the response time as a real-time and fast system was the goal in this study's outlook. Thus, the Interlink's FSR possessing the lowest response time in the data sheet (see Table 1) and in experiments [16] was chosen to perform the FMG data collection in this work. A typical Interlink Electronics's FSR sensor consists of a top carbon-based ink layer and a bottom conductive substrate layer with a spacer adhesive located in the middle of the two layers [17]. Therefore, during FMG collection, as the hand exerts a force, the corresponding muscles on the arm produce a deformation on the skin's surface. These deformations apply pressure to the surface of the top layer of the FSR, changing its resistance. These changes in resistance can be translated into corresponding changes in voltage by a voltage divider structure resulting in the FMG distinct patterns that could be used for hand gesture recognition with the best sensitivity, which is ensured by a reference resistance of 100 kΩ in the voltage divider [17,18].

2.3. Hand Gesture Recognition Based on FMG Sensors

FSRs have been used for hand gesture recognition often in recent years, sometimes alone [19], sometimes in combination with sEMG [20] or other sensors [21]. In these studies, the sensors were mostly worn on the forearm or the wrist [14]. In some rare cases, the sensors were worn while attached to a glove [20]. Moreover, FSR-based hand gesture recognition studies have practically focused on grasping [22], upper-arm activities such as pinching or rotations [21], and robotic hand [23] or prostheses control [24]. Moreover, there have been studies comparing grasp vs. nongrasp gestures [19]. However, sign language recognition have rarely been investigated with FMG signals and have never been the focus of any published scientific work except a few [18,25,26], where the feasibility of sign language recognition by FMG-based systems and investigations about the measurement system and the recognition with classic classification methods were conducted. Many studies have shown the advantages of FMG over EMG signals [19,27]. For example, FMG does not require much skin preparation and is less affected by skin impedance or sweat. Furthermore, FMG is characterized by its stability and robustness to external electrical noise; in addition, it does not necessitate the same amount of signal processing, and feature extraction as EMG [19]. Thus, all of these factors were the main reason for making the implementation of FMG in wearable devices more reliable in terms of cost and equipment. The oldest research discussing FMG features is from 2017 [28], while most research has implemented FMG as raw signals for gesture recognition. The discussed features for force myography are primarily used in grasping detection, robot hand control, and gait analysis [28–32]. Many researchers have achieved hand gesture recognition based on various machine learning methods. In addition, the hand gesture term includes a massive number of gestures with different levels of force and acceleration from sign language alphabets that generally cover postures and some slight motions to grasping and upper-arm activities that contain the interaction with objects and a high level of muscle contraction force. As for the different myography measurement techniques considered in this work, the high force level ensures a higher representation of the gesture. Most hand gesture recognition studies in the literature have focused on grasping and upper-arm activities. In contrast, sign language recognition is still an application where more investigations for features and classification methods are mandatory. Hence for the experimental part of this work, the application focus is on sign language recognition and, more specifically, American Sign Language (ASL) recognition. An overview of publications discussing American Sign Language recognition based on FMG as a standalone system or combining FMG and sEMG are listed in Table 2.

Table 2. State of the art for FMG-sensor-based ASL recognition.

Sensors	Features	Subjects	Gestures	Classifier	Accuracy
8 colocated sEMG FMG [33]	MAV, WL ZC, SSC	5	10	LDA	91.6%
8 FMG Self-produced [33]	MAV	5	10	LDA	80%
16 FMG [19]	RAW signal	12	16	LDA	96.70%
8 nanocomposite sensors [34]	min, max, mean, RMS, median, STD	10	10	ELM	93%

For FMG-based hand gesture recognition studies in the state of the art, the number of sensors is relatively high for portable and user-friendly systems. Moreover, the use of raw FMG signals in most of the studies limits the signal abilities and the machine learning methods' performance. In addition, the applications of FMG are mainly focused on grasping and robotic hand or prosthesis control where a significant muscle contraction force is included, and they are rarely investigated for sign language recognition. From Table 2, for force myography, only one publication presented the sign language recognition by FMG as a standalone system based on commercial sensors [19]. However, that previous study was based on raw FMG only. In our studies in [18,25,35], the feasibility of sign language recognition by a reduced number of FMG sensors up to four and the investigation of various features for the recognition from an FMG-based bracelet with classic classification methods were conducted. In [25], it was proved that for a low-level ambiguity in the gesture set that the ELM could recognize the signs with an accuracy of 89.65% based on six standard features extracted from signals collected by four commercial sensors. In [35], the ELM accuracy for sign recognition based on the same extracted features as in [25] from the FMG signals collected by six sensors was equal to 91.11%. In this work, an optimized classifier is proposed, and its adequate minimal number of sensors to recognize various sets of signs with different levels of ambiguities is investigated.

3. Proposed k-Tournament Grasshopper Extreme Learner

The ELM has been proven in the literature to outperform other algorithms in terms of accuracy, speed, and model size. Therefore, it is more suitable for embedded systems. However, the weights' random tuning remains a source of incertitude in terms of the optimal result this algorithm could reach. Researchers with different approaches proposed many optimizations of ELMs to reduce this effect. However, the used optimization methods were relatively old algorithms in the field. New optimization methods with good performances in various applications have been newly proposed and could give better results. Moreover, none of the proposed methods investigated the selection of randomly generated weights to optimize the architecture of an ELM without controlling its randomization process.

3.1. ELM Weights Selection

A neural network weight selection is one of the pruning types of network architecture, also named connection pruning, where the number of connections in the network is reduced. Another type is node pruning, where the number of hidden nodes is reduced by selecting the more significant hidden nodes [36]. For ELM pruning, researchers have proposed several methods for node pruning [8,37–42], but the weight pruning problem has not yet been studied. To cover this gap in the ELM architecture optimization strategies, a weight selection of the ELM is proposed in this paper as shown in Figure 1.

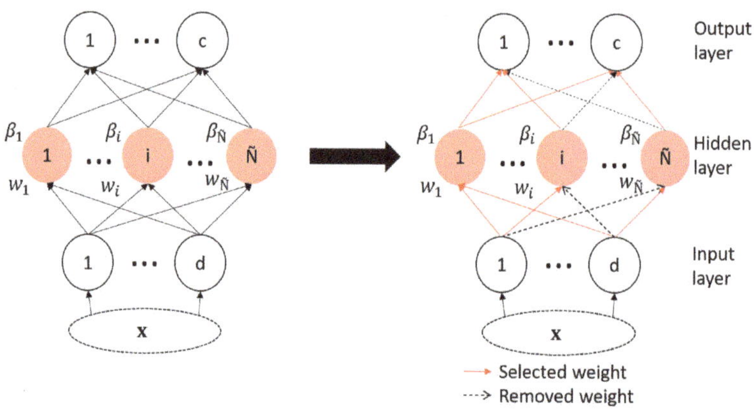

Figure 1. Proposed ELM architecture optimization strategy.

The selection of initially generated weights proposed in this work has the aim of keeping only the best subset of weights, which shares the same idea as other feature selection methods. In the latter methods, the goal, in general, is to define the best subset of features to improve the performance of the classification stage. Moreover, feature selection is important because the quality of the classification is variable along with the number of features. Hence, controlling this number is important because when it is too small, it may cause an overlap of data, which means it is not enough for the classification. However, if the number of features are too great, the dimension of the problem increases, and more complex classification algorithms are needed. Similar to the number of features, the number of weights in the ELM also impacts the overfitting, the model size, and the complexity. From there comes the inspiration to use a feature selection approach as a strategy for ELM weights' selection.

3.2. k-Tournament Grasshopper Extreme Learning Machine for Selection Problems

First, the tournament process is included in the grasshopper repositioning process, as shown in the pseudocode in Algorithm 2 by controlling the best population evaluation.

Algorithm 2: k-Tournament grasshopper optimization algorithm.

Initialization of CMin, CMax and MaxIteration;
Initialize the population of particles Xi;
Evaluate each solution in the population;
Set T as the best solution;
while $t < MaxIteration$ **do**
  Update c using the controlling parameter equation;
  for *each solution* **do**
    Normalize the distances between the grasshoppers in [1, 4];
    Update the step vector $\Delta X_i(i)$ of the current solution;
    Bring the current grasshopper back if it goes outside the boundaries;
  Conduct a K=2 tournament between the current solution and the rest of the population;
  Update T with the winners of the tournaments.;
  t = t + 1;
Return T

Furthermore, to perform the selection of this algorithm, the S-shaped transfer function is applied to the velocity of the search agents in the same way shown in the binary grasshopper optimization algorithm proposed in [43] presented by the pseudo-code in Algorithm 3 before combining it with the extreme learning machine shown in the Algorithm 1 as the wrapper's evaluation classifier.

Algorithm 3: Binary grasshopper optimization algorithm (BGOA) [43].

Initialization of CMin, CMax, and MaxIteration;
Initialize the population of particles Xi;
Evaluate each solution in the population;
Set T as the best solution;
while $t \leq MaxIteration$ **do**
  Update c using the controlling parameter equation;
  for *each solution* **do**
    Normalize the distances between the grasshoppers in [1, 4];
    Update the step vector ΔX_i of the current solution;
    for $i = 1 : dim$ **do**
      if $rand \geq T(\Delta X_{t+1})$ **then**
        $X_{t+1}(i) = 1$;
      else
        $X_{t+1}(i) = 0$;
  $t = t + 1$;
Return T;

For this wrapper, the ELM was chosen as the evaluation method of the selected subsets because it outperformed other classification methods customarily used for wrapper building, such as KNN and SVM, in terms of accuracy, speed, and minimal computation complexity [44–48]. Moreover, searching for the best feature subset in feature selection is a challenging problem, especially in wrapper-based methods. This is because the selected subset needs to be evaluated by the learning algorithm (e.g., classifier) at each individual optimization step. Hence, a proper optimization method is required to reduce the number of evaluations, which is ensured by the ELM's ability to solve multiclass problems in one iteration [49].

3.3. *k-Tournament Grasshopper Extreme Learner*

The final proposed KTGEL is shown in Algorithm 4. The proposed approach tends to optimize the extreme learning machine by selecting the most significant weights from the randomly generated ones during its initialization. The weight selection is integrated into the training process of the ELM. Moreover, the proposed KTGEL inherits the training procedure of the ELM, including the coupling between the input data and the input weights. Hence, the KTGEL is able to perform the feature selection within its training phase as an effect of the weight coupling relation with the input data during this phase. Each weight is coupled to one feature, but one feature is coupled to many weights, resulting in a feature being only eliminated if all its related weights are eliminated. Hence the proposed k-tournament grasshopper extreme learner is estimated to provide a better classification accuracy than the original ELM classifier on different biosignal databases for hand gesture recognition with a smaller model size as nonselected weights are replaced by zero so that no more computations are devoted to them.

Algorithm 4: Pseudocode of the proposed k-tournament grasshopper extreme learner.

Input: Given the training set $\mathbf{N} = \{(x_i, t_i) | x_i \in \mathbb{R}^n, t_i \in \mathbb{R}, i = 1 \cdots N\}$, the activation function g, and the number of hidden nodes $\widetilde{N}$
Output: The output weight matrix β and selected feature vector.
Assign random input weights w_i and biases b_i, for $i = 1 \cdots \widetilde{N}$;
Initialize the tournament size and maximal iterations;
Initialization of CMin, CMax; Initialize the population of grasshoppers Xi:(w_i,x_i);
Run "tournaments" among the k individuals chosen at random from the population;
The winner of each tournament is selected as the best solution;
while t < *MaxIteration+1* **do**
    Update c;
    for *each solution* **do**
        Normalize the distances between the grasshoppers in $[1, 4]$;
        Update the step vector ΔX_i of the current solution;
        for $i = 1 : dim$ **do**
            if $rand \geq T(\Delta X_{t+1})$ **then**
                $X_{t+1}(i) = 1$;
            else
                $X_{t+1}(i) = 0$;
    Conduct a k = 2 tournament between the current solution and the rest of the population;
    Update T with the winners of the tournaments;
    t = t + 1;
Return T: (w,x);
Calculate the hidden layer output matrix;
Calculate the output weight matrix: $\beta = \mathbf{H}^{\dagger}\mathbf{T}$ where $\mathbf{H}^{\dagger}$ is the Moore–Penrose generalized inverse of matrix H and $T = [t_1 \cdots t_N]^T$;

4. Experimental Investigations

In this section, three performance metrics were investigated: the accuracy in comparison with other works in the state of the art, reproducing the same set of gestures performed with the same number of subjects under as many similar conditions as possible, the influence of the number of sensors in relation with the number of subjects on the classification accuracy, and the influence of the ambiguity level in the set of gestures in comparison with an ELM after a feature selection step. The ELM and KTGEL were initialized with 3000 hidden nodes and compared on the data collected from eight FSR sensors with a total of 48 initial features in terms of accuracy, the final network architecture after weight selection by the KTGEL, the average sensitivity, and the average precision. To investigate the effect of the ambiguity level between gestures on their classification based on the FMG eight-sensor band, 40 participants from both genders in the age range between 20 and 32 years old participated in the collection of the 27 letters, the ASL numbers from 0 to 10, and the expression "I love you.". Each subject participated in collecting only 10 or 9 signs with ten repetitions for each. In total, the collected data included 39 signs from the ASL, with 100 observations for each one. From this database, two sets of gestures were exploited in this paper for the investigations of the ambiguity level influence on the KTGEL performance. As for the evaluation with the accuracy, both the micro precision and the micro recall are conventionally used for a multiclassification assessment, where TP_j, FP_j, FN_j are, respectively, the numbers of true positives, false positives, and false negatives of a class j, to show the overall classifier precision and sensitivity [50].

$$\text{micro-}P = \frac{\sum_{j=1}^{m} \text{TP}_j}{\sum_{j=1}^{m} \text{TP}_j + \text{FP}_j} \tag{2}$$

$$\text{micro-}R = \frac{\sum_{j=1}^{m} \text{TP}_j}{\sum_{j=1}^{m} \text{TP}_j + \text{FN}_j} \tag{3}$$

4.1. Comparison with the State of the Art of FMG-Based Gesture Recognition

In Table 3 a comparison between this work and the 2 studies from the state of the art was conducted to compare the algorithms' performance while keeping the number of sensors, observations, and subjects. In [33], the FMG signals were collected with eight self-produced sEMG-FMG colocated sensors placed on the forearm of the subjects, and in this work, eight commercial FSR sensors were integrated into a wristband. In [34], carbon-nanotube-based FMG sensors were customized to produce more sensitive sensors with a higher ability to detect signs than commercial FSR sensors.

Table 3. Performance of the proposed classifiers vs. the state of the art of ASL numbers' recognition by FMG.

Work	Hand Signs	Sensor No.	Sensor	Classifier	Accuracy in %	Observations	Subjects
[33]	10	8	Customized	LDA	80.00	50	5
This work	10	8	FSR	KTGEL	88.00	50	5
[34]	10	8	Customized	ELM	93.00	100	10
This work	10	8	FSR	KTGEL	98	100	10

For FMG, the comparison with [33] showed that the proposed FMG bracelet located on the wrist and commercial sensors could provide better accuracy for ASL numbers' classification. Moreover, the comparison with [34] was made with the exact same gestures proving that the KTGEL outperformed the ELM in terms of accuracy, even while implemented on data collected with commercial sensors. In contrast, the data in [34] were collected with optimized sensors that had been proved to outperform the commercial FSR sensors when the same signal processing was applied to data collected by both sensors.

4.2. Investigation of the Sensors and Subject Number Influence

Force myography is rarely used for sign language recognition, and it has only been used with raw signals. Thus, it has not been sufficiently investigated in the literature. That is why it was necessary to conduct tests and observe the results. Both the number of sensors and the convenient features should be examined in this part. The idea is to find the optimal number of sensors from the wrist sensor band since previous studies [19] confirmed that the wrist-positioned band had more sensitivity to ASL than forearm bands. To minimize the sensors' number and thereby ensure user comfort, two bands of six and eight commercial pressure sensors were designed, realized, and tested to find the band that led to the best accuracy for the ASL gesture recognition system. In the first band, eight sensors were placed with a gap of 2 cm around the wrist, while the second band had six sensors with a 2.25 cm gap between sensors. In all systems, Teensy boards with synchronized ADCs were employed as acquisition boards with a sampling frequency of 100 Hz. The two-band system was used to collect data during the performance of ASL signs according to the measurement protocol in Figure 2.

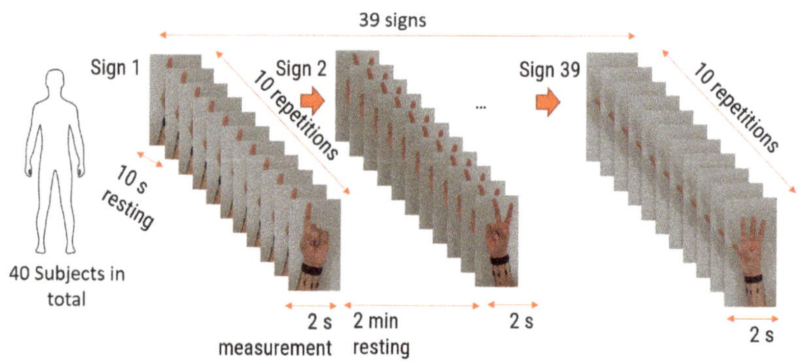

Figure 2. FMG signal collection protocol.

The first investigation aimed to test the feasibility of finger sign detection by the wrist FSR bands, including a small number of sensors compared to the state of the art, where the previous studies that implemented sign language included 16 commercial sensors [19] or 8 customized sensors [33,34]. Hence, only one person was asked to wear one of the two bands each time and perform the nine ALS numbers from one to nine shown in Figure 3 for twenty trials each.

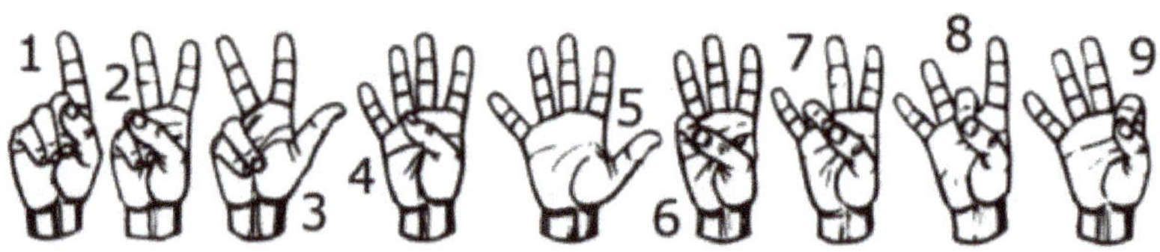

Figure 3. Performed ASL numbers from 1 to 9 .

Gestures have been performed with a resting of two minutes between every two gestures to avoid muscle fatigue. The collected signal seemed to have stationary behaviors for the different gestures, so it was estimated that even though features increased the performance of algorithms in comparison with raw data, there was no need for complicated features. Hence, six basic features, which were the min, max, RMS, var, STD, and mean, were extracted and normalized by the min–max method, and the KTGEL was used to classify the gestures. The classification accuracy was considered here as the evaluation criterion for the needed number of sensors for further data collection.

In this set of gestures, the ambiguity level between signs could be described as low since no dynamic gestures were considered, and the similarity between the gesture performance process was limited between the numbers six, seven, eight, and nine between all the possible combinations of the nine gestures. The collected data from only one person resulted in a total of 180 observations. In this investigation, 80% of the observations were used to train and validate the KTGEL using a fivefold cross-validation while saving a random 20% of each gesture's data to be used only as testing data. From the confusion matrices in Figures 4 and 5, it could be confirmed that for only one subject performing the gestures, both bracelets could detect and allow the classifier to predict the nine tested gestures correctly. It was proved by this investigation that the six sensors were sufficient to recognize gestures with a low ambiguity level collected from only one subject.

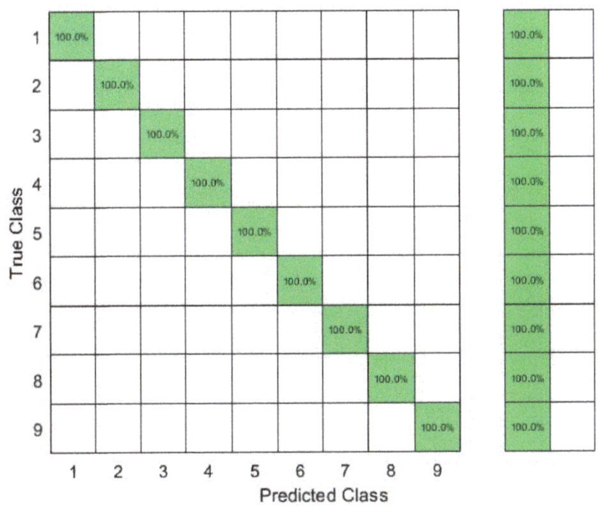

Figure 4. Confusion matrix of the KTGEL for one person for ASL numbers from 1 to 9:6 sensors.

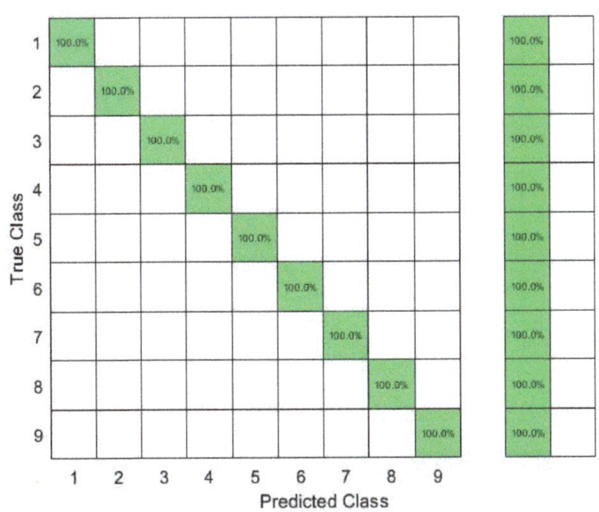

Figure 5. Confusion matrix of the KTGEL for one person for ASL numbers from 1 to 9:8 sensors.

The second investigation was to evaluate the system's stability and accuracy for the same gesture recognition while increasing the number of subjects to 10 subjects. However, in that investigation, each person was asked to perform each gesture only ten times. In total, 1000 observations were used in this implementation of the KTGEL, with 80% of the observations employed to train and validate the model using fivefold cross-validation while a random 20% of each gesture observations were safeguarded to be used only as testing data.

The results in Figure 6 show that with six sensors collecting nine gestures, the KTGEL had a test accuracy of 71%. Figure 7 shows that the eight sensors band collecting American Sign Language numbers could be recognized with an accuracy of 95%. These results confirmed that six sensors were not suitable enough for FMG-based gesture recognition with several subjects. The additional complexity in the signals induced by the physiological difference between the various subjects could not be canceled by the use of six sensors only. It is also observed in Figure 6 that the confusion between gestures could not be totally obvious from the gestures' nature, which led to the estimation that the collected data were not enough to differentiate the gestures. However, observing Figure 7, it could be seen that the confusions were limited, with the most relevant confusions being between gestures six, seven, and eight. Hence, this investigation showed that eight FSR sensors as the minimal number of sensors had an acceptable gesture recognition accuracy from the data collected from 10 subjects. In addition, to confirm the user's comfort with the used number of sensors, subjects were asked about their evaluation of the band. None of the subjects complained about the sensor band placement, but they announced that the material used for the actual band was not soft enough. Hence, the eight-sensor band was kept for further data collection as a possible standalone system for a future investigation of sign recognition with more features, and a modification of the bracelet material will be considered as an outlook of the system design.

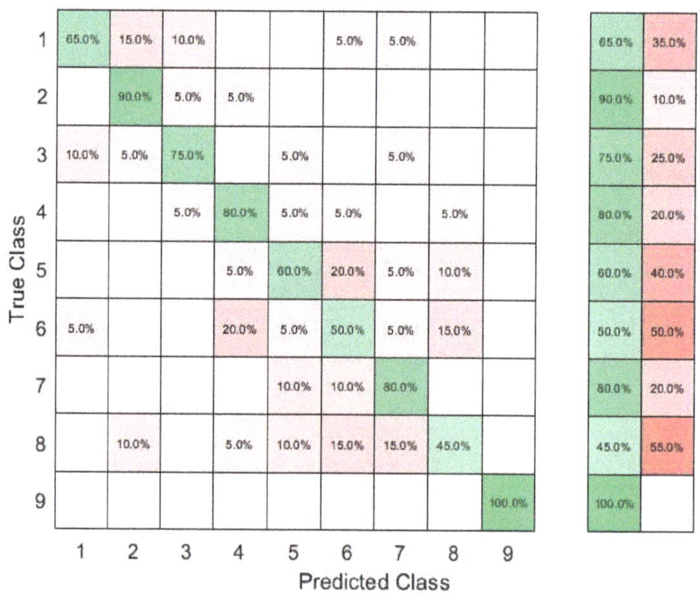

Figure 6. Confusion matrix of the KTGEL for ten person and nine numbers: 6 sensors.

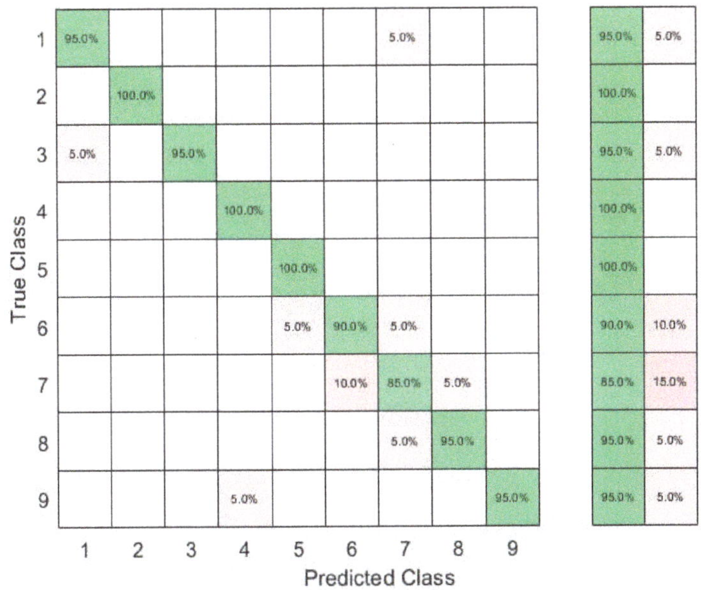

Figure 7. Confusion matrix of the KTGEL for ten person and nine numbers: 8 sensors.

4.3. Recognition of ASL Signs with a Middle Ambiguity Level

To investigate the influence of ambiguity on the KTGEL performance for American Sign Language recognition, the first ten alphabet letters from A to J were collected from 10 healthy subjects. During the data collection, subjects followed an informative video for ASL teaching. Gestures were collected as postures except for the letter J, which was a dynamic gesture including a rotation movement of the wrist as symbolized by the arrow in Figure 8. This set of gestures was considered to have a middle ambiguity level as it included a dynamic gesture and a similarity in the posture between the signs A, C, and E and the signs G and H.

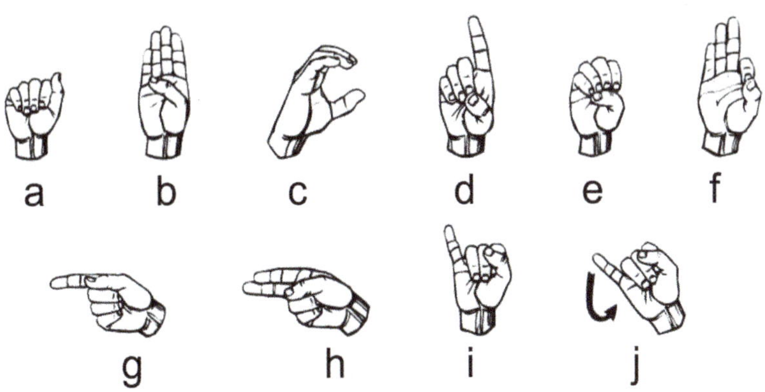

Figure 8. Ten ASL letters, A–J.

Implemented with 3000 hidden nodes after a feature selection by the KTGELM, the ELM had both a microaverage precision and a microaverage sensitivity of 97% when trained with only 13 selected features out of the original 48 features, as it is detailed in the comparison presented in Table 4. The KTGEL initialized with 3000 hidden nodes resulted in a trained model with only 1000 hidden nodes while it was given the full 48 features as inputs. The KTGEL reached the same total sensitivity with a reduced total precision by only 1% in comparison with the ELM after a separate feature selection stage.

Table 4. Comparison between the ELM and KTGEL in recognition of ASL signs with a middle ambiguity level.

	ELM	KTGEL
Additional feature selection algorithm	yes	no
Initial number of features	13	48
Initial number of hidden nodes	3000	3000
Final number of hidden nodes	3000	1000
Training time with feature selection in seconds	9.5	2.5
Testing time in seconds	0.22	0.04
Testing accuracy in %	95	94
Precision in %	97	96
Sensitivity in %	97	97

The confusion matrices in Figures 9 and 10 show that even though J was a dynamic gesture, it was 100% recognized using the FSR wrist band, which could be explained by the muscle deformation resulting from the rotation of the wrist which resulted in a stronger level of the signal in comparison with the other signs where the muscle movements in the wrist level were not visible.

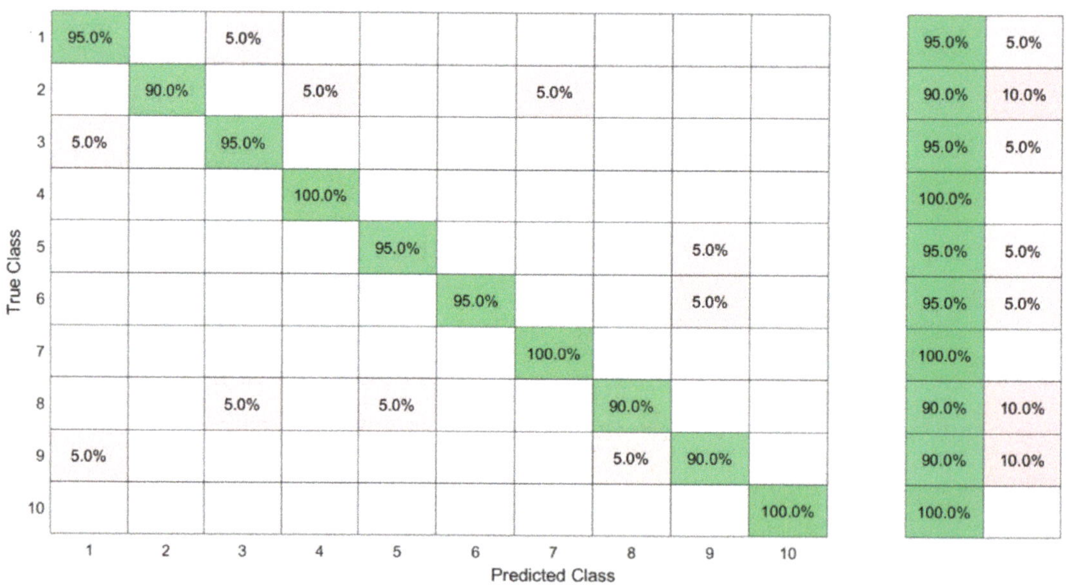

Figure 9. Ten ASL letters, A–J, detected with FMG and classified by the ELM after feature selection by the KTGELM.

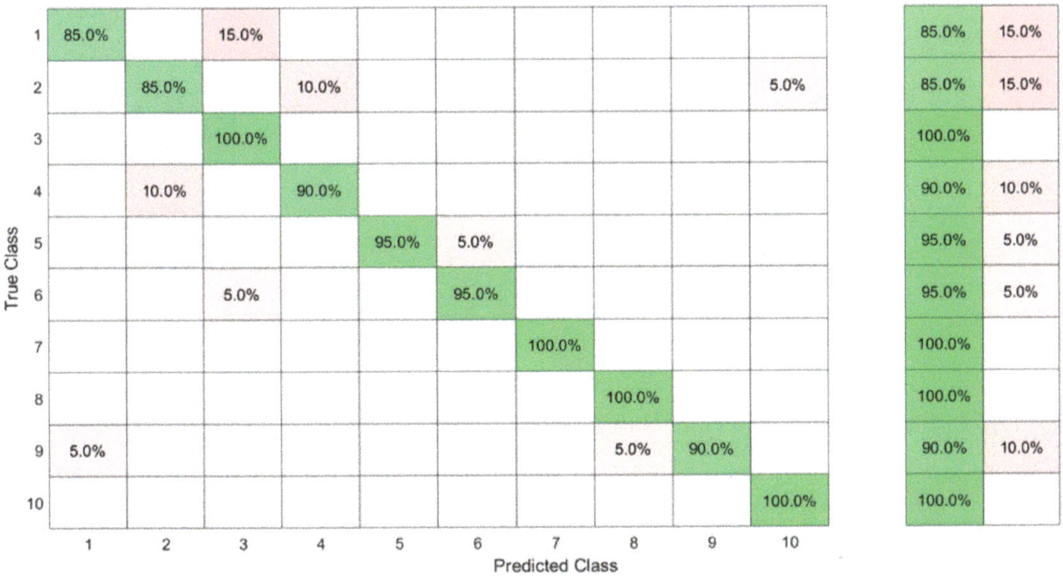

Figure 10. Ten ASL letters, A–J, detected with FMG and classified by the KTGEL.

4.4. Recognition of ASL Signs with a High Ambiguity Level

The used data set in this part included the 20 ASL letters shown in Figure 11 with 10 of the signs showing a big similarity, namely between "B" and "4", "M" and "N", "U" and "2", "6" and "W," "S" and "T", and a dynamic sign "Z", so the expected accuracy could be as low as 50% for this data set.

The same sign set was collected by 20 new subjects while wearing the eight-sensor band, and 100 observations of each sign were collected as FMG data. The data collected by the FMG sensor at the wrist level presented not only information about muscle contraction but also about the tendon state. As the sensors were distributed around the wrist, the FMG band could cover all the superficial muscles. Hence, more confusion between signs was noted due to the force transmission through the muscle fibers during the contraction and the influence of the deep muscle on the superficial ones. Therefore, different signs could have the same FMG response at the level of one or more sensors when signs shared an initial hand shape or the same performing fingers. For the ELM after the KTGELM feature selection shown in Figure 12, it could be seen that the signs "T" and "S", symbolized as classes 15 and 11, presented a source of confusion for the rest of the signs as not only the majority of their observations were misclassified, but also many other classes were mispredicted as signs "T" and "S". Based on the FMG data set, the ELM after feature selection by the KTGELM presented a classification microaveraged precision of 78% with a sensitivity of 80% among the 20 signs.

Using the same database, the KTGEL resulted a trained model with 1000 hidden nodes. The original data without feature selection are presented in the confusion matrix in Figure 13, where similar results to the ELM with feature selection could be seen with a micro-p of 77% and a micro-r of 80% and a thrice smaller model size. Evaluating the overall classification accuracy, it could be seen that the ELM after the KTGELM and the KTGEL had the same performance in most cases, with the second being less complicated as it had only 1000 hidden nodes and could do the feature selection and the classification in the same process.

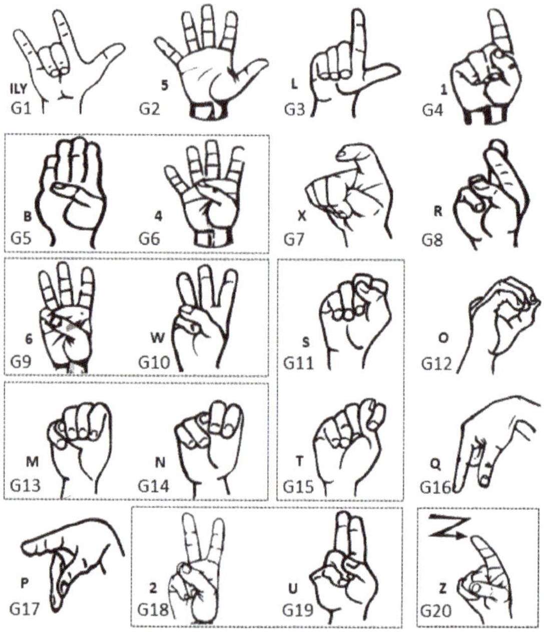

Figure 11. Data set of 20 ASL letters with expected high ambiguity, namely between "B" and "4", "M" and "N", "U" and "2", "6" and "W", "S" and "T", and a dynamic sign "Z".

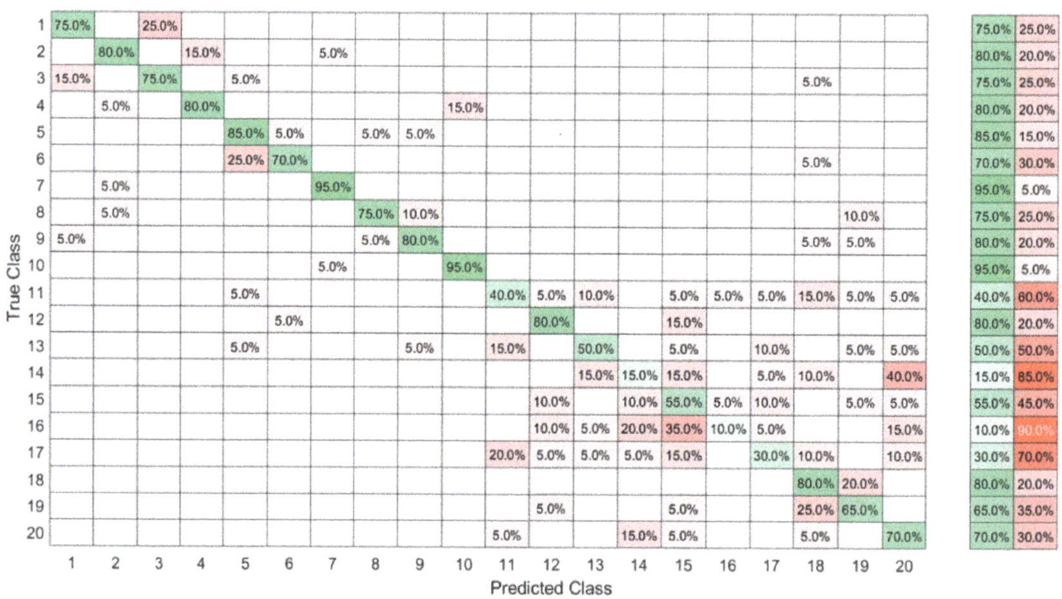

Figure 12. High-ambiguity data set classified by ELM after feature selection by the KTGELM.

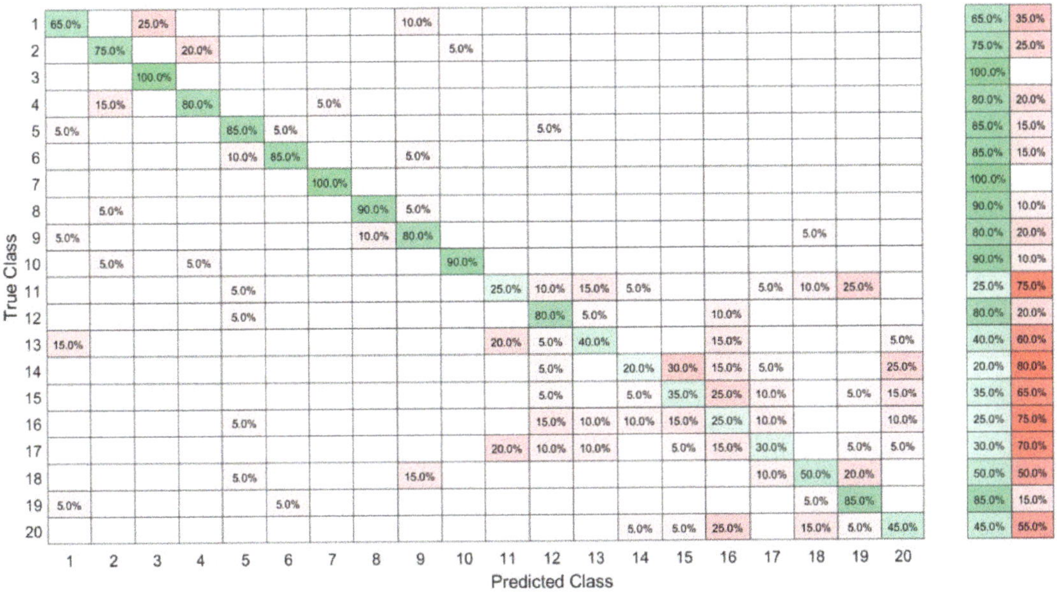

Figure 13. High-ambiguity data set classified by the KTGEL without a previous feature selection.

5. Conclusions

This work focused on recognizing American Sign Language based on commercial FMG sensors. We proposed to optimize an ELM by a weight-pruning method to optimize the network architecture and maintain the randomness of the initial weights. The pruning reduced the network size in the ELM by removing the weights, which were participating less in the classification result. We proposed to use the k-tournament grasshopper optimization algorithm (KTGOA) as the core of the ELM's weight-pruning process due to its fast convergence in multidimensional optimization spaces. A KTGOA was implemented to select the ELM weights. Thereby, a k-tournament grasshopper extreme learner (KTGEL) was proposed as a classifier with a reduced architecture, high performance, and internal feature selection. The influence of the number of FMG sensors and the number of subjects on the performance of the KTGEL was first investigated. It was proved in this paper that if only one subject was performing the data collection, a six-sensor bracelet was sufficient. However, with an increasing number of subjects, eight sensors were the minimal number needed to recognize the ASL numbers accurately. The investigation of the influence of the ambiguity level in the set of gestures on the performance of the KTGEL compared with the ELM showed that both had similar accuracy in the case of middle and high ambiguity levels. However, the ELM was trained with fewer features as it was preceded by a feature selection wrapper, while the KTGEL was trained with all the features. Moreover, in both tested cases, the KTGEL-trained model reduced the number of initially hidden nodes by two-thirds. The KTGEL also showed similar sensitivity and precision values with those of the ELM trained with selected features. The proposed KTGEL was created by the KTGOA that optimized the process of the best solution selection but inherited the linear behavior of the exploration–exploitation balancing coefficient from the original GOA. Similarly to the GOA, this linearity could lead to a trapping into a local optimum during the selection process of coupled features and weights, during the weight pruning in the KTGEL. Hence, in future work, the nonlinearization of the exploration–exploitation coefficient for the weight selection process will be investigated.

Author Contributions: Conceptualization, R.B. and O.K.; methodology, R.B.; software, R.B.; validation, R.B. and O.K.; formal analysis, R.B.; investigation, R.B.; resources, O.K.; data curation, R.B.; writing—original draft preparation, R.B.; writing—review and editing, R.B.; visualization, R.B.; supervision, O.K.; project administration, O.K.; funding acquisition, O.K. All authors have read and agreed to the published version of the manuscript.

Funding: This work was supported by the Deutsche Forschungsgemeinschaft (DFG, German Research Foundation)–Project-ID 416228727–SFB 1410, subproject A03.

Institutional Review Board Statement: The study was conducted according to the guidelines of the Declaration of Helsinki, and approved by the Institutional Ethics Committee of Technische Universitat Chemnitz, (reference: V-331-15-GJSensor-13052019).

Informed Consent Statement: Informed consent was obtained from all subjects involved in the study.

Conflicts of Interest: The authors declare no conflict of interest and the funders had no role in the design of the study; in the collection, analyses, or interpretation of data; in the writing of the manuscript, or in the decision to publish the results.

References

1. Albadr, M.A.A.; Tiun, S.; AL-Dhief, F.T.; Sammour, M.A.M. Spoken language identification based on the enhanced self-adjusting extreme learning machine approach. *PLoS ONE* **2018**, *13*, e0194770. [CrossRef] [PubMed]
2. Zhao, Z.; Chen, Z.; Chen, Y.; Wang, S.; Wang, H. A Class Incremental Extreme Learning Machine for Activity Recognition. *Cogn. Comput.* **2014**, *6*, 423–431. [CrossRef]
3. Song, S.; Wang, M.; Lin, Y. An improved algorithm for incremental extreme learning machine. *Syst. Sci. Control Eng.* **2020**, *8*, 308–317. [CrossRef]
4. Miche, Y.; Sorjamaa, A.; Bas, P.; Simula, O.; Jutten, C.; Lendasse, A. OP-ELM: Optimally Pruned Extreme Learning Machine. *IEEE Trans. Neural Netw.* **2010**, *21*, 158–162. [CrossRef]
5. Miche, Y.; Sorjamaa, A.; Lendasse, A. OP-ELM: theory, experiments and a toolbox. In Proceedings of the International Conference on Artificial Neural Networks, Prague, Czech Republic, 3–6 September 2008; Springer: Berlin/Heidelberg, Germany, 2008; pp. 145–154. [CrossRef]
6. Kaloop, M.R.; El-Badawy, S.M.; Ahn, J.; Sim, H.B.; Hu, J.W.; El-Hakim, R.T.A. A hybrid wavelet-optimally-pruned extreme learning machine model for the estimation of international roughness index of rigid pavements. *Int. J. Pavement Eng.* **2022**, *23*, 862–876. [CrossRef]
7. Pouzols, F.M.; Lendasse, A. Evolving fuzzy optimally pruned extreme learning machine for regression problems. *Evol. Syst.* **2010**, *1*, 43–58. [CrossRef]
8. Alencar, A.S.C.; Neto, A.R.R.; Gomes, J.P.P. A new pruning method for extreme learning machines via genetic algorithms. *Appl. Soft Comput.* **2016**, *44*, 101–107. [CrossRef]
9. Ahila, R.; Sadasivam, V.; Manimala, K. An integrated PSO for parameter determination and feature selection of ELM and its application in classification of power system disturbances. *Appl. Soft Comput.* **2015**, *32*, 23–37. [CrossRef]
10. Saremi, S.; Mirjalili, S.; Lewis, A. Grasshopper Optimisation Algorithm: Theory and application. *Adv. Eng. Softw.* **2017**, *105*, 30–47. [CrossRef]
11. Wang, J.; Tang, L.; Bronlund, J.E. Surface EMG Signal Amplification and Filtering. *Int. J. Comput. Appl.* **2013**, *82*, 15–22. [CrossRef]
12. Al-Betar, M.A.; Awadallah, M.A.; Faris, H.; Aljarah, I.; Hammouri, A.I. Natural selection methods for Grey Wolf Optimizer. *Expert Syst. Appl.* **2018**, *113*, 481–498. [CrossRef]
13. Amft, O.; Junker, H.; Lukowicz, P.; Troster, G.; Schuster, C. Sensing muscle activities with body-worn sensors. In Proceedings of the International Workshop on Wearable and Implantable Body Sensor Networks (BSN'06), Cambridge, MA, USA, 3–5 April 2006; pp. 4–141. ISSN: 2376-8886. [CrossRef]
14. Xiao, Z.G.; Menon, C. A Review of Force Myography Research and Development. *Sensors* **2019**, *19*, 4557. [CrossRef]
15. Vecchi, F.; Freschi, C.; Micera, S.; Sabatini, A.M.; Dario, P.; Sacchetti, R.; Vecchi, F.; Sabatini, P.R.; Sacchetti, R. Experimental evaluation of two commercial force sensors for applications in biomechanics and motor control. In *Proceedings of the IFESS 2000 and NP 2000 Proceedings: 5th Annual Conference of the International Functional Electrical Stimulation Society and 6th Triennial Conference "Neural Prostheses: Motor Systems;* Center for Sensory-Motor Interaction (SMI), Department of Health Science and Technology, Aalborg University: Aalborg, Denmark, 2000.
16. Hollinger, A.; Wanderley, M.M. Evaluation of Commercial Force-Sensing Resistors. In Proceedings of the International Conference on New Interfaces for Musical Expression NIME-06, Paris, France, 4–8 June 2006. [CrossRef]
17. Saadeh, M.Y.; Trabia, M.B. Identification of a force-sensing resistor for tactile applications. *J. Intell. Mater. Syst. Struct.* **2012**, *24*, 813–827. [CrossRef]
18. Barioul, R.; Ghribi, S.F.; Abbasi, M.B.; Fasih, A.; Jmeaa-Derbel, H.B.; Kanoun, O. Wrist Force Myography (FMG) Exploitation for Finger Signs Distinguishing. In Proceedings of the 2019 5th IEEE International Conference on Nanotechnology for Instrumentation and Measurement (NanofIM), Sfax, Tunisia, 30–31 October 2019. [CrossRef]

19. Jiang, X.; Merhi, L.K.; Xiao, Z.G.; Menon, C. Exploration of Force Myography and surface Electromyography in hand gesture classification. *Med. Eng. Phys.* **2017**, *41*, 63–73. [CrossRef] [PubMed]
20. Wan, B.; Wu, R.; Zhang, K.; Liu, L. A new subtle hand gestures recognition algorithm based on EMG and FSR. In Proceedings of the 2017 IEEE 21st International Conference on Computer Supported Cooperative Work in Design (CSCWD), Wellington, New Zealand, 26–28 April 2017; pp. 127–132. [CrossRef]
21. Dementyev, A.; Paradiso, J.A. WristFlex: Low-power Gesture Input with Wrist-worn Pressure Sensors. In Proceedings of the Proceedings of the 27th Annual ACM Symposium on User Interface Software and Technology, Honolulu, HI, USA, 5–8 October 2014; ACM: New York, NY, USA, 2014; UIST '14, pp. 161–166. [CrossRef]
22. Jiang, X.; Merhi, L.K.; Menon, C. Force Exertion Affects Grasp Classification Using Force Myography. *IEEE Trans. -Hum.-Mach. Syst.* **2018**, *48*, 219–226. [CrossRef]
23. Ferigo, D.; Merhi, L.K.; Pousett, B.; Xiao, Z.G.; Menon, C. A Case Study of a Force-myography Controlled Bionic Hand Mitigating Limb Position Effect. *J. Bionic Eng.* **2017**, *14*, 692–705. [CrossRef]
24. Cho, E.; Chen, R.; Merhi, L.K.; Xiao, Z.; Pousett, B.; Menon, C. Force Myography to Control Robotic Upper Extremity Prostheses: A Feasibility Study. *Front. Bioeng. Biotechnol.* **2016**, *4*, 18. [CrossRef] [PubMed]
25. Barioul, R.; Ghribi, S.F.; Derbel, H.B.J.; Kanoun, O. Four Sensors Bracelet for American Sign Language Recognition based on Wrist Force Myography. In Proceedings of the 2020 IEEE International Conference on Computational Intelligence and Virtual Environments for Measurement Systems and Applications (CIVEMSA), Tunis, Tunisia, 22–24 June 2020. [CrossRef]
26. Atitallah, B.B.; Abbasi, M.B.; Barioul, R.; Bouchaala, D.; Derbel, N.; Kanoun, O. Simultaneous Pressure Sensors Monitoring System for Hand Gestures Recognition. In Proceedings of the 2020 IEEE Sensors, Rotterdam, The Netherlands, 25–28 October 2020. [CrossRef]
27. Ahmadizadeh, C.; Merhi, L.K.; Pousett, B.; Sangha, S.; Menon, C. Toward Intuitive Prosthetic Control: Solving Common Issues Using Force Myography, Surface Electromyography, and Pattern Recognition in a Pilot Case Study. *IEEE Robot. Autom. Mag.* **2017**, *24*, 102–111. [CrossRef]
28. Sadarangani, G.P.; Menon, C. A preliminary investigation on the utility of temporal features of Force Myography in the two-class problem of grasp vs. no-grasp in the presence of upper-extremity movements. *Biomed. Eng. Online* **2017**, *16*, 59. [CrossRef]
29. Jiang, X.; Tory, L.; Khoshnam, M.; Chu, K.H.T.; Menon, C. Exploration of Gait Parameters Affecting the Accuracy of Force Myography-Based Gait Phase Detection. In Proceedings of the 2018 7th IEEE International Conference on Biomedical Robotics and Biomechatronics (Biorob), Enschede, The Netherlands, 26–29 August 2018. [CrossRef]
30. Islam, M.R.U.; Waris, A.; Kamavuako, E.N.; Bai, S. A comparative study of motion detection with FMG and sEMG methods for assistive applications. *J. Rehabil. Assist. Technol. Eng.* **2020**, *7*, 205566832093858. [CrossRef]
31. Godiyal, A.K.; Pandit, S.; Vimal, A.K.; Singh, U.; Anand, S.; Joshi, D. Locomotion mode classification using force myography. In Proceedings of the 2017 IEEE Life Sciences Conference (LSC), Sydney, Australia, 13–15 December 2017. [CrossRef]
32. Anvaripour, M.; Saif, M. Hand gesture recognition using force myography of the forearm activities and optimized features. In Proceedings of the 2018 IEEE International Conference on Industrial Technology (ICIT), Lyon, France, 19–22 February 2018; pp. 187–192. [CrossRef]
33. Jiang, S.; Gao, Q.; Liu, H.; Shull, P.B. A novel, co-located EMG-FMG-sensing wearable armband for hand gesture recognition. *Sens. Actuators A Phys.* **2020**, *301*, 111738. [CrossRef]
34. Ramalingame, R.; Barioul, R.; Li, X.; Sanseverino, G.; Krumm, D.; Odenwald, S.; Kanoun, O. Wearable Smart Band for American Sign Language Recognition With Polymer Carbon Nanocomposite-Based Pressure Sensors. *IEEE Sens. Lett.* **2021**, *5*, 1–4. [CrossRef]
35. Al-Hammouri, S.; Barioul, R.; Lweesy, K.; Ibbini, M.; Kanoun, O. Six Sensors Bracelet for Force Myography based American Sign Language Recognition. In Proceedings of the 2021 IEEE 18th International Multi-Conference on Systems, Signals & Devices (SSD), Monastir, Tunisia, 22–25 March 2021. [CrossRef]
36. Reed, R. Pruning algorithms-a survey. *IEEE Trans. Neural Netw.* **1993**, *4*, 740–747. [CrossRef]
37. Tian, Y.; Yu, Y. A new pruning algorithm for extreme learning machine. In Proceedings of the 2017 IEEE International Conference on Information and Automation (ICIA), Macau SAR, China, 18–20 July 2017. [CrossRef]
38. Freire, A.L.; Neto, A.R.R. A Robust and Optimally Pruned Extreme Learning Machine. In *Advances in Intelligent Systems and Computing*; Springer International Publishing: Cham, Switzerland, 2017; pp. 88–98. [CrossRef]
39. Cui, L.; Zhai, H.; Wang, B. An Improved Pruning Algorithm for ELM Based on the PCA. In Proceedings of the 2017 International Conference on Robotics and Artificial Intelligence-ICRAI, Shanghai, China, 29–31 December 2017; ACM Press: New York, NY, USA, 2017. [CrossRef]
40. de Campos Souza, P.V.; Araujo, V.S.; Guimaraes, A.J.; Araujo, V.J.S.; Rezende, T.S. Method of pruning the hidden layer of the extreme learning machine based on correlation coefficient. In Proceedings of the 2018 IEEE Latin American Conference on Computational Intelligence (LA-CCI), Gudalajara, Mexico, 7–9 November 2018. [CrossRef]
41. Li, X.; Wang, K.; Jia, C. Data-Driven Control of Ground-Granulated Blast-Furnace Slag Production Based on IOEM-ELM. *IEEE Access* **2019**, *7*, 60650–60660. [CrossRef]
42. de Campos Souza, P.V.; Torres, L.C.B.; Silva, G.R.L.; de Padua Braga, A.; Lughofer, E. An Advanced Pruning Method in the Architecture of Extreme Learning Machines Using L1-Regularization and Bootstrapping. *Electronics* **2020**, *9*, 811. [CrossRef]

43. Mafarja, M.; Aljarah, I.; Faris, H.; Hammouri, A.I.; Al-Zoubi, A.M.; Mirjalili, S. Binary grasshopper optimisation algorithm approaches for feature selection problems. *Expert Syst. Appl.* **2019**, *117*, 267–286. [CrossRef]
44. Huang, G.B. What are Extreme Learning Machines? Filling the Gap Between Frank Rosenblatt's Dream and John von Neumann's Puzzle. *Cogn. Comput.* **2015**, *7*, 263–278. [CrossRef]
45. Anam, K.; Al-Jumaily, A. Evaluation of extreme learning machine for classification of individual and combined finger movements using electromyography on amputees and non-amputees. *Neural Netw.* **2017**, *85*, 51–68. [CrossRef] [PubMed]
46. Chorowski, J.; Wang, J.; Zurada, J.M. Review and performance comparison of SVM- and ELM-based classifiers. *Neurocomputing* **2014**, *128*, 507–516. [CrossRef]
47. Ibrahim, H.T.; Mazher, W.J.; Ucan, O.N.; Bayat, O. A grasshopper optimizer approach for feature selection and optimizing SVM parameters utilizing real biomedical data sets. *Neural Comput. Appl.* **2018**, *31*, 5965–5974. [CrossRef]
48. Shi, J.; Cai, Y.; Zhu, J.; Zhong, J.; Wang, F. SEMG based hand motion recognition using cumulative residual entropy and extreme learning machine. *Med. Biol. Eng. Comput.* **2012**, *51*, 417–427. [CrossRef] [PubMed]
49. Huang, G.B.; Zhou, H.; Ding, X.; Zhang, R. Extreme Learning Machine for Regression and Multiclass Classification. *IEEE Trans. Syst. Man, Cybern. Part (Cybernetics)* **2012**, *42*, 513–529. [CrossRef] [PubMed]
50. Carrara, F.; Elias, P.; Sedmidubsky, J.; Zezula, P. LSTM-based real-time action detection and prediction in human motion streams. *Multimed. Tools Appl.* **2019**, *78*, 27309–27331. [CrossRef]

sensors

MDPI

Article

Hand Gesture Recognition Using EMG-IMU Signals and Deep Q-Networks

Juan Pablo Vásconez †,‡, Lorena Isabel Barona López †,‡, Ángel Leonardo Valdivieso Caraguay †,‡ and Marco E. Benalcázar *,†,‡

Artificial Intelligence and Computer Vision Research Lab, Escuela Politécnica Nacional, Quito 170517, Ecuador
* Correspondence: marco.benalcazar@epn.edu.ec; Tel.: +593-222976300 (ext. 4706)
† Current address: Ladrón de Guevara E11-253, Quito 170517, Ecuador.
‡ These authors contributed equally to this work.

Abstract: Hand gesture recognition systems (HGR) based on electromyography signals (EMGs) and inertial measurement unit signals (IMUs) have been studied for different applications in recent years. Most commonly, cutting-edge HGR methods are based on supervised machine learning methods. However, the potential benefits of reinforcement learning (RL) techniques have shown that these techniques could be a viable option for classifying EMGs. Methods based on RL have several advantages such as promising classification performance and online learning from experience. In this work, we developed an HGR system made up of the following stages: pre-processing, feature extraction, classification, and post-processing. For the classification stage, we built an RL-based agent capable of learning to classify and recognize eleven hand gestures—five static and six dynamic—using a deep Q-network (DQN) algorithm based on EMG and IMU information. The proposed system uses a feed-forward artificial neural network (ANN) for the representation of the agent policy. We carried out the same experiments with two different types of sensors to compare their performance, which are the Myo armband sensor and the G-force sensor. We performed experiments using training, validation, and test set distributions, and the results were evaluated for user-specific HGR models. The final accuracy results demonstrated that the best model was able to reach up to 97.50% ± 1.13% and 88.15% ± 2.84% for the classification and recognition, respectively, with regard to static gestures, and 98.95% ± 0.62% and 90.47% ± 4.57% for the classification and recognition, respectively, with regard to dynamic gestures with the Myo armband sensor. The results obtained in this work demonstrated that RL methods such as the DQN are capable of learning a policy from online experience to classify and recognize static and dynamic gestures using EMG and IMU signals.

Keywords: hand gesture recognition; electromyography; inertial measurement unit; reinforcement learning; deep Q-network

Citation: Vásconez, J.P.; Barona López, L.I.; Valdivieso Caraguay, Á.L.; Benalcázar, M.E. Hand Gesture Recognition Using EMG-IMU Signals and Deep Q-Networks. *Sensors* **2022**, 22, 9613. https://doi.org/10.3390/s22249613

Academic Editor: Georg Fischer

Received: 28 October 2022
Accepted: 19 November 2022
Published: 8 December 2022

1. Introduction

In recent years, the use of non-verbal communication techniques has proven useful for creating human–machine interfaces (HMIs). In particular, hand gesture recognition (HGR) systems have been used in applications such as sign language recognition, human–machine interfaces, muscle rehabilitation systems, prosthesis design, robotic applications, and augmented reality, among others [1–6]. However, designing HGR systems that are capable of determining with high accuracy the moment a certain gesture was performed is a challenging problem. This is due in part to the variability of the signals of each gesture between different users, as well as the similarities that the signals of different hand gestures may have.

Several HGR systems use vision-based methods, for example, Kinect [7] and Leap Motion Sensor [8]. On the other hand, sensor-based HGR systems typically use gloves with inertial measurement units (IMU) [9,10], as well as non-invasive surface electromyography

(EMG) methods for the detection of arm muscle activity, such as the G-force and Myo armband sensors [6]. However, the performance of vision-based method systems can be affected by occlusion and illumination issues, as well as the distance between the sensor and the hand. For this, sensor-based HGR systems based on EMG or IMU signals are preferred for different HGR applications. It is worth mentioning that EMG signals (EMG) are often selected when static gestures are used since the information from muscle activity is usually sufficient to characterize this type of hand gesture [1,4,5]. On the other hand, IMU signals (IMUs) are usually selected to characterize dynamic gestures since this type of gesture primarily depends on hand and arm movements [6]. Therefore, a combination of EMG and IMU signals to recognize static and dynamic hand gestures could increase the performance of HGR systems since more information is analyzed for each gesture [11]. However, this is still an open research problem [12,13].

EMG signals can be modeled as a stochastic process that depends on whether the muscle contraction is static or dynamic. However, to address these problems, machine learning (ML) and deep learning (DL) techniques have been commonly used to classify and recognize EMG signals instead of mathematical models since the latter have high design complexity and performance issues [1,14]. In particular, supervised methods, such as support vector machines (SVMs), k-nearest neighbors (K-NNs), artificial neural networks (ANNs), convolutional neural networks (CNNs), a fusion of the transformer model and the CNN model (transformer-CNN), and long short-term memory (LSTM) networks, have shown high-performance results for HGR systems (at least 80% classification accuracy and 300 ms processing time) [1,15–20]. However, these models still require a fully labeled dataset to be trained, which makes them unsuitable for learning using new experiences gained online when the user interacts with the system. On the contrary, reinforcement learning (RL) approaches can help build models that learn online from experience. These models could help improve the performance of the HGR system over time since the system can adapt to each user in an online manner after each interaction with the system, which helps reduce the problem of interpersonal variability. Reinforcement learning methods are based on the maximization of the accumulated reward that is obtained by trying to correctly predict a gesture from online experiences, which allows for finding an optimal policy for an agent to use to predict categories of signals in a given environment [16].

There have been a few attempts to use RL techniques for HGR and arm movement or hand gesture characterization using sensor-based systems. For example, in [21], the authors used the Myo armband sensor to extract 9-axis IMU and 8-axis EMG sensor information to classify dynamic hand gestures using a deep Q-network (DQN) model. The experiment consisted only of three different hand gestures based on drawing a circle, a rectangle, and a triangle in the air. Each of these three gestures had 30 training data and 20 test data. The agent was built using a CNN with and without LSTM layers and was demonstrated to obtain high classification performance. In [22], the authors used the UCI dataset, which contains EMG data from six users performing six different hand gestures. From this dataset, time-domain features were obtained using a CNN-based automatic feature extraction method. To learn a classification policy, a deep Q-learning dueling technique was used, which allows for the selection of the most relevant characteristics throughout the training. The base dataset was composed of a total of 2700 EMG signal samples for the six hand gestures. As this was a sparse dataset, the authors used data augmentation methods using Gaussian noise, random horizontal flipping, and vertical flipping on the EMG data to obtain 10,000 samples. The authors showed that CNN performed better than ANN for this dataset. In another work, the authors proposed a classifier based on the neural reinforcement learning (NRL) method to classify finger movements using only EMGs [23]. For this, the authors used four feature extraction methods, which were the variance, mean absolute value, zero crossing, and waveform length of seven different gesture classes. Then, they used a k-nearest neighbor classifier based on reinforcement learning to classify the extracted features using a trial-and-error approach. The authors performed experiments on 10 users with general and specific models, demonstrating that it was feasible for the NRL user to identify typing movements using EMG signals from the forearm. In [24], a reinforcement learning-based classifier capable of learning to classify arm and finger movements was designed. For this, a 26T System was used to obtain EMG signals from 10 subjects using 1, 2, and 3 electrodes, respectively, to compare their results. The temporal characteristics that were used were the length of the waveform, the mean absolute value, the variance, and the zero crossing. An algorithm based on Q-learning was used for the classification stage, where the agent was made up of an ANN to infer six classes of arm positions and four classes of finger movements. The authors used 144 training samples and 95 test samples to build specific models for each of the 10 subjects. Finally, we presented an approach to classify and recognize five different static hand gestures based only on the EMGs in [16]. For this, we used Q-learning with an ANN as a policy representation of the agent. However, we used only the EMG signals to recognize static gestures and data were obtained using only the Myo armband sensor. Although the results obtained were encouraging, it is still necessary to explore other types of gestures and sensor behaviors when using different RL-based methods. Moreover, the use of IMU is still key to recognizing dynamic gestures, and the combination of EMG-IMU signals still needs to be analyzed and compared to a case when only EMGs are used to develop HGR systems based on RL methods. In summary, the use of datasets with a considerable number of samples and participants for both dynamic and static gestures based on EMG and IMU information still needs to be explored for different RL-based methods and sensors. To the best of our knowledge, this work is the first attempt to use EMG-IMU signals from a large dataset from two different sensors (Myo armband and G-force) and compare the results with other methods.

Considering the literature review presented above, the main contributions of the present work are listed below:

- We use our large dataset composed of 85 users with information on 11 different hand gestures (5 static and 6 dynamic gestures) that contain EMG and IMU signals. The data were taken from two different armband sensors, the Myo armband and G-force sensors.
- We successfully combine the EMG-IMU signals with the deep Q-network (DQN) reinforcement learning algorithm. We propose an agent's policy representations based on artificial neural networks (ANN).
- We compare the results of the proposed method using both sensors, the Myo armband and G-force sensors. We also compare the results found in the present work, which uses EMG and IMU signals, with those of a method previously developed on a dataset that used only EMG signals and the Q-learning algorithm.

The rest of this work is organized as follows. In Section 2, the proposed method for an HGR system based on EMG-IMU signals and RL is presented and each stage is explained in detail. The classification and recognition results of the proposed method are presented in Section 3. The discussion section is in Section 4. Finally, the conclusions are provided in Section 5.

2. Hand Gesture Recognition Method

In this section, we present the proposed method for the HGR system based on EMG-IMU signals and RL (Figure 1). As can be observed, the proposed method is composed of data acquisition, pre-processing, feature extraction, classification (DQN), and post-processing stages. The data were taken from two different armband sensors to compare results, which are the Myo armband and G-force sensors. We combined the EMG-IMU signals with the deep Q-network (DQN) reinforcement learning algorithm to develop the proposed HGR system. Next, we explain in detail each stage.

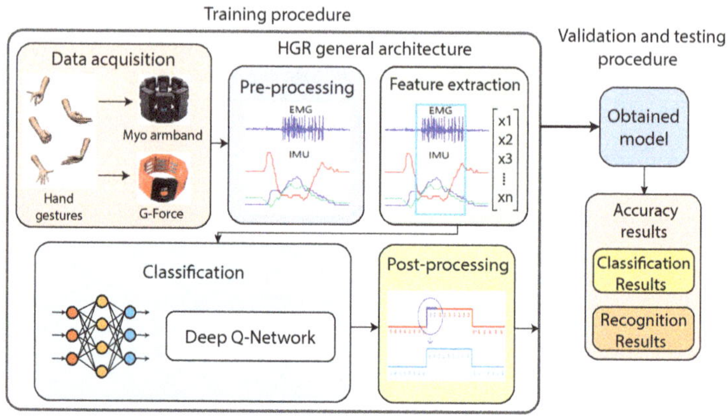

Figure 1. Hand gesture recognition method based on EMG-IMU and RL.

2.1. Data Acquisition

In this work, we use EMG-IMU data of 12 different hand gesture categories—11 different hand gestures and 1 relax gesture—in which 5 of them are static gestures—wave in, wave out, fist, open, and pinch—and the other 6 are dynamic gestures—up, down, left, right, forward, and backward. The data were collected using the Myo armband—a sensor with 8 channels at a sampling rate of 200 Hz—and the G-force armband—a sensor with 8 channels at a sampling rate of 1 kHz. The proposed dataset consists of 85 users, of whom 43 are used for training and validation to find the best possible hyperparameter configurations. From this group, 16 users are from the Myo armband sensor data and 27 from the G-force sensor data. On the other hand, 42 users are used for testing to evaluate overfitting and to calculate the final results. From this group, 16 users are from the Myo armband sensor data and 26 from the G-force sensor data. The data of each user in the training set is composed of 180 hand gesture repetitions—15 repetitions for each gesture—and the other 180 samples are for validation. This division of samples is similar to the test set. We summarize the dataset distribution for both the training and testing sets in Table 1. The dataset has been made public and is available at the following link https://laboratorio-ia.epn.edu.ec/en/resources/dataset/emg-imu-epn-100 accessed on 18 November 2022.

Table 1. Dataset distribution to evaluate user-specific models [25].

	User-Specific Model (One Model for Each of the 85 Users)			
	Number of Models	**Training**	**Validation**	**Test**
Training set	43 models trained (to find the best hyperparameters)	180 samples per user	180 samples per user	-
Testing set	42 models trained (to use the best of the found hyperparameters)	180 samples per user	-	180 samples per user

2.2. Pre-Processing

The preprocessing of each EMG sample consisted of using a sliding window on each sample to analyze it separately [1,14]. In this work, we chose a window length of 300 and a step of 40, where these values were selected based on experimentation to achieve high classification and recognition accuracy. Since we had two different sensors—Myo armband and G-force—with different sample frequencies—200 Hz and 1 kHz—a resampling was performed by applying an FIR antialiasing low-pass filter to the signals so that the EMGs and IMUs would have the same number of 1000 points for both sensors. However, only one window of 300 points was sent to the feature extraction stage to be evaluated at each time instant. Each EMG sensor had 8 channels, and to obtain the IMU signal, the 4 signals of the quaternions were used; thus, each EMG-IMU window information had a dimension of $[300, 12]$.

2.3. Feature Extraction

Feature extraction methods are used to extract relevant and non-redundant features from EMGs and IMUs. For this purpose, different domains can be used such as time, frequency, or time-frequency domains. In this work, five different features were extracted in the time domain over each step of the sliding window. The feature extraction functions used were root mean square (RMS), standard deviation (SD), energy (E), mean absolute value (MAV), and absolute envelope (AE), which are typically used to extract features of EMGs [1,14]. We used all these features in a feature vector since we obtained better results than when we used only one or a few of them. Since we had 5 feature extraction methods and an EMG-IMU window size of $[300, 12]$, a feature vector with a size of $[60, 1]$ was extracted from each of the EMG-IMU windows, which was made up of a feature vector with a size of $[40, 1]$ that corresponded to the EMGs and a vector with a size of $[20, 1]$ that corresponded to the quaternions obtained from the IMU.

2.4. Classification of EMGs

The objective of this stage is to identify the category of a hand gesture using an EMG-IMU signal among a set of categories with which the proposed algorithm was previously trained. In this work, we used an RL algorithm called deep Q-network (DQN), which is made up of a neural network to represent the agent's policy. In this section, we explain in detail the EMG-IMU signal sequential classification problem that can be modeled as a partially observable finite Markov decision process (POMDP).

2.4.1. Q-Learning

We can define the sliding window classification on an EMG-IMU signal sample during the development of a hand gesture as a sequential decision-making problem. In this problem, the actions correspond to the labels of the hand gestures to be inferred, whereas the states are the feature vectors corresponding to the observations of each window of an EMG-IMU sample. In this context, we can learn to estimate the optimal action for each state. For this purpose, we maximized the expected sum of future rewards by performing that action in the given states and then following an optimal policy [26]. Thus, considering a given policy π, the value of the action a taken in the initial state s can be defined as

$$Q_{\pi}(s,a) = \mathbb{E}_{\pi}[R_1 + \gamma R_2 + \gamma^2 R_3 + \ldots + \gamma^{n-1} R_n | S_0 = s, A_0 = a] \quad (1)$$

where R_i are the rewards or punishments that the agent receives at each state with $i = 1, 2, \ldots, n$, where n represents the number of states. The variable $\gamma \in [0,1]$ is the discount factor that determines how much future rewards affect the agent's learning process. Then, the optimal state-action value function can be expressed as $Q_*(s,a) = max_{\pi} Q_{\pi}(s,a)$. An optimal policy can be calculated from the optimal function $Q_*(s,a)$ by choosing the highest valued action at each state according to [27]. Typically, to estimate the optimal state-action values, we can use the Q-learning algorithm, which is an off-policy temporal difference RL method [26]. For any finite Markov decision process (MDP), the Q-learning algorithm can find an optimal policy by maximizing the expected return function that we presented in Equation (1) given an initial state and an initial action [27]. However, it is important to consider that we assume that only the observations O_t are measured instead of the complete state information of the environment s_t. This is because there may be a discrepancy between the set of EMG-IMU window observations and the set of feature vectors [16]. For this reason, in this work, we considered the HGR problem using EMG-IMU as a partially observable Markov decision process (POMDP) [16].

The Q-Learning algorithm uses Q-values to iteratively improve the behavior of the learning agent. The Q-values are an estimation of the performance of a certain action A_t at the observation O_t. There are different ways to represent the Q-values such as polynomial functions, tables, or neural networks [27]. In the proposed method, we used a continuous observation space represented by the extracted EMG-IMU features and a discrete action space represented by the predicted hand gestures. Therefore, the Q-learning algorithm should be combined with a function approximation approach to learningc a parameterized value function $Q(O_t, A_t; \theta_t)$. A critic representation can be used to obtain high-performance results when using discrete action spaces and continuous observations [27]. For a given observation and action, a critic agent output returns the expected value of the cumulative long-term reward. The standard Q-learning algorithm updates the parameters θ_t after taking action A_t in observation O_t, obtaining the reward R_{t+1} in O_{t+1}, described as follows:

$$\theta_{t+1} = \theta_t + \alpha\left(Y_t^Q - Q(O_t, A_t; \theta_t)\right) \cdot \nabla_{\theta_t} Q(O_t, A_t; \theta_t) \quad (2)$$

Here, θ_{t+1} and θ_t are the updated and the previous parameters, respectively, and α is the learning rate. Finally, the target function Y_t^Q is defined as

$$Y_t^Q \equiv R_{t+1} + \gamma \cdot \underset{a}{max}[Q(O_{t+1}, a; \theta_t)] \quad (3)$$

where the term $\underset{a}{max}[Q(O_{t+1}, a)]$ is the estimated optimal future Q value. The term γ is the discount factor, and a reward R_{t+1} is received by the agent when moving from the observation O_t by taking the action A_t to the next observation O_{t+1}.

2.4.2. Deep Q-Networks (DQN)

In this work, we use a deep Q-network (DQN) agent representation, which is composed of an artificial neural network (ANN) as a function approximation method to learn a parameterized value function. Thus, for a given observation O_t, a DQN returns a vector of action values $Q(O_t, \cdot\,; \theta)$, where θ are the parameters of the neural network [24,26,27]. The number of inputs of the network is the same as the dimension of the feature vector that represents an observation composed of the extracted EMG-IMU features $[60, 1]$, and the number of neurons at the output layer is the same as the number of possible actions that the agent can perform. According to [26,28], there are two key characteristics to consider in the DQN algorithm that are not considered in the standard Q-learning algorithm. The first is the use of a target network Y_t^{DQN} that is used in Equation (4), which has parameters θ^- that are updated periodically every τ steps from the online network in Equation (2), with the parameters θ_t. The rest of the time, the parameters θ^- remain fixed until the next update after τ steps. This helps to remove correlations with the target [26,28].

$$Y_t^{DQN} \equiv R_{t+1} + \gamma \cdot \underset{a}{max}\left[Q\left(O_{t+1}, a, \theta_t^-\right)\right] \quad (4)$$

The second important consideration is the use of experience replay, which randomly samples the data to remove correlations in the sequences of observations, which accelerates the training of the agent. For this purpose, the tuple $E_t = (O_t, A_t, R_t, S_{t+1})$ that represents the agent's experience at time t is saved in a pool of stored data sample transitions $\mathcal{D} = \{E_1, E_2, \cdots, E_T\}$. During learning, the parameters of the ANN are updated using Equations (2) and (4), with the mini-batches of experience drawn uniformly at random from $\mathcal{D}$ [28,29]. The use of the target network with parameters θ^- and the experience replay approach help to significantly improve the performance of the DQN algorithm compared to the standard Q-learning algorithm [26,28]. The pseudo-code for the DQN algorithm is presented in Algorithm 1.

Algorithm 1 DQN with Experience Replay

Initialize action-value function Q with random weights
Initialize replay memory $\mathcal{D}$ to capacity N
for episode = 1, M **do**
  Initialize agent in observation O_t
  for t = 1, T **do**
    With probability ϵ select a random action A_t
    otherwise, select $\underset{a}{max}\left[Q\left(O_{t+1}, A, \theta_t^-\right)\right]$
    store transition $E_t = (O_t, A_t, R_t, S_{t+1})$ in $\mathcal{D}$
    Sample random mini-batch of transitions (O_t, A_t, R_t, S_{t+1}) in $\mathcal{D}$

$$Y_t^{DQN} = \begin{cases} R_{t+1} & for \quad terminal \quad O_t \\ R_{t+1} + \gamma \cdot \underset{a}{max}\left[Q\left(O_{t+1}, a, \theta_t^-\right)\right] & for \quad non-terminal \quad O_t \end{cases}$$

    Perform gradient descent to update $\theta_{t+1} = \theta_t + \alpha\left(Y_t^Q - Q(O_t, A_t; \theta_t)\right) \cdot \nabla_{\theta_t} Q(O_t, A_t; \theta_t)$

  end for
end for

2.4.3. DQN for EMG-IMU Classification

The proposed method modeled as a partially observable Markov decision process (POMDP) that we use in this work uses DQN the algorithm to learn an optimal policy, which allows an agent to learn to classify and recognize hand gestures from EMG-IMU signals. A figure that represents the interaction between the DQN agent representation and the proposed environment for the EMG-IMU classification is illustrated in Figure 2. We briefly explain each part of Figure 2 below.

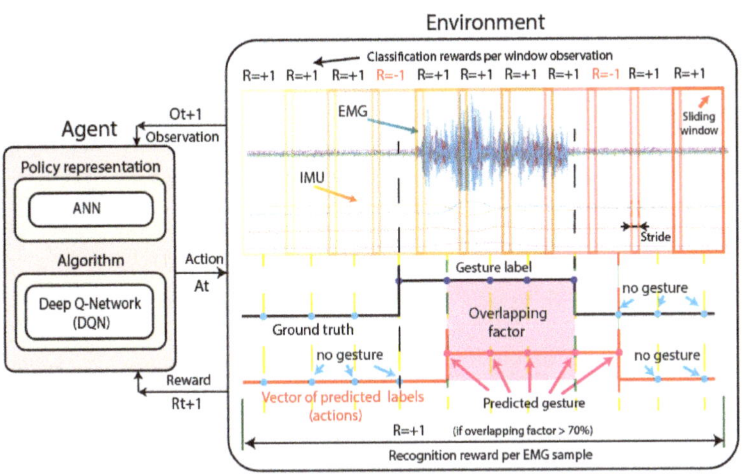

Figure 2. Scheme of the interaction between the DQN agent representation and the proposed environment for the EMG-IMU classification.

Agent: The agent is made up of the DQN algorithm and an artificial neural network ANN as the policy representation. During training, the agent learns a policy that maximizes the total sum of rewards using the DQN algorithm. The inputs of the neural network are the features extracted from each window of the EMG-IMU signals (observations), and as its output, the network returns the values of the predicted gestures (actions). In this way, the agent learns to classify window observations from EMG-IMU signals. Each EMG-IMU signal sample is considered an independent episode, and each sliding window step is considered an observation during that episode.

Observation: The observation O_t for a given unknown state S_t is defined as the feature vector obtained from each EMG-IMU signal window. This vector is composed of RMS, SD, E, MAV, and AE information. The end of an episode occurs when the agent reaches the last sliding window observation of an EMG-IMU sample.

Action: An action A_t is defined as the category of the gesture that the agent predicts to go from the current observation O_t to the observation O_{t+1}, after which it receives a reward R_{t+1}. The categories of gestures used for this work are: wave in, wave out, fist, open, pinch, and relax (static gestures), and up, down, left, right, forward, and backward (dynamic gestures).

Environment: The environment is the defined environment within which the agent performs an action to move from one observation to the next, which returns a reward. In this case, we define the environment from the sliding window information—feature vectors and labels—extracted from each EMG-IMU signal and the ground truth (vector of known labels) of the EMG-IMU signal.

Reward: The agent receives a positive or negative reward depending on whether during its interaction with the environment it was able to correctly predict a gesture for a given observation. We define two different rewards, one for ranking and one for recognition. An illustration of the rewards that the agent obtains is presented in Figure 2. The agent can receive a positive reward $R_t = +1$ or a negative reward $R_t = -1$ depending on whether or not it correctly predicts the label of a window gesture. Once an episode ends, the vector of the known labels—ground-truth—is compared with the vector of the predicted labels, and if the overlapping factor between these vectors is greater than 70%, then recognition is considered successful and the agent receives a reward $R_t = +1$. If the recognition fails, the agent is penalized with $R_t = -1$.

2.5. Post-Processing

Once an EMG-IMU sample is processed and the vector of the predicted labels is obtained, we use post-processing to remove false labels and improve the accuracy of the proposed HGR system. There are several ways to perform post-processing such as using filters, majority voting, and heuristics, among others [1,16]. In this work, based on experimentation, we obtained the best results by calculating the mode on the vector of the predicted labels that are different from the relax labels. Then all the labels in those vectors that are different from the mode are replaced with it. The post-processing step is key to improving the classification and especially the recognition results since a single erroneous label in an EMG-IMU window can cause the recognition prediction to fail.

3. Results

In this section, we present the validation and testing results for the proposed HGR user-specific method for both the Myo armband and G-force sensors with regard to static and dynamic gestures. First, to find the best possible hyperparameters, we perform a validation procedure, and the best model results found during the validation are presented. Then, we present the final testing results with the previously found best hyperparameters. The validation and testing results for the Myo armband and G-force sensors are analyzed to compare their performance, considering separately static and dynamic gestures. Finally, we briefly compare the proposed method using the EMG-IMU signals with a similar method that uses only EMG.

3.1. Validation Results

For the validation results, we trained and tested different user-specific models based on an agent that uses neural networks as policy representations with the DQN algorithm that we presented previously in Section 2.4. For each model, we evaluated different hyperparameters such as the learning rate and mini-batch size to evaluate the classification and recognition results. Appendix A contains a summary of several of the tests performed to find the best hyperparameters. The best hyperparameter values found for the proposed method are summarized in Table 2.

Table 2. Best hyperparameters found during validation procedure.

Hyperparameter Name	Hyperparameter Values
Activation function between layers	Relu
Target Smooth Factor	5×10^{-3}
Experience buffer length	1×10^{6}
Learn rate (α)	0.3×10^{-3}
Epsilon initial value	1
Epsilon greedy epsilon decay	1×10^{-4}
Discount factor	0.99
Training set replay per user	15 times
Sliding window size	300 points
Stride size	40 points
Mini-batch size	64
Optimizer	Adam
Gradient decay factor	0.9
L2 regularization factor	0.0001
Number of neurons for layer	60, 50, 50, 7 for the input layer, hidden layer 1, hidden layer 2, and output layer, respectively

A training sample illustration of the average reward versus the number of episodes is illustrated in Figure 3. As can be observed, the curve in the figure shows satisfactory growth and convergence to the maximum average reward as the number of episodes increased. It is worth mentioning that this figure varied slightly depending on the data of each user. However, for all users, the same trend of convergence to the maximum average reward value was observed.

We present the classification and recognition results per user for the Myo armband sensor for static and dynamic gestures in Figure 4. Likewise, we present the classification and recognition results per user for the G-force sensor for static and dynamic gestures in Figure 5. Moreover, we present a summary of the best classification and recognition results of the user-specific HGR models obtained during validation in Table 3. It can be observed that for the validation results, the DQN-based model with the Myo armband sensor achieved slightly better results than the same model with the G-force sensor. There was a 6.5% classification accuracy difference between the Myo armband and G-force sensors for static gestures and a 4.3% difference for dynamic gestures. Moreover, the standard deviation was also lower for the Myo armband sensor, which was only 2.78% compared to a value of 9.04% for the G-force sensor. On the other hand, for dynamic gestures, the Myo obtained slightly better results. For example, for the Myo armband sensor, we obtained a 4.3% higher efficiency in the classification when using dynamic gestures with a standard deviation of only 1.37% compared to a value of 7.20% for the G-force sensor. The same analysis applied to the recognition accuracy metrics, demonstrating that the Myo armband sensor obtained slightly better results using this metric.

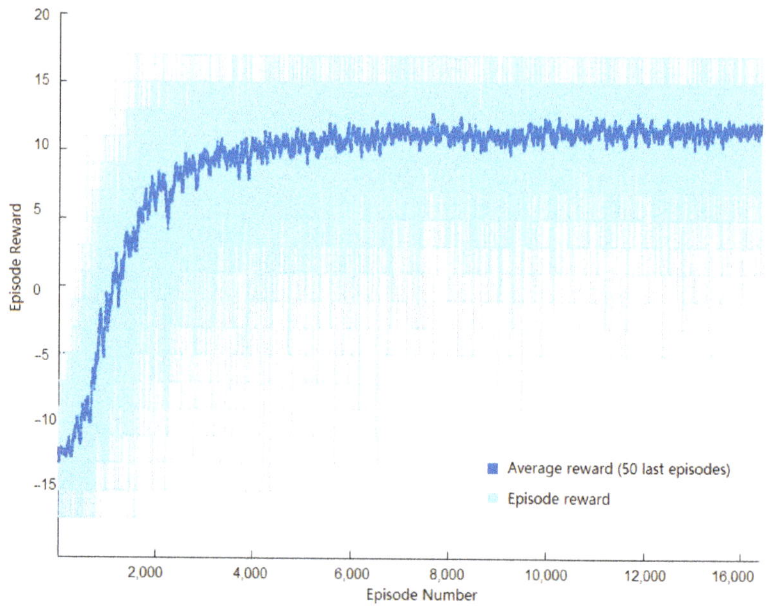

Figure 3. Sample of episode rewards versus episode numbers during the training of one user.

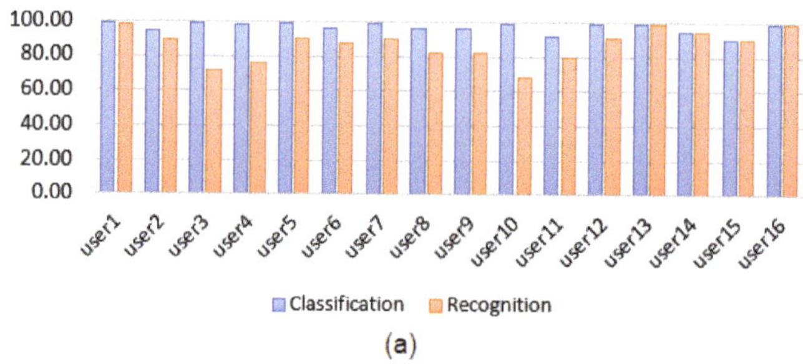

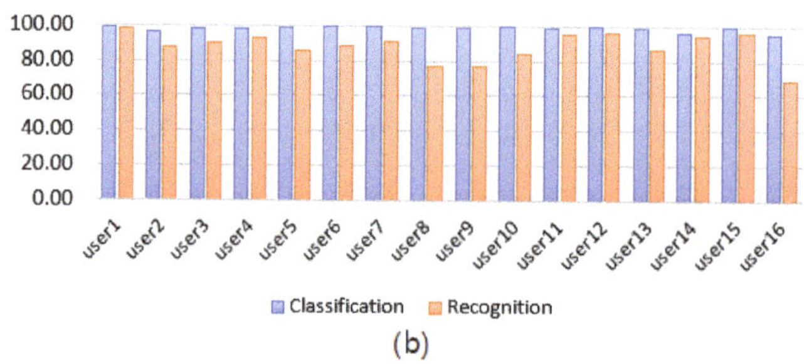

Figure 4. User-specific HGR model classification and recognition accuracy results for the Myo armband sensor using DQN. (**a**) Static gestures. (**b**) Dynamic gestures.

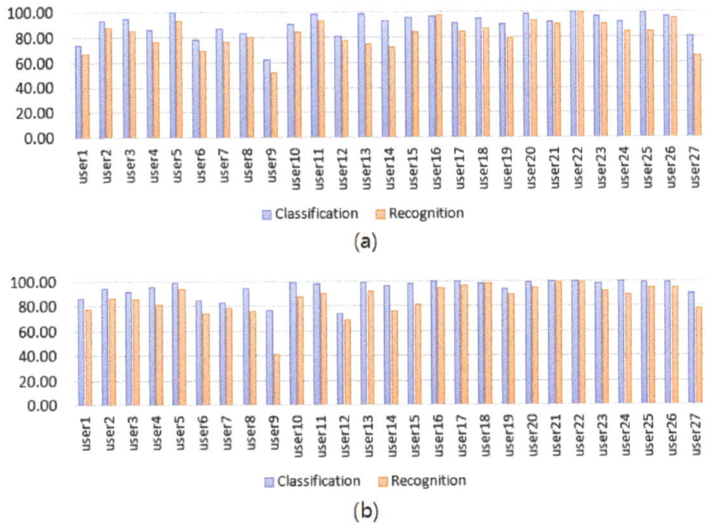

Figure 5. User-specific HGR model classification and recognition accuracy results for the G-force sensor using DQN. (**a**) Static gestures. (**b**) Dynamic gestures.

Table 3. User-specific validation: best results for Myo armband and G-force sensors.

Sensor	Classification Accuracy	Recognition Accuracy
Myo armband (Static gestures)	96.9% ± 2.78%	87.0% ± 9.36%
Myo armband (Dynamic gestures)	98.6% ± 1.37%	88.2% ± 8.28%
G-force (Static gestures)	90.4% ± 9.04%	82.2% ± 10.98%
G-force (Dynamic gestures)	94.3% ± 7.20%	85.5% ± 12.3%

3.2. Testing Results

To present the testing results, we performed experiments on the test set based on the best hyperparameters previously found during the validation procedure presented in Section 3.1. This procedure helped us to evaluate our models with different data and analyze overfitting. We summarized the test results for 306 users with the best-found hyperparameters in Table 4. The classification results were similar for the two sensors, with the Myo-armband sensor obtaining slightly better results, with differences of 4.26% for static gestures and 1.82% for dynamic gestures compared to the G-force sensor. On the other hand, the recognition accuracy was similar for both sensors for the testing results compared with the validation results, with the exception of the G-force sensor, in which the recognition values were 56.45% ± 8.12% and 70.57% ± 11.99% for static and dynamic gestures, respectively. Overall, the testing classification results were similar to the validation results, demonstrating that the proposed models are robust to the effect of overfitting in terms of the classification of the proposed dataset distribution. Only for static gestures of the G-force sensor were the recognition results slightly lower. This is explained by the different distribution of the data and the variability of the users, as well as the fact that the hyperparameters were calibrated only for the validation dataset and not for the testing dataset.

We also present the confusion matrices that represent the classification results on the test set of the Myo armband sensor for static gestures in Figure 6 and dynamic gestures in Figure 7, as well as for the G-force sensor for static gestures in Figure 8 and dynamic gestures in Figure 9. In these figures, the results for each hand gesture can be observed in detail, which include both static and dynamic gestures for both sensors. It is worth mentioning that the processing time of each window observation was, on average, 33 ms.

Table 4. User-specific testing results for Myo armband and G-force sensors.

Sensor	Classification Accuracy	Recognition Accuracy
Myo armband (Static gestures)	97.50% ± 1.13%	88.15% ± 2.84%
Myo armband (Dynamic gestures)	98.95% ± 0.62%	90.47% ± 4.57%
G-force (Static gestures)	93.24% ± 3.43%	56.45% ± 8.12%
G-force (Dynamic gestures)	97.13% ± 2.04%	70.57% ± 11.99%

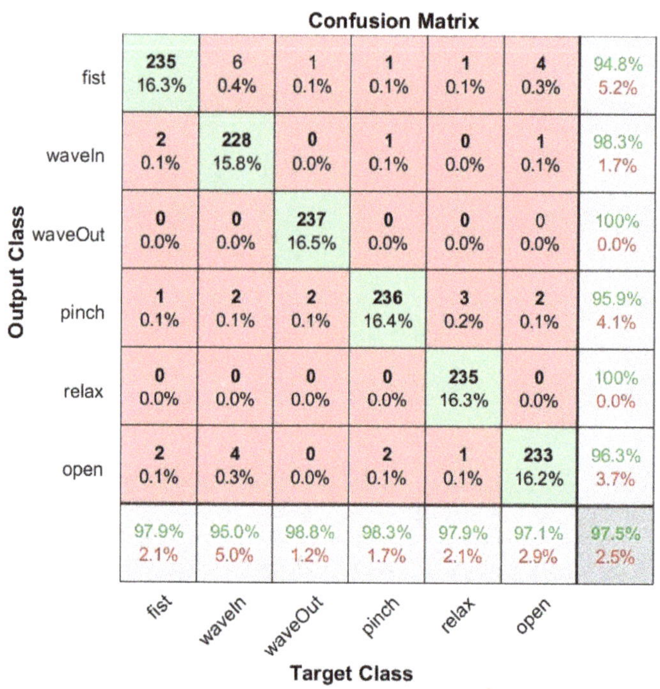

Figure 6. User-specific HGR model confusion matrix for 16 users from the test set with the best hyperparameter configuration for the Myo armband sensor for static gestures.

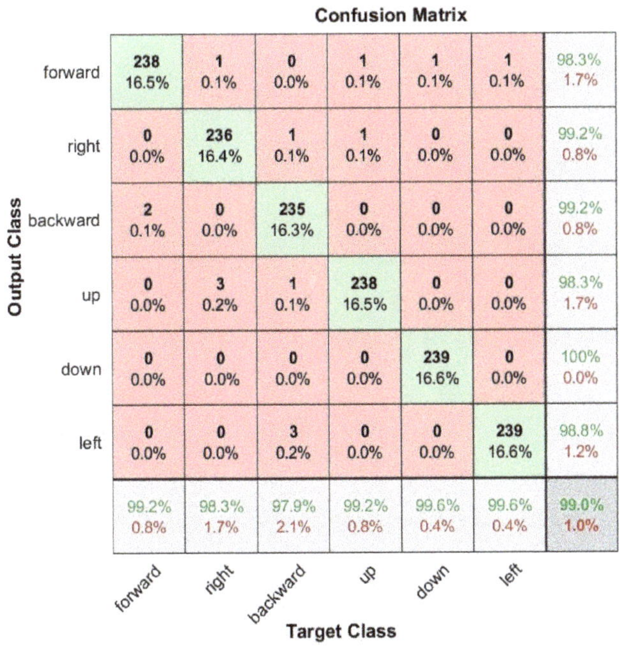

Figure 7. User-specific HGR model confusion matrix for 16 users from the test set with the best hyperparameter configuration for the G-force sensor for dynamic gestures.

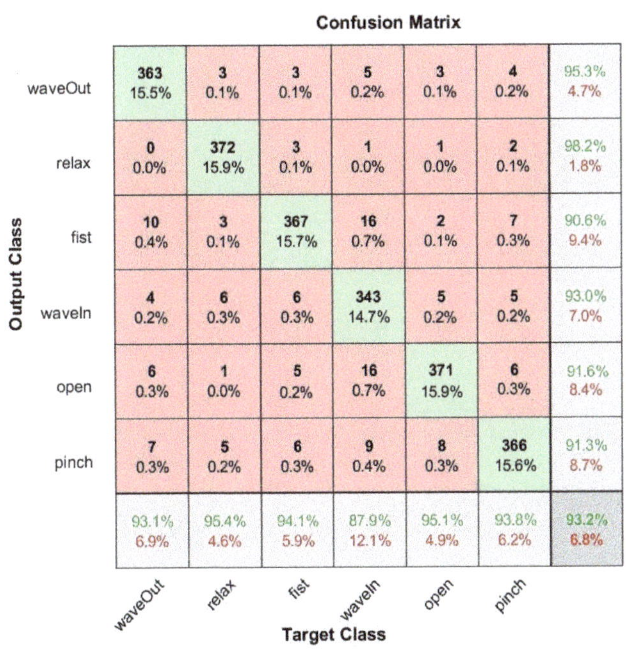

Figure 8. User-specific HGR model confusion matrix for 26 users from the test set with the best hyperparameter configuration for the Myo armband sensor for static gestures.

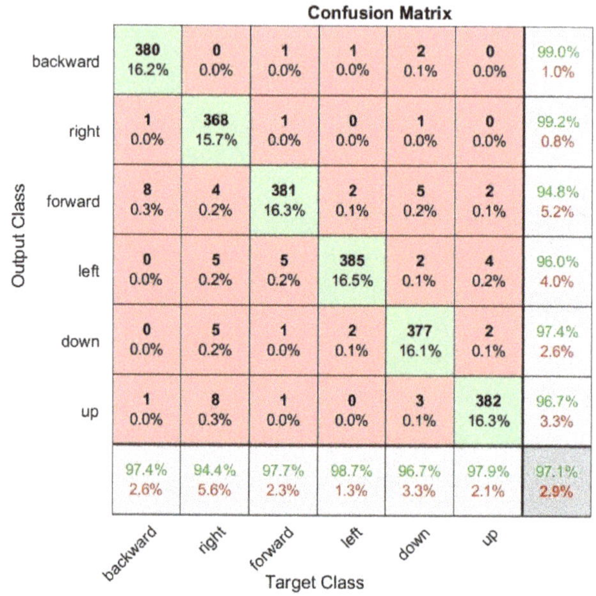

Figure 9. User-specific HGR model confusion matrix for 26 users from the test set with the best hyperparameter configuration for the G-force sensor for dynamic gestures.

3.3. Comparison with Other Methods

We implemented two additional tests for our proposed dataset and method, but the classification stage was based on supervised learning methods such as k-nearest neighbor (KNN) and a convolutional neural network (CNN). We also compared the results found in the present work, which uses EMG and IMU signals, with methods previously developed using the same sensor, with a similar dataset distribution with similar method stages that work with supervised and reinforcement learning [16,25]. These comparisons were useful for evaluating the effect of using EMG-IMU signals with respect to using EMG signals only, as well as comparing supervised and reinforcement learning methods for the proposed dataset. The selection criteria for the selected articles were based first on the type of sensor and its location on the user's arm, which needs to be consistent with what we proposed in this work. Another important point that we considered is that we found that in the works based on EMGs only, the HGR models were trained to recognize static gestures only. To successfully recognize dynamic gestures, it was necessary to use IMU signals or a combination of IMU and EMG signals. This is because dynamic gestures are highly dependent on the user's arm movements, which can be analyzed using information obtained from the IMU. We searched for approaches using similar methods that contained pre-processing, feature extraction, classification, and post-processing to fairly and objectively assess the effect of using EMG with IMU signals instead of just using EMG signals to develop HGR systems. The results using EMG and IMU signals that we obtained in this work for static gestures using the Myo armband sensor can be seen in Table 5, where we obtained 97.5% $\pm$ 1.13% and 88.15% $\pm$ 2.84% for the classification and recognition, respectively. On the other hand, another approach that used only EMG signals and Q-learning obtained 90.47% $\pm$ 14.24% and 87.51% $\pm$ 14.1% for the classification and recognition, respectively [16]. The approach that used EMG and IMU signals with supervised learning based on KNN obtained 80.04% and 66.12% for the classification and recognition, respectively, whereas the approach based on a CNN classifier obtained 84.49% $\pm$ 7.10% for the classification and 70.02% $\pm$ 8.21% for the recognition. Finally, another approach that used only EMG signals and a supervised learning approach based on a support vector machine obtained 95% and 81.6% for the classification and recognition, respectively [25]. As can be observed, using EMG and IMU signals helped to improve the classification and recognition results for static gestures when considering models based on reinforcement and supervised learning. Moreover, it can be observed that our model based on reinforcement learning with EMG and IMU signals presented the best results for this application.

Table 5. Comparison of classification and recognition accuracy results on the test set of the proposed model compared with other methods.

Learning Method	Type of Signal	Classification	Recognition
Reinforcement learning (this work)	EMG + IMU	97.5% ± 1.13%	88.15% ± 2.84%
Reinforcement learning [16]	EMG	90.47% ± 14.24%	87.51% ± 14.1%
Supervised learning—KNN classifier	EMG + IMU	80.04% ± 13.66%	66.12% ± 18.30%
Supervised learning—CNN classifier	EMG + IMU	84.49% ± 7.10%	70.02% ± 8.21%
Supervised learning [25]	EMG	95%	81.6%

4. Discussion

- According to the test results, the best classification accuracies were obtained for static gestures using the Myo armband sensor and were 97.50% ± 1.13% and 88.15% ± 2.84% for the classification and recognition, respectively. On the other hand, for dynamic gestures using the Myo armband sensor, the accuracies were 98.95% ± 0.62% and 90.47% ± 4.57% for the classification and recognition, respectively. The accuracies of the test results for static gestures using the G-force sensor were 93.24% ± 3.43% and 56.45% ± 8.12% for the classification and recognition, respectively. On the other hand, for dynamic gestures using the G-force sensor, the accuracies were 97.13% ± 2.04% and 70.57% ± 11.99% for the classification and recognition, respectively. This indicates that the method based on a DQN for the Myo armband sensor obtained slightly better results than the method based on a DQN for the G-force sensor.
- We compared the proposed method that used EMG and IMU signals with respect to other similar works where the same sensor was used with only EMG signals for static gestures. We obtained accuracies of 97.5% ± 1.13% and 88.15% ± 2.84% for the classification and recognition, respectively, using both EMG and IMU signals versus accuracies of 90.47% ± 14.24% and 87.51% ± 14.1% for the classification and recognition, respectively, using only EMG signals. This indicates the benefits of using EMG-IMU signals over using EMGs alone. This represents a 7% and 1% improvement in the classification and recognition, as well as a substantial reduction of more than 10% in the standard deviation of these metrics when using EMG-IMU signals instead of EMG signals alone. This also indicates the benefits of using EMG-IMU signals over using EMGs alone. Moreover, it can be seen that we are the first study to use RL with EMG-IMU signals to obtain better results compared to using only EMG signals with RL. Our results also outperformed those obtained with methods that use EMG or EMG-IMU with supervised learning.
- In general, the difference between the results of the validation and testing with regard to the classification and recognition was less than 5%. This difference is small so it can be said that the proposed method is robust and does not suffer from the effects of overfitting for the proposed dataset distribution.

- The processing time of each window observation was, on average, 33 ms for both sensors. Since this is less than 300 ms, we can consider that both models work in real time for the proposed application.
- Although the proposed results are encouraging, it is important to mention that in future works we will focus on the convenience and comfort that users experience when using static or dynamic gestures. User preference data can impact the development of HGR architectures so we will study this in depth in future work.

5. Conclusions

In this work, we proposed an HGR system based on the DQN algorithm for the classification of 11 different hand gestures including static and dynamic gestures. We tested and compared the results of two different sensors, the Myo armband and G-force sensors, from which we used the EMG and IMU signals to obtain the feature vectors. The proposed models were validated on 43 users and tested on 42 different users. The best classification accuracy was obtained for the Myo armband sensor, reaching up to 97.50% $\pm$ 1.13% and 88.15% $\pm$ 2.84% for the classification and recognition, respectively, with regard to static gestures, and 98.95% $\pm$ 0.62% and 90.47% $\pm$ 4.57% for the classification and recognition, respectively, with regard to dynamic gestures. The results obtained in this work showed that the DQN was able to learn a policy from online experience to classify and recognize gestures based on EMG and IMU signals, significantly improving the results obtained by similar methods using only EMG. It was also observed that the use of the Myo armband sensor compared to the G-force sensor obtained better accuracy for this application and data distribution. Future work includes testing other feature extraction methods and reinforcement learning algorithms to evaluate the proposed dataset.

Author Contributions: Conceptualization, L.I.B.L., Á.L.V.C. and M.E.B.; Methodology, J.P.V., L.I.B.L. and M.E.B.; Software, J.P.V., L.I.B.L., Á.L.V.C. and M.E.B.; Validation, J.P.V.; Formal analysis, J.P.V., L.I.B.L., Á.L.V.C. and M.E.B.; Investigation, J.P.V., L.I.B.L., Á.L.V.C. and M.E.B.; Resources, M.E.B.; Data curation, J.P.V., L.I.B.L., Á.L.V.C. and M.E.B.; Writing—review & editing, J.P.V., L.I.B.L., Á.L.V.C. and M.E.B.; Visualization, J.P.V.; Supervision, M.E.B.; Project administration, Á.L.V.C.; Funding acquisition, M.E.B. All authors have read and agreed to the published version of the manuscript.

Funding: This research received no external funding.

Informed Consent Statement: Informed consent was obtained from all subjects involved in the study.

Data Availability Statement: The dataset is available at https://laboratorio-ia.epn.edu.ec/en/resources/dataset/emg-imu-epn-100 accessed on 18 November 2022.

Conflicts of Interest: The authors declare no conflict of interest.

Abbreviations

The following abbreviations are used in this manuscript:

HGR	hand gesture recognition systems
EMG	electromyography
EMGs	electromyography signals
IMU	inertial measurement unit
IMUs	inertial measurement unit signals
ML	machine learning
RL	reinforcement learning
CNN	convolutional neural network
ANN	artificial neural network
DQN	deep Q-network

Appendix A

A summary of the validation results from changing the learning rate (alpha) parameter is presented in Table A1.

Table A1. User-specific validation results for Myo armband and G-force sensors.

Alpha	Classification Accuracy	Recognition Accuracy
0.07	39.2% ± 16.52%	25.0% ± 16.86%
0.05	38.9% ± 17.5%	22.0% ± 17.86%
0.03	45.9% ± 15.79%	37.0% ± 15.55%
0.01	51.9% ± 16.45%	47.0% ± 14.36%
0.007	54.2% ± 15.58%	48.6% ± 16.57%
0.005	70.5% ± 9.58%	57.2% ± 10.45%
0.003	71.4% ± 10.25%	58.2% ± 13.33%
0.001	77.3% ± 6.78%	73.4% ± 11.56%
0.0007	87.3% ± 4.11%	83.2% ± 12.22%
0.0005	89.2% ± 3.58%	75.2% ± 10.12%
0.0003	**96.9% ± 2.78%**	**87.0% ± 9.36%**
0.0001	93.2% ± 4.51%	83.5% ± 9.78%
0.00007	94.3% ± 3.58%	82.1% ± 8.35%
0.00005	88.3% ± 4.58%	80.1% ± 8.89%
0.00003	83.5% ± 6.52%	77.0% ± 13.48%
0.00001	85.3% ± 5.86%	81.0% ± 12.89%

References

1. Jaramillo-Yánez, A.; Benalcázar, M.E.; Mena-Maldonado, E. Real-Time Hand Gesture Recognition Using Surface Electromyography and Machine Learning: A Systematic Literature Review. *Sensors* **2020**, *20*, 2467. [CrossRef] [PubMed]
2. Kim, J.; Yang, S.; Koo, B.; Lee, S.; Park, S.; Kim, S.; Cho, K.H.; Kim, Y. sEMG-Based Hand Posture Recognition and Visual Feedback Training for the Forearm Amputee. *Sensors* **2022**, *22*, 7984. [CrossRef] [PubMed]
3. Lin, W.; Li, C.; Zhang, Y. Interactive Application of Data Glove Based on Emotion Recognition and Judgment System. *Sensors* **2022**, *22*, 6327. [CrossRef] [PubMed]
4. Chico, A.; Cruz, P.J.; Vásconez, J.P.; Benalcázar, M.E.; Álvarez, R.; Barona, L.; Valdivieso, Á.L. Hand Gesture Recognition and Tracking Control for a Virtual UR5 Robot Manipulator. In Proceedings of the 2021 IEEE Fifth Ecuador Technical Chapters Meeting (ETCM), Cuenca, Ecuador, 12–15 October 2021; pp. 1–6.
5. Romero, R.; Cruz, P.J.; Vásconez, J.P.; Benalcázar, M.; Álvarez, R.; Barona, L.; Valdivieso, Á.L. Hand Gesture and Arm Movement Recognition for Multimodal Control of a 3-DOF Helicopter. In *International Conference on Robot Intelligence Technology and Applications*; Springer: Cham, Switzerland, 2022; pp. 363–377.
6. Benalcázar, M.E.; Jaramillo, A.G.; Zea, A.; Páez, A.; Andaluz, V.H. Hand gesture recognition using machine learning and the Myo armband. In Proceedings of the 2017 25th European Signal Processing Conference (EUSIPCO), Kos, Greece, 28 August–2 September 2017; pp. 1040–1044.
7. Nuzzi, C.; Pasinetti, S.; Lancini, M.; Docchio, F.; Sansoni, G. Deep learning based machine vision: First steps towards a hand gesture recognition set up for collaborative robots. In Proceedings of the 2018 Workshop on Metrology for Industry 4.0 and IoT, Brescia, Italy, 16–18 April 2018; pp. 28–33.
8. Yang, L.; Chen, J.; Zhu, W. Dynamic hand gesture recognition based on a leap motion controller and two-layer bidirectional recurrent neural network. *Sensors* **2020**, *20*, 2106. [CrossRef] [PubMed]
9. Kim, M.; Cho, J.; Lee, S.; Jung, Y. IMU sensor-based hand gesture recognition for human-machine interfaces. *Sensors* **2019**, *19*, 3827. [CrossRef] [PubMed]
10. Wen, F.; Sun, Z.; He, T.; Shi, Q.; Zhu, M.; Zhang, Z.; Li, L.; Zhang, T.; Lee, C. Machine learning glove using self-powered conductive superhydrophobic triboelectric textile for gesture recognition in VR/AR applications. *Adv. Sci.* **2020**, *7*, 2000261. [CrossRef] [PubMed]

11. Kundu, A.S.; Mazumder, O.; Lenka, P.K.; Bhaumik, S. Hand gesture recognition based omnidirectional wheelchair control using IMU and EMG sensors. *J. Intell. Robot. Syst.* **2018**, *91*, 529–541. [CrossRef]
12. Zhang, X.; Yang, Z.; Chen, T.; Chen, D.; Huang, M.C. Cooperative sensing and wearable computing for sequential hand gesture recognition. *IEEE Sens. J.* **2019**, *19*, 5775–5783. [CrossRef]
13. Jiang, S.; Lv, B.; Guo, W.; Zhang, C.; Wang, H.; Sheng, X.; Shull, P.B. Feasibility of wrist-worn, real-time hand, and surface gesture recognition via sEMG and IMU sensing. *IEEE Trans. Ind. Inform.* **2017**, *14*, 3376–3385. [CrossRef]
14. Benalcázar, M.E.; Motoche, C.; Zea, J.A.; Jaramillo, A.G.; Anchundia, C.E.; Zambrano, P.; Segura, M.; Palacios, F.B.; Pérez, M. Real-time hand gesture recognition using the Myo armband and muscle activity detection. In Proceedings of the 2017 IEEE Second Ecuador Technical Chapters Meeting (ETCM), Salinas, Ecuador, 16–20 October 2017; pp. 1–6.
15. Englehart, K.; Hudgins, B. A robust, real-time control scheme for multifunction myoelectric control. *IEEE Trans. Biomed. Eng.* **2003**, *50*, 848–854. [CrossRef] [PubMed]
16. Vásconez, J.P.; López, L.I.B.; Caraguay, Á.L.V.; Cruz, P.J.; Álvarez, R.; Benalcázar, M.E. A Hand Gesture Recognition System Using EMG and Reinforcement Learning: A Q-Learning Approach. In *International Conference on Artificial Neural Networks*; Springer: Cham, Switzerland, 2021; pp. 580–591.
17. Zhang, C.; Wang, Z.; An, Q.; Li, S.; Hoorfar, A.; Kou, C. Clustering-Driven DGS-Based Micro-Doppler Feature Extraction for Automatic Dynamic Hand Gesture Recognition. *Sensors* **2022**, *22*, 8535. [CrossRef] [PubMed]
18. Jiang, Y.; Song, L.; Zhang, J.; Song, Y.; Yan, M. Multi-Category Gesture Recognition Modeling Based on sEMG and IMU Signals. *Sensors* **2022**, *22*, 5855. [CrossRef] [PubMed]
19. Pan, T.Y.; Tsai, W.L.; Chang, C.Y.; Yeh, C.W.; Hu, M.C. A hierarchical hand gesture recognition framework for sports referee training-based EMG and accelerometer sensors. *IEEE Trans. Cybern.* **2022**, *52*, 3172–3183. [CrossRef] [PubMed]
20. Colli Alfaro, J.G.; Trejos, A.L. User-Independent Hand Gesture Recognition Classification Models Using Sensor Fusion. *Sensors* **2022**, *22*, 1321. [CrossRef] [PubMed]
21. Seok, W.; Kim, Y.; Park, C. Pattern recognition of human arm movement using deep reinforcement learning. In Proceedings of the 2018 International Conference on Information Networking (ICOIN), Chiang Mai, Thailand, 10–12 January 2018; pp. 917–919.
22. Song, C.; Chen, C.; Li, Y.; Wu, X. Deep Reinforcement Learning Apply in Electromyography Data Classification. In Proceedings of the 2018 IEEE International Conference on Cyborg and Bionic Systems (CBS), Shenzhen, China, 25–27 October 2018; pp. 505–510.
23. Sharma, R.; Kukker, A. Neural Reinforcement Learning based Identifier for Typing Keys using Forearm EMG Signals. In Proceedings of the 9th International Conference on Signal Processing Systems, Auckland, New Zealand, 27–30 November 2017; pp. 225–229.
24. Kukker, A.; Sharma, R. Neural reinforcement learning classifier for elbow, finger and hand movements. *J. Intell. Fuzzy Syst.* **2018**, *35*, 5111–5121. [CrossRef]
25. Barona López, L.I.; Valdivieso Caraguay, Á.L.; Vimos, V.H.; Zea, J.A.; Vásconez, J.P.; Álvarez, M.; Benalcázar, M.E. An Energy-Based Method for Orientation Correction of EMG Bracelet Sensors in Hand Gesture Recognition Systems. *Sensors* **2020**, *20*, 6327. [CrossRef] [PubMed]
26. Van Hasselt, H.; Guez, A.; Silver, D. Deep reinforcement learning with double q-learning. In Proceedings of the AAAI Conference on Artificial Intelligence, Phoenix, AZ, USA, 12–17 February 2016; Volume 30.
27. Sutton, R.S.; Barto, A.G. *Reinforcement Learning: An Introduction*; MIT Press: Cambridge, MA, USA, 2018.
28. Mnih, V.; Kavukcuoglu, K.; Silver, D.; Rusu, A.A.; Veness, J.; Bellemare, M.G.; Graves, A.; Riedmiller, M.; Fidjeland, A.K.; Ostrovski, G.; et al. Human-level control through deep reinforcement learning. *Nature* **2015**, *518*, 529–533. [CrossRef] [PubMed]
29. Kapturowski, S.; Ostrovski, G.; Quan, J.; Munos, R.; Dabney, W. Recurrent experience replay in distributed reinforcement learning. In Proceedings of the International Conference on Learning Representations, Vancouver, BC, Canada, 30 April–3 May 2018.

sensors MDPI

Article

sEMG-Based Hand Posture Recognition and Visual Feedback Training for the Forearm Amputee

Jongman Kim [1], Sumin Yang [1], Bummo Koo [1], Seunghee Lee [1], Sehoon Park [2], Seunggi Kim [2], Kang Hee Cho [3] and Youngho Kim [1,*]

1 Department of Biomedical Engineering and Institute of Medical Engineering, Yonsei University, Wonju 26493, Korea
2 Korea Orthopedics and Rehabilitation Engineering Center, Incheon 21417, Korea
3 Department of Rehabilitation Medicine, Chungnam National University College of Medicine, Daejeon 35015, Korea
* Correspondence: younghokim@yonsei.ac.kr; Tel.: +82-33-760-2859

Abstract: sEMG-based gesture recognition is useful for human–computer interactions, especially for technology supporting rehabilitation training and the control of electric prostheses. However, high variability in the sEMG signals of untrained users degrades the performance of gesture recognition algorithms. In this study, the hand posture recognition algorithm and radar plot-based visual feedback training were developed using multichannel sEMG sensors. Ten healthy adults and one bilateral forearm amputee participated by repeating twelve hand postures ten times. The visual feedback training was performed for two days and five days in healthy adults and a forearm amputee, respectively. Artificial neural network classifiers were trained with two types of feature vectors: a single feature vector and a combination of feature vectors. The classification accuracy of the forearm amputee increased significantly after three days of hand posture training. These results indicate that the visual feedback training efficiently improved the performance of sEMG-based hand posture recognition by reducing variability in the sEMG signal. Furthermore, a bilateral forearm amputee was able to participate in the rehabilitation training by using a radar plot, and the radar plot-based visual feedback training would help the amputees to control various electric prostheses.

Keywords: surface electromyography; forearm amputee; hand posture; visual feedback training; pattern recognition; artificial neural network

Citation: Kim, J.; Yang, S.; Koo, B.; Lee, S.; Park, S.; Kim, S.; Cho, K.H.; Kim, Y. sEMG-Based Hand Posture Recognition and Visual Feedback Training for the Forearm Amputee. *Sensors* **2022**, 22, 7984. https://doi.org/10.3390/s22207984

Academic Editor: Giovanni Saggio

Received: 14 September 2022
Accepted: 17 October 2022
Published: 19 October 2022

1. Introduction

Surface electromyography (sEMG) records the electrical biosignals generated by the action potentials that occur during the contraction of muscle fibers [1]. Various information in sEMG signals has been used to estimate and diagnose a user's condition or recognize a user's motion and intention [2]. In particular, sEMG-based gesture recognition was suggested to be a promising technology for human–computer interactions (HCIs) [3]. Indeed, sEMG-based gesture recognition technology has already been applied to both healthy adults and various patients for rehabilitation training and control of electric prostheses [4].

Electric prostheses have been developed to improve patients' quality of life following a limb amputation with the importance of their control [5]. An sEMG-based control system is the most direct protocol for controlling electric prostheses, and there exist two different types: non-pattern recognition algorithms and pattern recognition algorithms [6,7]. Non-pattern recognition algorithms using the magnitude of the sEMG signal and threshold values have the advantages of ease of use and fast response time, but they work for only a few hand gestures. As the number of recognized gestures for a non-pattern recognition algorithm increases, it becomes increasingly slow and difficult to use due to its complexity and the multiple stages involved in muscle contractions [8]. Therefore, many previous studies have developed pattern recognition algorithms to classify various gestures. However, it has been reported that the muscles of the amputees were lost or weakened, depending on the surgery and the period of amputation [9]. The differences in the amputees' muscles increase the variability in sEMG signals, i.e., different signal patterns appear even when the same gestures are repeated, and that variability in sEMG signals critically decreases the classification performance [10]. These results indicate that user training is as important as optimizing the recognition system.

1.1. Related Work

1.1.1. sEMG-Based Gesture Recognition

Many studies have been performed to recognize hand gestures using multichannel sEMG sensors. Emayavaramban et al. developed a recognition algorithm for twelve hand gestures by using five sEMG sensors on the forearm [11]. sEMG signals were measured in ten healthy adults, and the best classification accuracy (95.1%) appeared with a pattern net neural network classifier and an autoregressive Burg feature vector. Shi et al. used two-channel sEMG sensors to measure signals from thirteen healthy adults and develop a recognition algorithm for four hand gestures to control a bionic hand [12]. MAV and WL were selected as the feature vectors with the best classification accuracy (93.8%) with the k-nearest neighbor (k-NN) classifier. However, it was difficult to apply those algorithms to amputees because the sEMG signals were measured in healthy adults. Adewuyi et al. developed a hand gesture recognition algorithm by using multichannel sEMG sensors on sixteen healthy adults and four partial hand amputees [13]. Four classifiers (linear discriminant analysis [LDA], quadratic discriminant analysis, linear neural network, and multilayer perceptron artificial neural network [MLPANN]) and five feature sets (time domain and autoregressive, time domain, sequential forward searching [SFS], separability index, and all feature vectors) were used to recognize the hand gestures. The healthy adults showed fewer classification errors than the amputees, and the combination of the MLPANN classifier and SFS feature vector was the best option for recognizing the hand gestures of all subjects. Betthauser et al. measured sEMG signals using eight sEMG sensors on eight healthy adults and two forearm amputees to recognize five hand and wrist gestures [14]. Seven classifiers (LDA, artificial neural network [ANN], regularized LDA, support vector machine [SVM], non-negative least squares, sparse representation classification [SRC], and extreme learning machine with adaptive SRC [EASRC]) were trained with three feature sets. The classification performances of the healthy adults were higher than those of amputees, and the EASRC classifier showed the fewest classification errors. Most previous studies suggested that the classifier and feature vectors be optimized using multichannel sEMG sensors to improve gesture recognition. In addition, the classification performance in the previous studies was better in healthy adults than in amputees. Variability in the sEMG signal was increased by muscle loss in amputees, which is a critical factor that decreases the performance of sEMG-based gesture recognition [10]. For these reasons, rehabilitation and user training are as important to patients as improvements in the hardware and software of sEMG-based gesture recognition devices.

1.1.2. Rehabilitation Training for the Amputees

Previous clinical research used two types of rehabilitation training for amputees: (1) mirror therapy, which trains both the amputated side and the intact side at the same time; and (2) mental imagery, in which the amputee imagines movements without actually moving the residual limb [15]. However, neither of those procedures allows the subjects to check their movements themselves in real time. Few studies have quantitatively examined the effect of rehabilitation using mirror therapy or mental imagery [16]. In addition, patients with bilateral amputations cannot participate in rehabilitation with mirror therapy because they lack an intact side. Powell et al. tested repetitive training with sixteen sEMG sensors on four amputees to improve the consistency and distinguishability of nine hand and wrist gestures [17]. The amputees repeated the gestures in a random order by following the image of a virtual prosthesis on a screen. That study reported that classification accuracy for the amputees improved from 77.5% to 94.4% during ten days of training. Rehabilitation training with a screen could be used for both unilateral and bilateral amputees because only the amputated side was involved in the rehabilitation. However, that rehabilitation training was still inefficient, so a way of training that improves the gestures on the amputated side is still needed.

In this study, sEMG-based ANN classifiers were developed to recognize the hand postures for the control of myoelectric prostheses. In addition, the radar plot-based visual feedback training was suggested to improve the performance of hand posture recognition considering the bilateral forearm amputee. The sEMG signals of healthy adults and a bilateral forearm amputee were measured by multichannel sEMG sensors. Radar plot-based visual feedback training, which can be applied to bilateral amputees, was performed by the healthy adults for two days and by the forearm amputee for five days, respectively. Those sEMG signals were then used to develop ANN classifiers that could be used with two types of feature vectors. t-distributed stochastic neighbor embedding (t-SNE) and the silhouette coefficient (SC) were used to analyze changes in the variability of the sEMG signals during posture training. In addition, classification accuracy was determined according to the type of feature vector and the hand postures. The classification accuracies of the healthy adults and a forearm amputee increased by the visual feedback training and optimized feature vectors. In particular, the visual feedback training was more effective than the optimization of the feature vectors to improve the classification performance of the forearm amputee.

2. Materials and Methods

2.1. Participants

Ten healthy adults (HA, seven males and three females, 24.1 ± 1.2 years) and one bilateral forearm amputee (FA, male, 45 years) were recruited to participate in this study. The healthy adults had no neurological or musculoskeletal disorders. The amputee had no cognitive problems and had lost both his left and right forearms 21 years before participation in this study. The forearm amputee used a cosmetic prosthesis on the right forearm, which was shorter than the left side, and an electric prosthesis on the left forearm. At the time of this study, he had used a three-finger electric prosthesis with two degrees of freedom (DoFs) for 20 years and a five-finger electric prosthesis with multiple DoFs for 18 months. All participants were fully informed of the risks associated with the experiments, and they gave their written consent to participate in this study. The experimental procedures for healthy adults and a forearm amputee were approved by the Yonsei University Mirae Institutional Review Board (1041849-202002-BM-018-02) and the Institutional Review Board of the Korea Orthopedics & Rehabilitation Engineering Center (RERI-IRB-210915-2), respectively.

2.2. Equipment

A commercial sEMG system, Delsys Trigno wireless sEMG system (Delsys Inc., Natick, MA, USA), was used to measure sEMG signals at a sampling rate of 1926 Hz with the amplification factor of 909 in the analog mode (Figure 1a) [18]. Baseline hand dynamometers (Fabrication Enterprises, Inc., White Plains, NY, USA) were used to minimize the effects of muscle fatigue and the confounding factor of grasp force (Figure 1b) [19]. The bilateral forearm amputee, who could not use the hand dynamometers, performed the hand postures with their preferred power, and the radar plot from the sEMG signal was used to monitor their present power.

The forearm muscles used for sEMG-based hand posture recognition were selected from previous studies [12,20–24]. Nine sEMG sensors were positioned on the healthy adults' muscles: flexor digitorum superficialis (FDS), extensor digitorum (ED), extensor digitorum minimi (EDM), extensor pollicis (EP), flexor carpi radialis (FCR), flexor carpi ulnaris (FCU), extensor carpi radialis (ECR), extensor carpi ulnaris (ECU), and brachioradialis (BR). Magnetom Skyra MRI (Siemens Healthineers AG, Erlangen, Germany) recording and 3D reconstruction (Mimics Research 20.0, Materialise NV, Leuven, Belgium) were performed at Chungnam National University Hospital to analyze the residual muscles of the amputee, and eight forearm muscles were selected on the amputee: BR, FCR, ECR, ED, ECU, flexor digitorum profundus (FDP), FDS, and FCU (Figure 2). The muscle bellies were found for the right place of the electrodes based on the human anatomy, the amputee's 3D reconstruction data, and palpation.

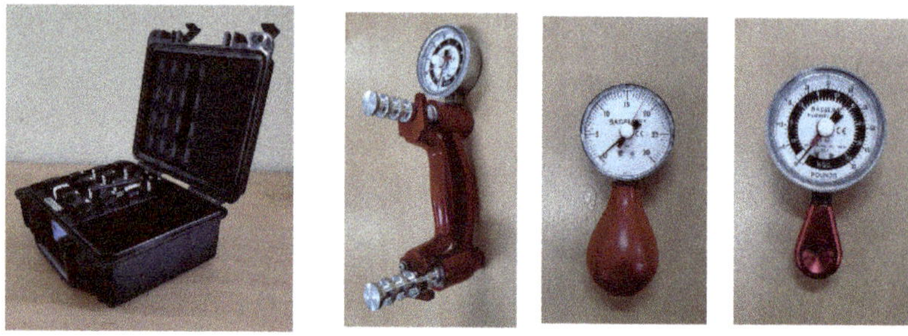

(a) (b)

Figure 1. Experimental equipment: (**a**) multichannel sEMG system, (**b**) hand dynamometers.

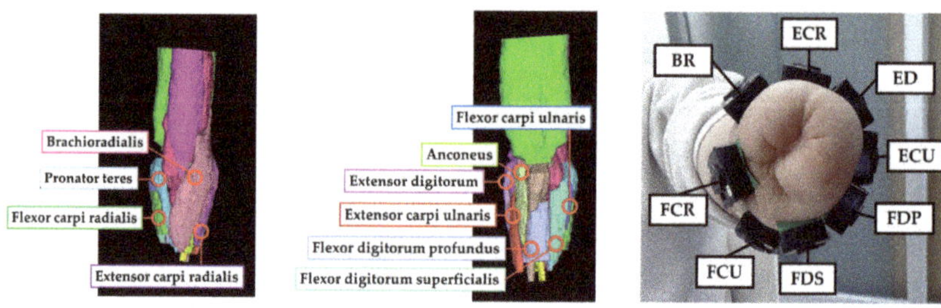

(a) (b)

Figure 2. The amputated limb of the forearm amputee: (**a**) 3D reconstruction data of the forearm amputee, (**b**) position of the sEMG sensors on the forearm amputee.

A graphic user interface (GUI) was developed using LabVIEW (National Instruments Corp., Austin, TX, USA) for real-time monitoring and recording of the sEMG signals. The GUI was designed with a radar plot (Figure 3), and the radar plot was useful to directly visualize the patterns of sEMG signals. The participants controlled their muscle contractions by following the displayed sEMG patterns for each hand posture.

2.3. Experimental Protocol

Twelve hand postures (Figure 4) were suggested in the previous study considering the hand function and the frequency of use in daily life [25–36]. All participants performed each hand posture for five seconds in a random order during one session. The sessions were repeated ten times each training day. The healthy adults used hand dynamometers to maintain 20% of their maximum voluntary contraction, and the experiments were performed for two days. On the first day of the experiment (the untrained session), the participants performed the postures without visual feedback training. On the second day of the experiment (the trained session), the sEMG signals were measured during the sessions with the radar plot-based visual feedback training. Participants tried to control the patterns in the sEMG signals to match those on the radar plot. The forearm amputee participated in the experiments for five days because they needed more time for hand posture training to control the pattern of the sEMG signal. For the amputee, the first day of the experiment was defined as the untrained session, and the other days of the experiment were defined as trained sessions.

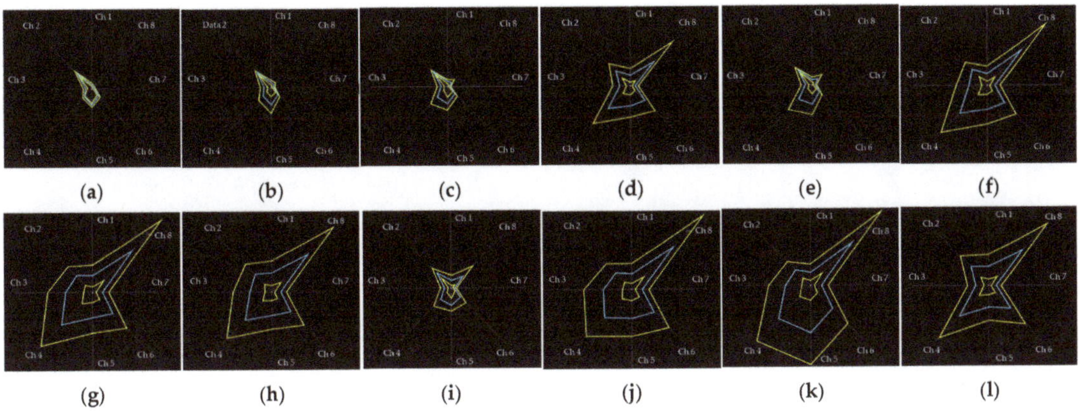

Figure 3. The radar plots of the forearm amputee on LabVIEW GUI: (**a**) rest, (**b**) spread, (**c**) finger pointing, (**d**) scissor sign, (**e**) V sign, (**f**) O.K. sign, (**g**) thumb up (hook), (**h**) cylindrical grasp, (**i**) spherical grasp, (**j**) lateral pinch, (**k**) palmar pinch, (**l**) tip pinch.

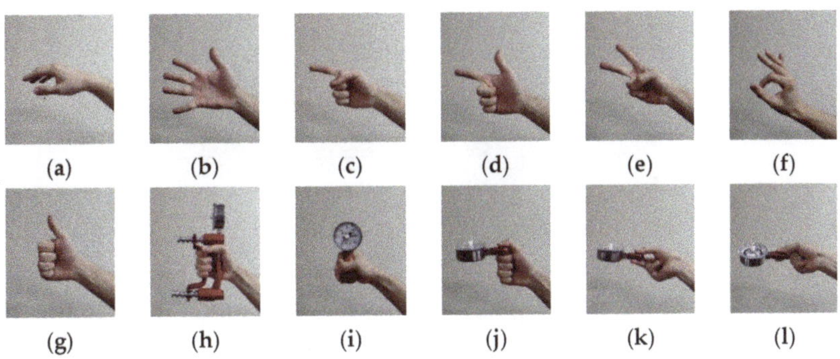

Figure 4. Hand postures for the sEMG-based posture recognition: (**a**) rest, (**b**) spread, (**c**) finger pointing, (**d**) scissor sign, (**e**) V sign, (**f**) O.K. sign, (**g**) thumb up (hook), (**h**) cylindrical grasp, (**i**) spherical grasp, (**j**) lateral pinch, (**k**) palmar pinch, (**l**) tip pinch.

2.4. Feature Vectors and Classifier

sEMG signals were filtered using the fourth-order Butterworth bandpass filter with a bandwith of 10–500 Hz, and the filtered sEMG signals were used to calculate the feature vectors. As suggested in a previous study [37], the mean absolute value (MAV) and Hudgins' set (MAV, waveform length [WL], zero crossing [ZC], and slope sign change [SSC]) were selected as the time-domain feature vectors. The previous studies reported that these feature vectors were useful to provide various information, such as MAV and WL for amplitude information and ZC and SSC for frequency information in the time domain for the pattern recognition algorithms [25,37–43]. The threshold values used to calculate the ZC and SSC feature vectors were selected following the optimization method of a previous study [21]. Table 1 shows the formulas for the feature vectors.

The ANN classifiers were developed using the Matlab software (Mathworks, Inc., Natick, MA, USA). Ten session data of each participant were divided into the training sessions and the testing sessions. The ANN classifiers were trained and validated using the automatically partitioned data within the training session data (yellow boxes in Figure 5) in the Deep Learning Toolbox of Matlab. The recognition performances of the ANN classifiers were evaluated following the ten-fold cross-testing protocol with the remained session data (blue boxes in Figure 5). The number of training data ranged from one session (TRN1) to nine sessions (TRN9) among the ten session data, and the remaining session data were used for the testing of the classifier.

Table 1. Formulas for the feature vectors.

$N:$ window size, $i:$ data sample, EMG_i: sEMG signal	
$MAV = \frac{1}{N}\sum_{i=1}^{N}\lvert EMG_i\rvert$	$ZC = \sum_{i=1}^{N-1}[f(x_i \times x_{i+1}) \cap \lvert x_i - x_{i+1}\rvert \geq threshold]$ $SSC = \sum_{i=2}^{N-1}[f[(x_i - x_{i-1}) \times (x_i - x_{i+1})]]$
$WL = \sum_{i=1}^{N-1}\lvert EMG_{i+1} - EMG_i\rvert$	$f(x) = \begin{cases} 1, & if\ x \geq threshold \\ 0, & otherwise \end{cases}$
$Threhold\ value = R \times RMS_{sEMG\ at\ rest}$, R = 0.0:0.5:10.0	

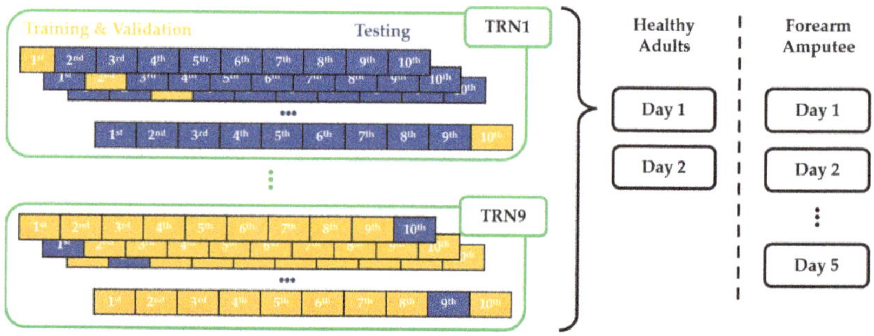

Figure 5. Ten-fold cross-testing of the ANN classifiers.

2.5. Performance Evaluation

t-SNE and the *SC* were used to analyze changes in the sEMG signals according to the radar plot-based visual feedback training. Most previous studies used a principal component analysis (PCA) to reduce the dimensions of the data or feature vectors [44–47]. A PCA is an unsupervised linear transformation algorithm that provides new features by determining the maximum variance of the data, and it can visualize data as a scatterplot [48]. However, a PCA is difficult to apply to nonlinear data processing and is affected by the scale of data when selecting the maximum variance axis [49]. For these reasons, some previous studies suggested using t-SNE, which uses Student's t distribution to compute the similarity between two points in a low-dimensional space, to solve the problems of the PCA [50]. t-SNE is effective for nonlinear data processing and shows better visualization results than a PCA. In the sEMG signals, the number of dimensions was decided by the number of channels in the sEMG system. Furthermore, sEMG signals depend on the muscle size and power. Therefore, in this study, the t-SNE function in Matlab software was used to reduce the dimensions of multichannel sEMG data and to visualize clusters of sEMG signals. In addition, the *SC* was calculated to quantify changes in the sEMG signal clusters according to the visual feedback training.

The *SC* quantifies data clustering by comparing inter- and intracluster similarity [51]. In this study, the Mahalanobis distance was used to calculate the similarity of a cluster by considering the relationships within the multivariable data [52]. The calculation of the *SC* is as follows:

$$a(i) = \frac{1}{|C_I| - 1}\sum_{j\in C_I,\ j\neq i} d(i,\ j)\ ;\ b(i) = \min_{J\neq I}\frac{1}{|C_J|}\sum_{j\in C_J} d(i,\ j) \tag{1}$$

$$s(i) = \begin{cases} \frac{b(i)-a(i)}{\max\{a(i),\ b(i)\}}, & if\ |C_I| > 1 \\ 0, & if\ |C_I| = 1 \end{cases} \tag{2}$$

$$SC = \max_{1 \le J \le K} \check{s}(J) \tag{3}$$

where $a(i)$ is the average distance between data points within a cluster (intracluster similarity). C_I is the number of sample data points in the Ith cluster, and $d(i, j)$ was the distance between the ith data point and the jth data point. $b(i)$ was the average distance between cluster C_I and cluster C_J and indicates intercluster similarity. $s(i)$ was the Silhouette value for the specific data in a cluster, and $\check{s}(J)$ was the average Silhouette value for the Jth cluster. The SC was defined as the maximum Silhouette value in each cluster. A high SC indicates good clustering, with high intracluster similarity and low intercluster similarity, and the SC range is from -1 to 1.

The performance of sEMG-based hand posture recognition in healthy adults and a forearm amputee was evaluated using classification accuracy and confusion matrixes. Significant differences ($p < 0.05$) between the classification performance results were statistically analyzed using the Kruskal–Wallis H test and pairwise comparison in IBM SPSS Statistics (IBM, Corp., Armonk, NY, USA).

3. Results

3.1. t-SNE and SC with Visual Feedback Training

In this study, the effects of radar plot-based visual feedback training on variability in the sEMG signal were visually analyzed using t-SNE and quantified using the SC.

The t-SNE results show that the clusters of both the healthy adults and the forearm amputee were improved by the visual feedback training (Figure 6, Figure 7 and Figures S1–S9). In particular, the sEMG signals of the forearm amputee were well-clustered after Day 3, compared with those from Days 1 and 2.

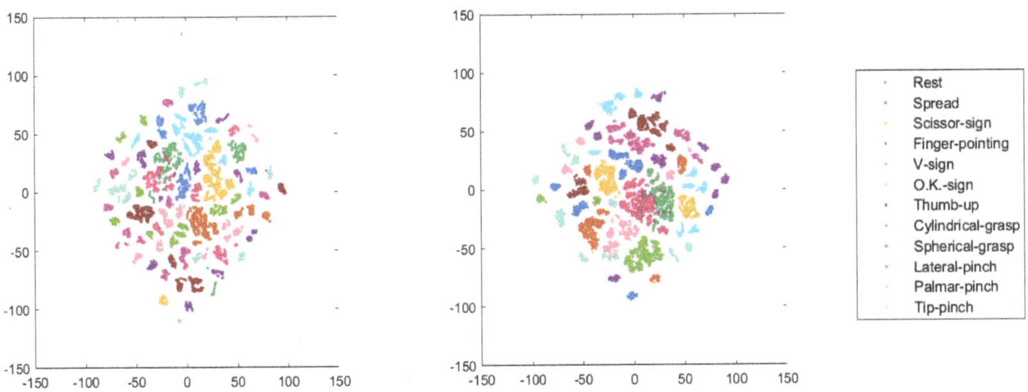

(a) (b)

Figure 6. t-SNE visualization of variability in the sEMG signals of a healthy adult (subject 1): (**a**) Day 1, (**b**) Day 2.

The SCs of all participants increased with the visual feedback training, and these results correlate well with the t-SNE visualizations (Figure 8). Most of the healthy adults had SCs higher than zero before the visual feedback training (Day 1: 0.000021 ± 0.000115), and those SCs were improved by the hand posture training (Day 2: 0.0001 ± 0.000159). In the forearm amputee, the SCs were higher than zero after Day 3 (Day 1: −0.000198, Day 2: −0.000033, Day 3: 0.000004, Day 4: 0.000018, Day 5: 0.000010).

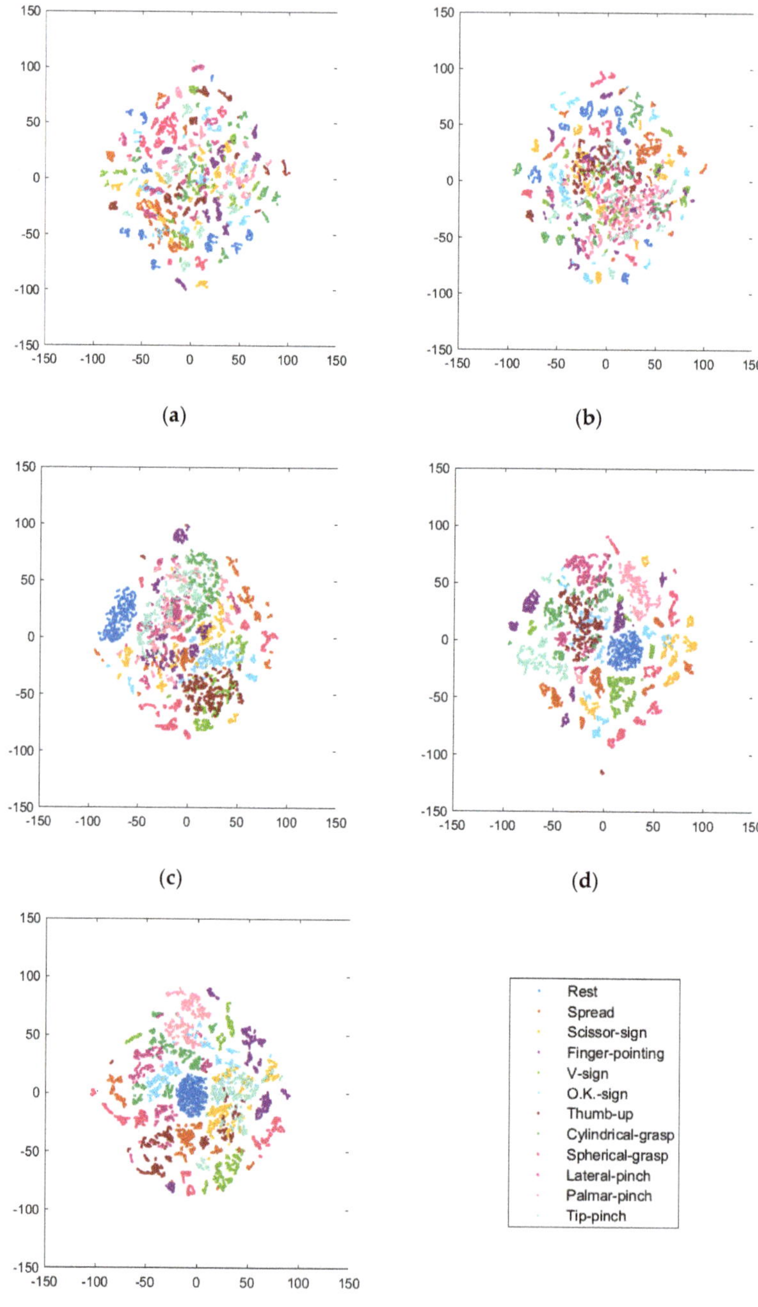

(a) (b) (c) (d) (e)

Figure 7. t-SNE visualization of variability in the forearm amputee's sEMG signals: (**a**–**e**) Day 1–Day 5.

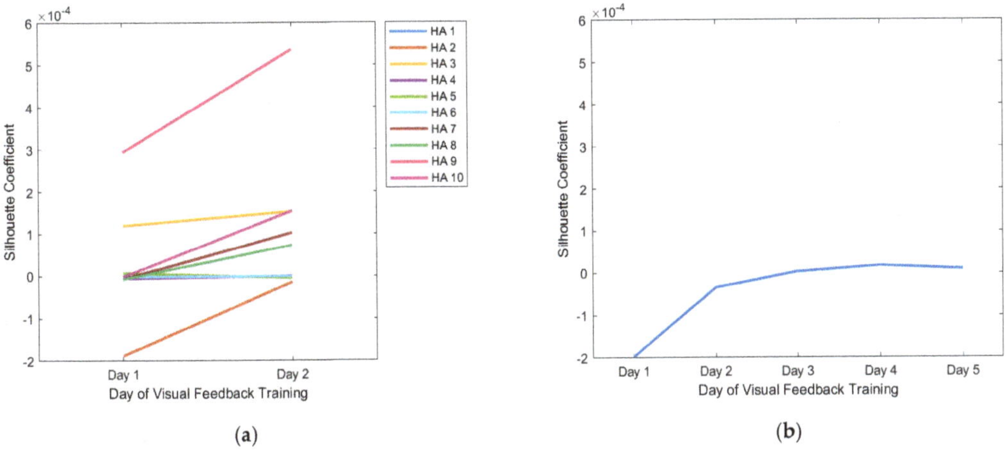

Figure 8. *SC* quantification of variability in the sEMG signals: (**a**) healthy adults, (**b**) forearm amputee.

3.2. Classification Accuracy

Tables 2 and 3 show the classification accuracy with MAV only and Hudgins' feature vector set. The classification accuracies of both the healthy adults and the forearm amputee improved as the number of training sessions increased. Significant improvements in the classification performance appeared after six and five training sessions in the healthy adults and forearm amputee, respectively.

Table 2. Classification accuracy with visual feedback training using MAV only (bold: $p < 0.05$).

		Classification Accuracy (%): Mean (Standard Deviation)								
		TRN1	TRN2	TRN3	TRN4	TRN5	TRN6	TRN7	TRN8	TRN9
Healthy Adults	Day 1	70.6 (7.7)	77.2 (6.9)	80.1 (6.3)	82.0 (6.2)	83.7 (5.7)	**85.4 (5.5)**	**85.9 (6.1)**	**86.8 (6.3)**	**87.7 (6.5)**
	Day 2	75.0 (6.8)	81.7 (6.8)	84.8 (6.8)	86.4 (6.4)	87.6 (6.0)	**88.4 (5.9)**	**89.1 (5.7)**	**89.9 (5.4)**	**90.3 (4.7)**
Forearm Amputee	Day 1	28.1 (4.3)	30.4 (4.6)	31.4 (3.0)	30.8 (2.1)	**31.8 (2.3)**	**31.3 (2.7)**	**30.7 (3.7)**	**31.2 (6.0)**	**32.8 (5.7)**
	Day 2	34.5 (4.9)	36.9 (5.0)	40.3 (4.2)	40.3 (2.6)	**42.0 (2.1)**	**43.6 (3.6)**	**44.5 (4.8)**	**44.5 (9.1)**	**48.3 (9.5)**
	Day 3	45.3 (3.8)	48.6 (4.7)	50.0 (3.9)	49.2 (5.4)	**49.4 (4.6)**	**51.8 (4.8)**	**54.1 (2.2)**	**56.4 (3.9)**	**59.7 (10.6)**
	Day 4	67.0 (3.1)	70.0 (2.3)	68.3 (3.0)	71.9 (3.4)	**72.1 (2.6)**	**74.0 (3.2)**	**75.5 (4.2)**	**78.4 (5.4)**	**80.7 (11.9)**
	Day 5	58.5 (5.0)	61.3 (5.5)	61.7 (4.8)	62.2 (6.1)	**63.9 (2.9)**	**65.2 (4.1)**	**70.3 (3.2)**	**72.0 (5.4)**	**76.5 (11.1)**

Table 3. Classification accuracy with visual feedback training using Hudgins' set (bold: $p < 0.05$).

		Classification Accuracy (%): Mean (Standard Deviation)								
		TRN1	TRN2	TRN3	TRN4	TRN5	TRN6	TRN7	TRN8	TRN9
Healthy Adults	Day 1	75.2 (7.1)	81.5 (5.8)	84.4 (4.8)	86.1 (4.8)	87.5 (4.6)	**88.9 (4.4)**	**89.7 (4.4)**	**90.9 (4.3)**	**91.2 (4.3)**
	Day 2	82.5 (6.7)	87.4 (5.7)	89.7 (5.1)	91.2 (4.8)	92.1 (4.6)	**92.9 (4.3)**	**93.5 (4.3)**	**94.3 (4.0)**	**95.1 (3.4)**
Forearm Amputee	Day 1	29.4 (4.2)	31.2 (4.5)	32.2 (3.6)	32.2 (2.6)	**32.1 (2.4)**	**32.5 (2.4)**	**31.2 (3.5)**	**32.0 (7.2)**	**30.9 (8.9)**
	Day 2	36.3 (4.2)	41.0 (4.5)	43.0 (3.5)	45.2 (2.7)	**45.5 (1.8)**	**46.9 (4.7)**	**47.7 (5.0)**	**47.3 (7.3)**	**49.5 (9.0)**
	Day 3	46.3 (3.9)	50.8 (4.9)	52.7 (3.9)	53.2 (4.1)	**55.8 (2.6)**	**56.5 (3.9)**	**58.8 (3.9)**	**60.2 (3.1)**	**64.3 (9.8)**
	Day 4	71.0 (4.4)	72.5 (3.0)	72.7 (2.7)	74.1 (3.9)	**75.3 (3.1)**	**75.9 (3.4)**	**78.2 (4.0)**	**81.1 (5.7)**	**85.5 (9.8)**
	Day 5	64.8 (4.0)	68.2 (3.5)	69.3 (4.2)	70.6 (5.8)	**72.6 (6.0)**	**74.9 (5.3)**	**77.9 (3.1)**	**80.1 (5.7)**	**84.2 (6.7)**

The radar plot-based visual feedback training effectively increased the classification accuracy of all participants (Figure 9). In the healthy adults, visual feedback training improved the accuracy of the ANN classifiers from 87.7 ± 6.5% to 91.2 ± 4.3% and from 90.3 ± 4.7% to 95.1 ± 3.4% when using MAV only and Hudgins' set, respectively. However, the classification accuracy did not differ significantly between Day 1 and Day 2. For the forearm amputee, the classification accuracy changed from 32.8 ± 5.7% (Day 1) to 76.5 ± 11.1% (Day 5) with MAV only and from 30.9 ± 8.9% (Day 1) to 84.2 ± 6.7% (Day 5) with Hudgins' set. The forearm amputee showed significant improvements in classification accuracy on Day 3 and Day 4 with MAV only and Hudgins' set, respectively. In addition, most of the classification results, excluding Day 1 of the forearm amputee, show that the classification accuracies with Hudgins' set were higher than those with MAV only. However, a significant difference on Day 2 occurred only for the healthy adults. The classification results with MAV only and Hudgins' set did not differ significantly for the forearm amputee.

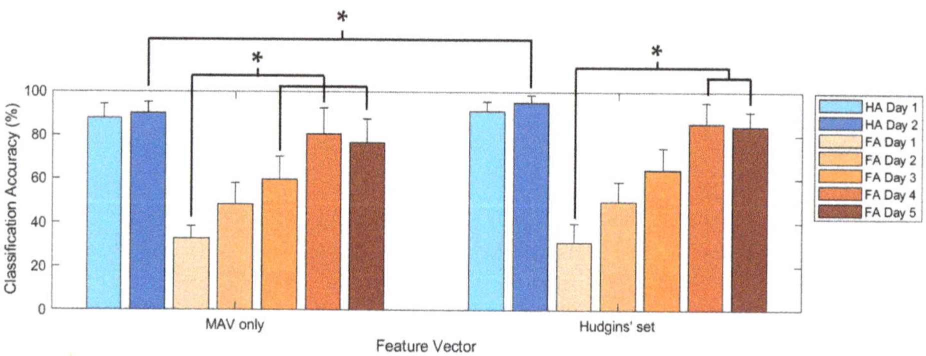

Figure 9. Classification accuracy according to the feature vectors and visual feedback training (*: $p < 0.05$).

3.3. *Confusion Matrix*

Figures 10 and 11 are the confusion matrixes showing the classification accuracy for each hand posture in the healthy adults and forearm amputee, respectively. With MAV only, the healthy adults showed high misclassification rates for cylindrical grasp vs. spherical grasp (14.5 ± 0.2%) and palmar pinch vs. tip pinch (15.0 ± 0.3%). Those misclassifications improved when Hudgins' set was applied (cylindrical grasp vs. spherical grasp: 13.1 ± 1.7%, palmar pinch vs. tip pinch: 10.2 ± 1.8%) and following visual feedback training (cylindrical grasp vs. spherical grasp: 11.7 ± 0.8%, palmar pinch vs. tip pinch: 11.7 ± 1.1%). The fewest misclassifications (cylindrical grasp vs. spherical grasp: 6.4 ± 0.9%, palmar pinch vs. tip pinch: 5.7 ± 0.7%) were found when Hudgins' set and visual feedback training were used together.

In the forearm amputee, only the hand postures of rest (MAV: 95.1%, Hudgins' set: 99.8%) and spherical grasp (MAV: 94.1%, Hudgins' set: 86.0%) were well-recognized with either feature vector on Day 1, which was the experiment before visual feedback training. The classification accuracies of each hand posture were improved by the visual feedback training, and most of the hand postures were recognized with a classification accuracy of higher than 70.0% on Day 4. In the data from Day 5, the last day of visual feedback training, MAV only showed many misclassifications of scissor sign vs. tip pinch (32.3 ± 6.2%) and cylindrical grasp vs. lateral pinch (21.7 ± 1.1%). Those misclassifications remained high with Hudgins' set (scissor sign vs. tip pinch: 22.3 ± 2.0%, cylindrical grasp vs. lateral pinch: 11.1 ± 0.8%).

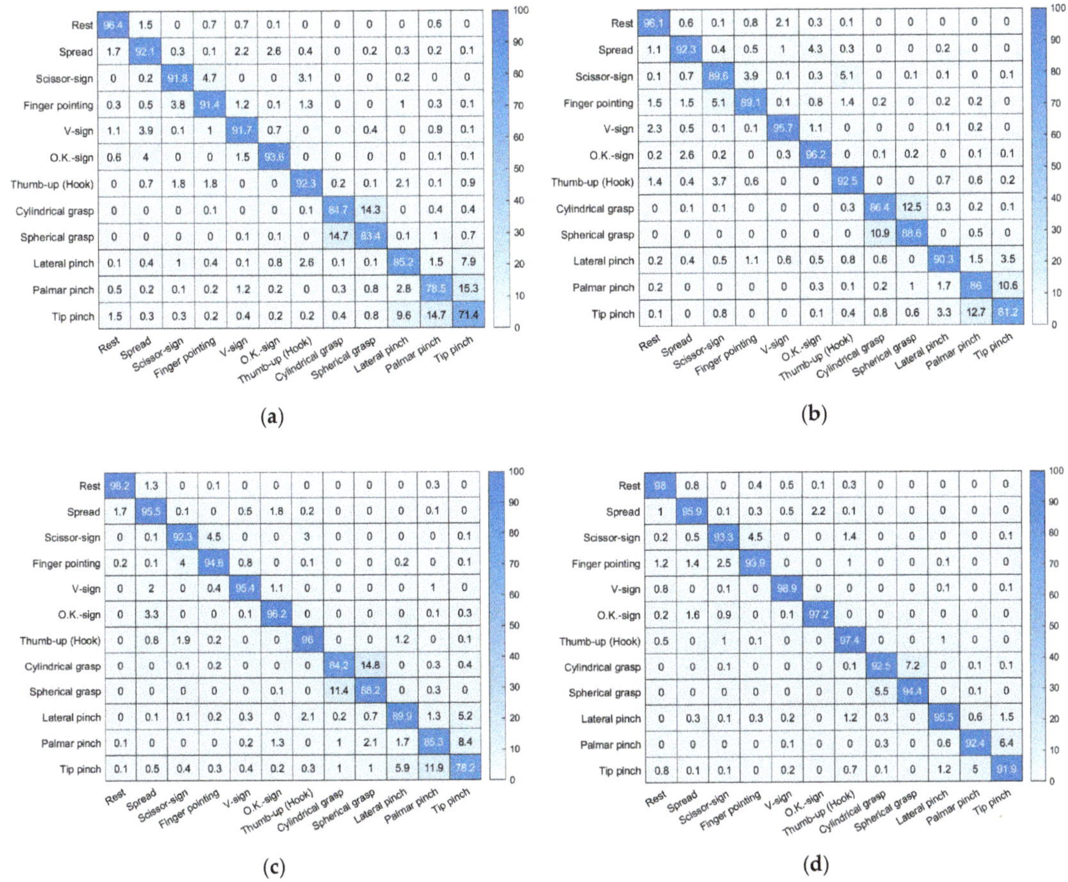

Figure 10. Confusion matrixes for the healthy adults: (**a**) Day 1 with MAV only, (**b**) Day 2 with MAV only, (**c**) Day 1 with Hudgins' set, (**d**) Day 2 with Hudgins' set.

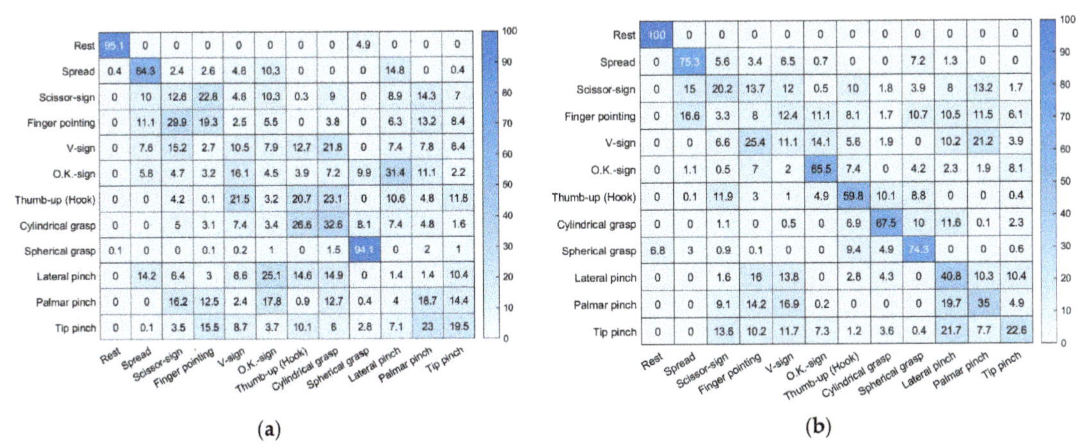

Figure 11. *Cont.*

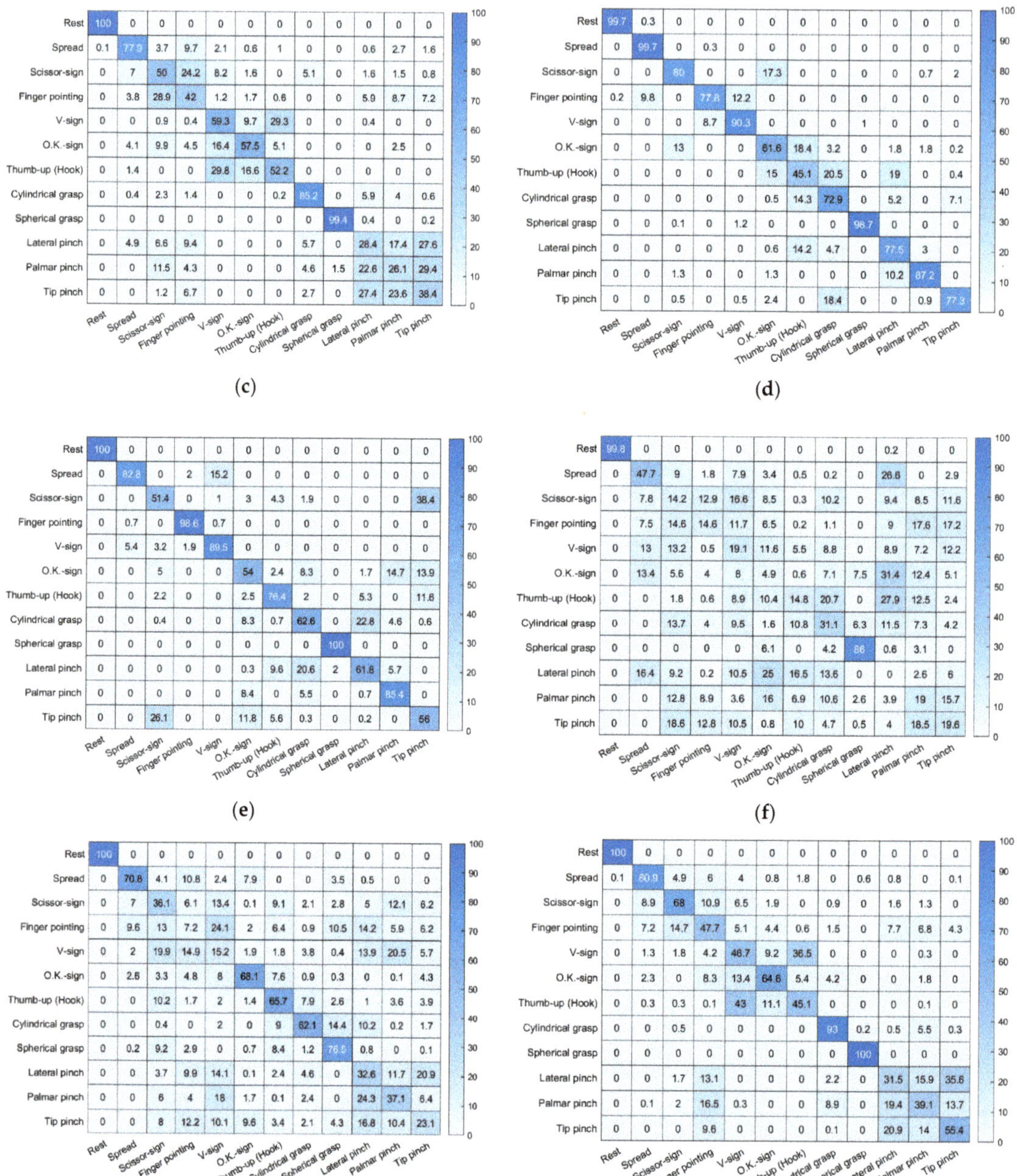

Figure 11. *Cont.*

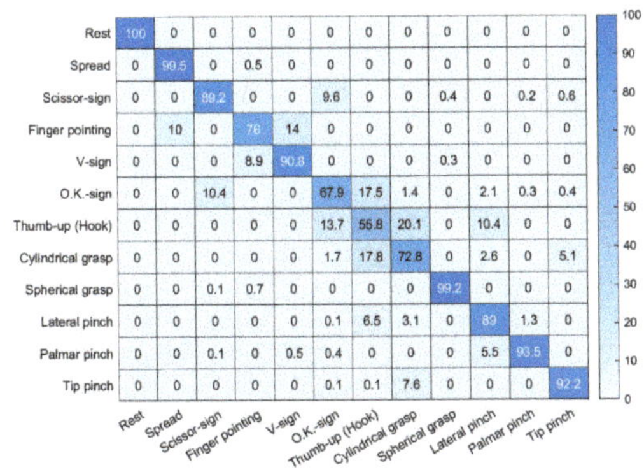

(i)

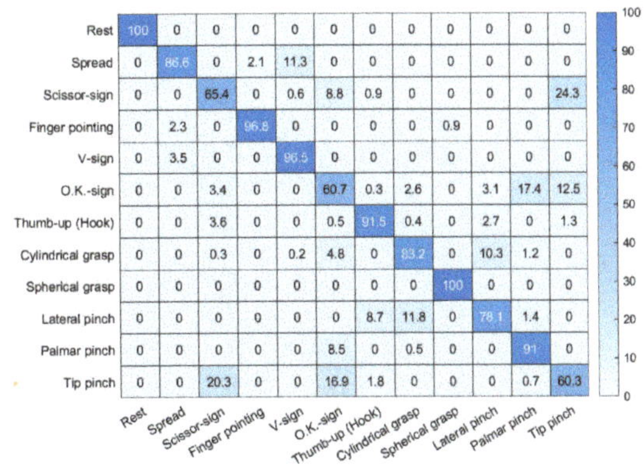

(j)

Figure 11. Confusion matrixes for the forearm amputee: **(a–e)** Day 1 to Day 5 with MAV only, **(f–j)** Day 1 to Day 5 with Hudgins' set.

4. Discussion

An electric prosthesis can perform some of the functions of a lost limb by using electromechanical motors and structures; it is an essential device for improving the quality of life for amputees [53]. The hand is an especially important body part that is used to perform many gestures in daily life, so amputees who lose a hand need a multiple DoF electric hand prosthesis. Many previous studies have reported the development of various electric hand prostheses with improved motors or newly designed structures [53–57]. However, few studies have designed algorithms to control electric prostheses. Therefore, despite advances in the hardware of electric prostheses, hand amputees have had access to only a few functions because the control algorithms have recognized only a few hand gestures [19].

This study was performed to develop a multichannel sEMG-based gesture recognition algorithm for twelve hand postures using data from healthy adults and a bilateral forearm amputee. In addition, it reports the design of a radar plot-based visual feedback training protocol that was usable by all subjects, even the bilateral amputee, to reduce variability in the sEMG signals. The visual feedback training effectively improved classification performance with data from both the healthy adults and the forearm amputee. These findings could help to efficiently improve sEMG-based gesture recognition for amputee rehabilitation and the control of electric prostheses.

Various training protocols have been tested for amputee rehabilitation in the previous studies. However, most of them show low training effects due to a lack of feedback, and few studies have quantified the effects of rehabilitation [15,16]. Furthermore, most published rehabilitation training protocols involve comparison with an intact side, which excludes bilateral amputees. Powell et al. suggested a rehabilitation protocol that uses only the amputated side with sixteen-channel sEMG sensors and a virtual electric prosthesis on the screen [17]. After ten days of rehabilitation training, the classification accuracy for data from the amputees increased from 77.5% to 94.4%, and that performance was maintained beyond the end of training. However, Powell's training protocol still lacked real-time feedback to suggest methods for improving the gestures. Fang et al. used sixteen sEMG sensors to measure signals for nine hand gestures in twelve healthy adults, and they analyzed the effects of visual feedback training on sEMG-based gesture recognition [58]. Their training protocols were divided into three types: no feedback, label feedback, and clustering feedback. No feedback was the only repetition without any feedback option, and its classification accuracy was 74.3%. Label feedback involved gesture repetition with the classification results provided as feedback, and it had a classification accuracy of 75.1%. The clustering feedback used a PCA algorithm to provide the visualized sEMG pattern, and it had the highest classification accuracy of 82.6%. These results indicate that visual feedback that includes real-time changes in the sEMG pattern improved the classification performance more effectively than the label feedback training. Therefore, in this study, a radar plot visualizing the sEMG pattern was used in the visual feedback training, and the effects of that radar plot-based visual feedback training were analyzed in both healthy adults and a bilateral forearm amputee.

t-SNE was a well-known visualization method with the dimension reduction, and *SC* was useful to quantify the clustering of sEMG signals. Zhang et al., measured sEMG signals in twelve healthy adults by using an armband-type sEMG sensor to recognize five hand gestures [59]. The feature vectors were calculated with various window sizes in sEMG signals. The best classification accuracy of 98.7% appeared with the selection of window size based on the cluster of feature vectors through t-SNE. Those results indicate that dimension reduction and data visualization through t-SNE were suitable for improving sEMG-based gesture recognition algorithms. In this study, t-SNE was used to analyze the cluster of sEMG signals in visualizations with dimension reduction. The visualized clusters of each sEMG signal correlated well with the classification accuracies, which were themselves improved by the visual feedback training. Likewise, the *SC* quantitatively showed that the sEMG signals of healthy adults and the forearm amputee were well clustered by the visual feedback training. The sEMG signals analyzed by t-SNE and the *SC* in this study seem to have lower clustering than reported in previous studies because of the characteristics of the muscles and sEMG sensors considered here. The sEMG signals visualized by t-SNE showed dispersed clusters even after the visual feedback training, and the *SC*s were only slightly higher than zero. These results were caused by the cocontractions of various muscles required by the hand gestures used in this study, such as agonist muscles for the main activity, antagonist muscles for the balance of tension with resistance, and synergist muscles to assist in the activity [60]. Because the movements required complex muscle activation, sEMG signals of all the muscles were measured during the movements, which caused dispersed clusters of sEMG signals to appear. In addition, the crosstalk among the sEMG sensors, which indicates that each sEMG sensor also measured signals from other muscles through the skin, also increased complexity and variability in the sEMG signals [61]. Nevertheless, the improved results in the t-SNE and *SC*s following the visual feedback training show that the training effectively reduced variability in the sEMG signals and improved the data clustering.

The previous studies measured the amputee's sEMG signals to practically improve an sEMG-based gesture recognition algorithm for the control of myoelectric prostheses. Benatti et al., used four-channel sEMG sensors and an SVM classifier to develop a recognition algorithm with four hand gestures for the control of multijoint prostheses [62]. They reported a classification accuracy of 89.1% for four amputees. Ahmadizadeh et al., used five force-sensitive resistor (FSR) sensors and two sEMG sensors to control a commercially available bebionic hand (Ottobock SE & Co. KGaA, Duderstadt, Germany) [63]. Their gesture recognition algorithms were developed based on k-NN, SVM, and LDA, and an amputee participated in the training and testing of each one. The classification accuracies were reported as 75.2%, 78.5%, and 81.6% for ten, six, and three hand gestures, respectively. Most previous studies that enrolled amputees reported low classification accuracy when recognizing various hand gestures, and they improved classification accuracy by applying fewer hand gestures to the recognition system. However, the recognition of four or fewer hand gestures significantly limits the control of a multijoint prosthesis, and non-pattern recognition algorithms are more efficient when the number of hand gestures is small. In this study, sEMG signals from forearm muscles were measured using nine and eight sEMG sensors on the healthy adults and forearm amputee, respectively. ANN classifiers were then developed to recognize twelve hand postures by using two types of feature vectors, MAV only and Hudgins' set. The healthy adults showed classification accuracies of 87.7% with MAV only and 90.3% with Hudgins' set. The classification accuracies for the forearm amputee were 32.8% with MAV only and 30.9% with Hudgins' set. Thus, the optimized feature vectors (Hudgins' set in this study) improved the classification performance in healthy adults, which agrees with the results of previous studies [11–14,19]. However, the classification accuracies for the forearm amputee decreased when using the optimized feature vectors because of high variability and low consistency in the sEMG signals. Those problems were solved by the radar plot-based visual feedback training. After the visual feedback training, the healthy adults showed classification accuracies of 91.2% with MAV only and 95.1% with Hudgins' set, and the forearm amputee showed classification accuracies of 76.5% with MAV only and 84.2% with Hudgins' set. Thus, the radar plot-based visual feedback training successfully improved both classification accuracy and the effect of the optimized feature vectors by reducing variability in the sEMG signals. For these reasons, reducing variability in the sEMG signals was more important for the amputee than the advanced hardware and software in the sEMG-based gesture recognition system.

The classification accuracies for each hand posture are shown as confusion matrixes in this paper. Misclassifications appeared mainly for cylindrical grasp vs. spherical grasp and palmar pinch vs. tip pinch, which had misclassification rates of 14.5% and 15.0%, respectively. Those misclassifications occurred because those gestures are similar and require cocontractions of the same muscles. We reported similar results in our previous study of armband-type sEMG sensors [25]. Some of the misclassifications in our previous studies, which appeared in palmar pinch vs. lateral pinch, finger pointing vs. scissor sign, and thumb up (hook) vs. scissor sign, did not occur in this study because we minimized the effects of crosstalk in the sEMG system by positioning the sEMG sensors on specific muscles. In addition, misclassification of the movements of healthy adults was improved by the optimized feature vectors (Hudgins' set) and visual feedback training, with the misclassification rates after applying both Hudgins' set and visual feedback training reduced to 6.4% and 5.7% for cylindrical grasp vs. spherical grasp and palmar pinch vs. tip pinch, respectively. In the forearm amputee, only the hand postures of rest and spherical grasp were well-recognized, with classification accuracies of 95.1% and 94.1%, respectively. The radar plot-based visual feedback training improved the classification accuracies of most hand postures to be higher than 70.0%. However, misclassifications persisted for scissor sign vs. tip pinch and cylindrical grasp vs. lateral pinch, which had misclassification rates of 32.3% and 21.7%, respectively, even after five days of hand posture training. In the healthy adults, misclassifications appeared between similar gestures, whereas the forearm amputee showed misclassifications between dissimilar gestures because of muscle loss. The forearm amputee had lost his extensor digitorum minimi and extensor pollicis muscles on the amputated side. In particular, the loss of the extensor pollicis, which contracts to move the thumb, caused information loss in the sEMG patterns that decreased classification accuracy. Misclassifications for the forearm amputee remained high, even when the feature vectors were optimized—22.3% in scissor sign vs. tip pinch and 11.1% in cylindrical grasp vs. lateral pinch. These results indicate that the optimized feature vectors effectively reinforced the consistency of the sEMG pattern in healthy adults, but they were not effective for the forearm amputee because of information loss in the sEMG patterns. Therefore, training for users, such as visual feedback training, would improve the classification performance for amputees more effectively than optimizing classifiers or feature vectors.

Many previous studies have suggested optimized classifiers and feature vectors and advanced hardware to improve the performance of sEMG-based gesture recognition. However, other optimization is required to successfully increase the number of recognized gestures or change users. In this study, the classification performance was efficiently improved by reducing variability in the sEMG signals through visual feedback training. Our method will not only reduce the time and cost of system optimization but also improve the user accessibility of future systems.

This study has three limitations. The first is that only one forearm amputee participated in the experiment. Amputees have larger individual differences in their sEMG patterns than healthy adults because of variations in the size of their residual limbs and periods of amputation. Specifically, misclassifications will differ for each amputee depending on which muscles have been lost. The second limitation is that the period for the hand posture training was shorter than the rehabilitation periods reported in previous studies [15,16]. Typically, the amputee rehabilitation programs lasted for several months in previous clinical research, whereas the visual feedback training in this study lasted for only five days. Our visual feedback training was useful to improve the classification performance of the bilateral forearm amputee dramatically within a short period. However, it is also important to analyze whether the number of recognizable gestures could be increased by reinforcing muscles through continuous posture training and whether the improved classification performance would be maintained after the end of training. The third limitation is the number of sEMG sensors used to recognize hand postures. A small number of sEMG sensors is more efficient in rehabilitation protocols and the control of electric prostheses. Specifically, the eight-channel sEMG sensors used in this study and their positions would be difficult to apply to an electric prosthesis because of the size of the socket on the amputated limb.

5. Conclusions

An sEMG-based hand posture recognition algorithm and radar plot-based visual feedback training were developed for the control of myoelectric prostheses and the amputee's rehabilitation in this paper. The classification accuracies for the healthy adults and a forearm amputee were improved by the visual feedback training and optimized feature vectors. The visual feedback training improved the classification performance of the healthy adults and a forearm amputee by 2.6% and 43.7%, respectively. The optimization of feature vectors (Hudgins' set) increased the classification accuracy by 4.8% more in trained healthy adults and 7.7% more in a trained forearm amputee, respectively. t-SNE and the *SC* both showed that the visual feedback training reduced variability in the sEMG signals in both healthy adults and the forearm amputee. The radar plot-based visual feedback training was very effective to improve the classification performance of the bilateral forearm amputee by the real-time monitoring of activation patterns of sEMG in the residual limb.

These findings could be used to improve the performance of sEMG-based hand posture recognition, not only in rehabilitation and the control of electric prostheses for amputees, but also in HCI systems for healthy adults. In future work, the measurement of sEMG signals and visual feedback training will be performed with various forearm amputees, and the number and positions of the sEMG sensors will be analyzed to develop an efficient sEMG-based hand posture recognition algorithm.

Supplementary Materials: The following supporting information can be downloaded at: https://www.mdpi.com/article/10.3390/s22207984/s1, Figure S1: t-SNE visualization of variability in the sEMG signals of a healthy adult (subject 2) (a) Day 1, (b) Day 2; Figure S2: t-SNE visualization of variability in the sEMG signals of a healthy adult (subject 3) (a) Day 1, (b) Day 2; Figure S3: t-SNE visualization of variability in the sEMG signals of a healthy adult (subject 4) (a) Day 1, (b) Day 2; Figure S4: t-SNE visualization of variability in the sEMG signals of a healthy adult (subject 5) (a) Day 1, (b) Day 2; Figure S5: t-SNE visualization of variability in the sEMG signals of a healthy adult (subject 6) (a) Day 1, (b) Day 2; Figure S6: t-SNE visualization of variability in the sEMG signals of a healthy adult (subject 7) (a) Day 1, (b) Day 2; Figure S7: t-SNE visualization of variability in the sEMG signals of a healthy adult (subject 8) (a) Day 1, (b) Day 2; Figure S8: t-SNE visualization of variability in the sEMG signals of a healthy adult (subject 9) (a) Day 1, (b) Day 2; Figure S9: t-SNE visualization of variability in the sEMG signals of a healthy adult (subject 10) (a) Day 1, (b) Day 2.

Author Contributions: Conceptualization, J.K. and Y.K.; data curation, J.K., S.Y., B.K. and S.L.; formal analysis, J.K.; funding acquisition, Y.K.; investigation, J.K., S.Y. and K.H.C.; methodology, J.K., S.Y., B.K., S.L., S.P. and S.K.; project administration, J.K. and Y.K.; resources, Y.K.; software, J.K., B.K. and S.P.; supervision, Y.K.; validation, J.K.; visualization, J.K.; writing—original draft, J.K. and Y.K.; writing—review and editing, J.K. and Y.K. All authors have read and agreed to the published version of the manuscript.

Funding: This research was financially supported by a major project of the Korea Institute of Machinery and Materials (Project ID: NK238F) and the Institute of Civil Military Technology Cooperation funded by the Defense Acquisition Program Administration and Ministry of Trade, Industry and Energy of Korean government under grant No. 21-SF-GU-07.

Institutional Review Board Statement: This study was conducted according to the guidelines of the Declaration of Helsinki and was approved by the Yonsei University Mirae Institutional Review Board (1041849-202002-BM-018-02; approval date: 18 March 2020) and the Institutional Review Board of the Korea Orthopedics & Rehabilitation Engineering Center (RERI-IRB-210915-2; approval date: 15 September 2021).

Informed Consent Statement: Informed consent was obtained from all subjects involved in the study.

Data Availability Statement: The data presented in this study are available upon request from the corresponding author. The data are not publicly available because the authors are continuing the study.

Conflicts of Interest: The authors declare no conflict of interest. The funders had no role in the design of the study; in the collection, analysis, or interpretation of data; in the writing of the manuscript; or in the decision to publish the results.

References

1. Farina, D.; Merletti, R.; Enoka, R.M. The Extraction of Neural Strategies from the Surface EMG: An Update. *J. Appl. Physiol.* **2014**, *117*, 1215–1230. [CrossRef] [PubMed]
2. Xu, H.; Xiong, A. Advances and Disturbances in sEMG-Based Intentions and Movements Recognition: A Review. *IEEE Sens. J.* **2021**, *21*, 13019–13028. [CrossRef]
3. Nam, Y.; Koo, B.; Cichocki, A.; Choi, S. GOM-Face: GKP, EOG, and EMG-Based Multimodal Interface with Application to Humanoid Robot Control. *IEEE. Trans. Biomed. Eng.* **2013**, *61*, 453–462. [CrossRef] [PubMed]
4. Ahmadizadeh, C.; Khoshnam, M.; Menon, C. Human Machine Interfaces in Upper-Limb Prosthesis Control: A Survey of Techniques for Preprocessing and Processing of Biosignals. *IEEE Signal. Process. Mag.* **2021**, *38*, 12–22. [CrossRef]
5. Parajuli, N.; Sreenivasan, N.; Bifulco, P.; Cesarelli, M.; Savino, S.; Niola, V.; Esposito, D.; Hamilton, T.J.; Naik, G.R.; Gunawardana, U.; et al. Real-Time EMG Based Pattern Recognition Control for Hand Prostheses: A Review on Existing Methods, Challenges and Future Implementation. *Sensors* **2019**, *19*, 4596. [CrossRef] [PubMed]
6. Witteveen, H.J.B.; Droog, E.A.; Rietman, J.S.; Veltink, P.H. Vibro- and Electrotactile User Feedback on Hand Opening for Myoelectric Forearm Prostheses. *IEEE Trans. Biomed. Eng.* **2012**, *59*, 2219–2226. [CrossRef]
7. Van Der Linde, H.; Hofstad, C.J.; Geurts, A.C.; Postema, K.; Geertzen, J.H.; van Limbeek, J. A Systematic Literature Review of the Effect of Different Prosthetic Components on Human Functioning with a Lower-Limb Prosthesis. *J. Rehabil. Res. Dev.* **2004**, *41*, 555–570. [CrossRef] [PubMed]
8. Samuel, O.W.; Asogbon, M.G.; Geng, Y.; Al-Timemy, A.H.; Pirbhulal, S.; Ji, N.; Chen, S.; Fang, P.; Li, G. Intelligent EMG Pattern Recognition Control Method for Upper-Limb Multifunctional Prostheses: Advances, Current Challenges, and Future Prospects. *IEEE Access* **2019**, *7*, 10150–10165. [CrossRef]
9. Klarich, J.; Brueckner, I. Amputee Rehabilitation and Preprosthetic Care. *Phys. Med. Rehabil Clin. N. Am.* **2014**, *25*, 75–91. [CrossRef]
10. Phinyomark, A.; Scheme, E. EMG Pattern Recognition in the Era of Big Data and Deep Learning. *Big Data Cogn. Comput.* **2018**, *2*, 21. [CrossRef]
11. Emayavaramban, G.; Amudha, A. Identifying Hand Gestures Using sEMG for Human-Machine Interaction. *ARPN J. Eng. Appl. Sci.* **2016**, *11*, 12777–12785.
12. Shi, W.T.; Lyu, Z.J.; Tang, S.T.; Chia, T.L.; Yang, C.Y. A Bionic Hand Controlled by Hand Gesture Recognition Based on Surface EMG Signals: A Preliminary Study. *Biocybern. Biomed. Eng.* **2018**, *38*, 126–135. [CrossRef]
13. Adewuyi, A.A.; Hargrove, L.J.; Kuiken, T.A. Evaluating EMG Feature and Classifier Selection for Application to Partial-Hand Prosthesis Control. *Front. Neurorobot.* **2016**, *10*, 15. [CrossRef]
14. Betthauser, J.L.; Hunt, C.L.; Osborn, L.E.; Masters, M.R.; Lévay, G.; Kaliki, R.R.; Thakor, N.V. Limb Position Tolerant Pattern Recognition for Myoelectric Prosthesis Control with Adaptive Sparse Representations from Extreme Learning. *IEEE Trans. Biomed. Eng.* **2017**, *65*, 770–778. [CrossRef] [PubMed]
15. Mallik, A.K.; Pandey, S.K.; Srivastava, A.; Kumar, S.; Kumar, A. Comparison of Relative Benefits of Mirror Therapy and Mental Imagery in Phantom Limb Pain in Amputee Patients at a Tertiary Care Center. *Arch. Rehabil. Res. Clin. Transl.* **2020**, *2*, 100081. [CrossRef] [PubMed]
16. Barbin, J.; Seetha, V.; Casillas, J.M.; Paysant, J.; Perennou, D. The Effects of Mirror Therapy on Pain and Motor Control of Phantom Limb in Amputees: A Systematic Review. *Ann. Phys. Rehabil. Med.* **2016**, *59*, 270–275. [CrossRef] [PubMed]

17. Powell, M.A.; Kaliki, R.R.; Thakor, N.V. User Training for Pattern Recognition-Based Myoelectric Prostheses: Improving Phantom Limb Movement Consistency and Distinguishability. *IEEE Trans. Neural Syst. Rehabil. Eng.* **2013**, *22*, 522–532. [CrossRef] [PubMed]
18. DELSYS. Trigno Wireless Biofeedback System—User's Guide. Available online: https://www.delsys.com/downloads/USERSGUIDE/trigno/wireless-biofeedback-system.pdf (accessed on 3 October 2022).
19. Kim, S.; Kim, J.; Koo, B.; Kim, T.; Jung, H.; Park, S.; Kim, S.; Kim, Y. Development of an Armband EMG Module and a Pattern Recognition Algorithm for the 5-Finger Myoelectric Hand Prosthesis. *Int. J. Precis. Eng. Manuf.* **2019**, *20*, 1997–2006. [CrossRef]
20. Phinyomark, A.; Phukpattaranont, P.; Limsakul, C. Feature Reduction and Selection for EMG Signal Classification. *Expert Syst. Appl.* **2012**, *39*, 7420–7431. [CrossRef]
21. Kamavuako, E.N.; Scheme, E.J.; Englehart, K.B. Determination of Optimum Threshold Values for EMG Time Domain Features; A Multi-Dataset Investigation. *J. Neural Eng.* **2016**, *13*, 046011. [CrossRef] [PubMed]
22. Asif, A.R.; Waris, A.; Gilani, S.O.; Jamil, M.; Ashraf, H.; Shafique, M.; Niazi, I.K. Performance Evaluation of Convolutional Neural Network for Hand Gesture Recognition Using EMG. *Sensors* **2020**, *20*, 1642. [CrossRef] [PubMed]
23. Tang, X.; Liu, Y.; Lv, C.; Sun, D. Hand Motion Classification Using a Multi-Channel Surface Electromyography Sensor. *Sensors* **2012**, *12*, 1130–1147. [CrossRef] [PubMed]
24. Laksono, P.W.; Matsushita, K.; Suhaimi, M.S.A.B.; Kitamura, T.; Njeri, W.; Muguro, J.; Sasaki, M. Mapping Three Electromyography Signals Generated by Human Elbow and Shoulder Movements to Two Degree of Freedom Upper-Limb Robot Control. *Robotics* **2020**, *9*, 83. [CrossRef]
25. Kim, J.; Koo, B.; Nam, Y.; Kim, Y. sEMG-Based Hand Posture Recognition Considering Electrode Shift, Feature Vectors, and Posture Groups. *Sensors* **2021**, *21*, 7681. [CrossRef]
26. Jiang, S.; Lv, B.; Guo, W.; Zhang, C.; Wang, H.; Sheng, X.; Shull, P.B. Feasibility of Wrist-Worn, Real-Time Hand, and Surface Gesturerecognition Via sEMG and IMU Sensing. *IEEE Trans. Ind. Inf.* **2017**, *14*, 3376–3385. [CrossRef]
27. De Andrade, F.H.C.; Pereira, F.G.; Resende, C.Z.; Cavalieri, D.C. Improving sEMG-based hand gesture recognition using maximal overlap discrete wavelet transform and an autoencoder neural network. In Proceedings of the 16th Brazilian Congress on Biomedical Engineering, Armação dos Búzios, Brazil, 21–25 October 2019; pp. 271–279.
28. Suarez, J.; Murphy, R.R. Hand gesture recognition with depth images: A review. In Proceedings of the 2012 IEEE RO-MAN: The 21st IEEE International Symposium on Robot and Human Interactive Communication, Paris, France, 9–13 September 2012; pp. 411–417.
29. Murthy, G.R.S.; Jadon, R.S. Hand gesture recognition using neural networks. In Proceedings of the 2010 IEEE 2nd International Advance Computing Conference (IACC), Patiala, India, 19–20 February 2010; pp. 134–138.
30. Li, W.J.; Hsieh, C.Y.; Lin, L.F.; Chu, W.C. Hand gesture recognition for post-stroke rehabilitation using leap motion. In Proceedings of the 2017 International Conference on Applied System Innovation (ICASI), Sapporo, Japan, 13–17 May 2017; pp. 386–388.
31. Chonbodeechalermroong, A.; Chalidabhongse, T.H. Dynamic contour matching for hand gesture recognition from monocular image. In Proceedings of the 2015 12th International Joint Conference on Computer Science and Software Engineering (JCSSE), Hatyai, Thailand, 22–24 July 2015; pp. 47–51.
32. Ren, Z.; Meng, J.; Yuan, J.; Zhang, Z. Robust hand gesture recognition with Kinect sensor. In Proceedings of the 19th ACM International Conference on Multimedia, New York, NY, USA, 1–28 December 2011; pp. 759–760.
33. Sayin, F.S.; Ozen, S.; Baspinar, U. Hand gesture recognition by using sEMG signals for human machine interaction applications. In Proceedings of the 2018 Signal Processing: Algorithms, Architectures, Arrangements, and Applications (SPA), Poznan, Poland, 19–21 September 2018; pp. 27–30.
34. Yang, Y.; Fermuller, C.; Li, Y.; Aloimonos, Y. Grasp type revisited: A modern perspective on a classical feature for vision. In Proceedings of the IEEE Conference on Computer Vision and Pattern Recognition, Boston, MA, USA, 7–12 June 2015; pp. 400–408.
35. Plouffe, G.; Cretu, A.M. Static and Dynamic Hand Gesture Recognition in Depth Data Using Dynamic Time Warping. *IEEE Trans. Instrum. Meas.* **2015**, *65*, 305–316. [CrossRef]
36. Apostol, B.; Mihalache, C.R.; Manta, V. Using spin images for hand gesture recognition in 3D point clouds. In Proceedings of the 2014 18th International Conference on System Theory, Control and Computing (ICSTCC), Sinaia, Romania, 17–19 October 2014; pp. 544–549.
37. Hudgins, B.; Parker, P.; Scott, R.N. A New Strategy for Multifunction Myoelectric Control. *IEEE. Trans. Biomed. Eng.* **1993**, *40*, 82–94. [CrossRef] [PubMed]
38. Oskoei, M.A.; Hu, H. Myoelectric Control Systems—A Survey. *Biomed. Signal. Process Control* **2007**, *2*, 275–294.
39. Hargrove, L.J.; Englehart, K.; Hudgins, B. A Comparison of Surface and Intramuscular Myoelectric Signal Classification. *IEEE Trans. Biomed. Eng.* **2007**, *54*, 847–853. [CrossRef]
40. Kuiken, T.A.; Li, G.; Lock, B.A.; Lipschutz, R.D.; Miller, L.A.; Stubblefield, K.A.; Englehart, K.B. Targeted Muscle Reinnervation for Real-Time Myoelectric Control of Multifunction Artificial Arms. *JAMA* **2009**, *301*, 619–628. [CrossRef]
41. Tkach, D.; Huang, H.; Kuiken, T.A. Study of Stability of Time-Domain Features for Electromyographic Pattern Recognition. *J. NeuroEng. Rehabil.* **2010**, *7*, 1–13. [CrossRef] [PubMed]
42. Phinyomark, A.; Quaine, F.; Charbonnier, S.; Serviere, C.; Tarpin-Bernard, F.; Laurillau, Y. EMG Feature Evaluation for Improving Myoelectric Pattern Recognition Robustness. *Expert Syst. Appl.* **2013**, *40*, 4832–4840. [CrossRef]

43. Samuel, O.W.; Zhou, H.; Li, X.; Wang, H.; Zhang, H.; Sangaiah, A.K.; Li, G. Pattern Recognition of Electromyography Signals Based on Novel Time Domain Features for Amputees Limb Motion Classification. *Comput. Electr. Eng.* **2018**, *67*, 646–655. [CrossRef]
44. Zhang, T.; Yang, B. Big data dimension reduction using PCA. In Proceedings of the 2016 IEEE International Conference on Smart Cloud (SmartCloud), New York, NY, USA, 18–20 November 2016.
45. Kambhatla, N.; Leen, T.K. Dimension Reduction by Local Principal Component Analysis. *Neural Comput.* **1997**, *9*, 1493–1516. [CrossRef]
46. Huang, D.; Jiang, F.; Li, K.; Tong, G.; Zhou, G. Scaled PCA: A New Approach to Dimension Reduction. *Manag. Sci.* **2022**, *68*, 1678–1695. [CrossRef]
47. Farrell, M.D.; Mersereau, R.M. On the Impact of PCA Dimension Reduction for Hyperspectral Detection of Difficult Targets. *IEEE Geosci. Remote. Sens.* **2005**, *2*, 192–195. [CrossRef]
48. Hotelling, H. Analysis of a Complex of Statistical Variables into Principal Components. *J. Educ. Psychol.* **1933**, *24*, 417. [CrossRef]
49. Anowar, F.; Sadaoui, S.; Selim, B. Conceptual and Empirical Comparison of Dimensionality Reduction Algorithms (pca, kpca, lda, mds, svd, lle, isomap, le, ica, t-sne). *Comput. Sci. Rev.* **2021**, *40*, 100378. [CrossRef]
50. Van der Maaten, L.; Hinton, G. Visualizing Data Using t-SNE. *J. Mach. Learn. Res.* **2008**, *9*, 2579–2605.
51. Rousseeuw, P.J. Silhouettes: A Graphical Aid to the Interpretation and Validation of Cluster Analysis. *J. Comput. Appl. Math.* **1987**, *20*, 53–65. [CrossRef]
52. Chandra, M.P. On the Generalised Distance in Statistics. *Proc. Natl. Inst. Sci. India* **1936**, *2*, 49–55.
53. Jung, S.Y.; Kim, S.G.; Kim, J.H.; Park, S.H. Development of Multifunctional Myoelectric Hand Prosthesis System with Easy and Effective Mode Change Control Method Based on the Thumb Position and State. *Appl. Sci.* **2021**, *11*, 7295. [CrossRef]
54. Belter, J.T.; Segil, J.L.; Dollar, A.M.; Weir, R.F. Mechanical Design and Performance Specifications of Anthropomorphic Prosthetic Hands: A Review. *J. Rehabil. Res. Dev.* **2013**, *50*, 599–618. [CrossRef] [PubMed]
55. Bennett, D.A.; Dalley, S.A.; Truex, D.; Goldfarb, M. A Multigrasp Hand Prosthesis for Providing Precision and Conformal Grasps. *IEEE/ASME Trans. Mechatron.* **2014**, *20*, 1697–1704. [CrossRef]
56. Ryu, W.; Choi, Y.; Choi, Y.J.; Lee, Y.G.; Lee, S. Development of an Anthropomorphic Prosthetic Hand with Underactuated Mechanism. *Appl. Sci.* **2020**, *10*, 4384. [CrossRef]
57. Controzzi, M.; Clemente, F.; Barone, D.; Ghionzoli, A.; Cipriani, C. The SSSA-MyHand: A Dexterous Lightweight Myoelectric Hand Prosthesis. *IEEE Trans. Neural. Syst. Rehabil. Eng.* **2016**, *25*, 459–468. [CrossRef]
58. Fang, Y.; Zhou, D.; Li, K.; Liu, H. Interface Prostheses with Classifier-Feedback-Based User Training. *IEEE. Trans. Biomed. Eng.* **2016**, *64*, 2575–2583.
59. Zhang, Z.; Yang, K.; Qian, J.; Zhang, L. Real-Time Surface EMG Pattern Recognition for Hand Gestures Based on an Artificial Neural Network. *Sensors* **2019**, *19*, 3170. [CrossRef]
60. Neumann, D.A. *Essential Topics of Kinesiology. Kinesiology of the Musculoskeletal System: Foundation for Rehabilitation*, 2nd ed.; Elsevier: Mosby, MO, USA, 2010; pp. 1–115.
61. Winter, D.A.; Fuglevand, A.J.; Archer, S.E. Crosstalk in Surface Electromyography: Theoretical and Practical Estimates. *J. Electromyogr. Kinesiol.* **1994**, *4*, 15–26. [CrossRef]
62. Benatti, S.; Milosevic, B.; Farella, E.; Gruppioni, E.; Benini, L. A Prosthetic Hand Body Area Controller Based on Efficient Pattern Recognition Control Strategies. *Sensors* **2017**, *17*, 869. [CrossRef]
63. Ahmadizadeh, C.; Merhi, L.K.; Pousett, B.; Sangha, S.; Menon, C. Toward Intuitive Prosthetic Control: Solving Common Issues Using Force Myography, Surface Electromyography, and Pattern Recognition in a Pilot Case Study. *IEEE Robot. Autom. Mag.* **2017**, *24*, 102–111. [CrossRef]

sensors

MDPI

Article

Characterization of Infants' General Movements Using a Commercial RGB-Depth Sensor and a Deep Neural Network Tracking Processing Tool: An Exploratory Study

Diletta Balta [1,*,†], HsinHung Kuo [2,†], Jing Wang [2], Ilaria Giuseppina Porco [3], Olga Morozova [4], Manon Maitland Schladen [2,5], Andrea Cereatti [1], Peter Stanley Lum [2] and Ugo Della Croce [3]

1 Department of Electronics and Telecommunications, Politecnico di Torino, 10129 Torino, Italy
2 Department of Biomedical Engineering, The Catholic University of America, Washington, DC 20064, USA
3 Department of Biomedical Sciences, University of Sassari, 07100 Sassari, Italy
4 Children's National Hospital, Washington, DC 20010, USA
5 Department of Rehabilitation Medicine, Georgetown University Medical Center, Washington, DC 20057, USA
* Correspondence: diletta.balta@polito.it
† These authors contributed equally to this work.

Abstract: Cerebral palsy, the most common childhood neuromotor disorder, is often diagnosed through visual assessment of general movements (GM) in infancy. This skill requires extensive training and is thus difficult to implement on a large scale. Automated analysis of GM performed using low-cost instrumentation in the home may be used to estimate quantitative metrics predictive of movement disorders. This study explored if infants' GM may be successfully evaluated in a familiar environment by processing the 3D trajectories of points of interest (PoI) obtained from recordings of a single commercial RGB-D sensor. The RGB videos were processed using an open-source markerless motion tracking method which allowed the estimation of the 2D trajectories of the selected PoI and a purposely developed method which allowed the reconstruction of their 3D trajectories making use of the data recorded with the depth sensor. Eight infants' GM were recorded in the home at 3, 4, and 5 months of age. Eight GM metrics proposed in the literature in addition to a novel metric were estimated from the PoI trajectories at each timepoint. A pediatric neurologist and physiatrist provided an overall clinical evaluation from infants' video. Subsequently, a comparison between metrics and clinical evaluation was performed. The results demonstrated that GM metrics may be meaningfully estimated and potentially used for early identification of movement disorders.

Keywords: markerless; RGB-D; general movements; infant movement analysis; movement disorders

Citation: Balta, D.; Kuo, H.; Wang, J.; Porco, I.G.; Morozova, O.; Schladen, M.M.; Cereatti, A.; Lum, P.S.; Della Croce, U. Characterization of Infants' General Movements Using a Commercial RGB-Depth Sensor and a Deep Neural Network Tracking Processing Tool: An Exploratory Study. *Sensors* **2022**, 22, 7426. https://doi.org/10.3390/s22197426

Academic Editor: Giovanni Saggio

Received: 20 July 2022
Accepted: 26 September 2022
Published: 29 September 2022

1. Introduction

Cerebral palsy (CP) is the clinical description given to a constellation of neuromotor impairments stemming from perinatal brain injuries such as periventricular leukomalacia, intracerebral hemorrhage, infection, and infant stroke [1]. A systematic review and meta-analysis [2] estimated the worldwide prevalence of CP at 2.11 births per 1000. Subsequent studies of various populations in Africa [3] Asia [4], and North America [5] suggest that the prevalence of CP is on the rise, at a rate of more than three per 1000, a phenomenon which may be due to the increasing likelihood of survival of early, preterm infants [6]. The average age for diagnosis of CP is 12–24 months in high-income countries and as late as five years in less well-resourced countries [7]. There are many reasons for diagnostic delay: the lack of definitive biomarkers for CP and definitive signs on traditional clinical examination, reluctance to communicate what might be a false positive to parents and triggering grief, uncertainty, and stigma, as well as the absence of curative treatments [8]. Arguably, the greatest boon to early identification of infants with CP has been the validation and dissemination of the general movement assessment, GMA. This instrument came into being as the understanding of the significance of infants' spontaneous movements increased during the latter decades of the 20th century [9,10]. Two patterns in particular, predominantly cramped-synchronized movements and the absence of fidgety movements at three to five months of age reliably predict a later finding of CP [11]. Despite the power of the GMA in early identification, its key mechanism of gestalt pattern recognition (from video) requires a significant investment in assessor training and validation [12], making the GMA challenging to implement broadly across clinical practices. Early intervention depends on early identification, which suggests the need for a widely disseminated screening process, which is not found in the current approach to delivering CP care [8]. Families have been identified as the cornerstone of early intervention [13] and exploration of a more integral role for families in neurodevelopmental monitoring and therapy may result in earlier detection of developmental delays as well as earlier application of appropriate therapies. Engagement of families of infants with CP whose early signs of impairment are subtle may be particularly helpful given that these infants have been shown to be at greater risk for not receiving early diagnosis and intervention than are more profoundly affected infants [14]. Computationally assisted screening procedures likewise suggest a way to manage the increased clinical workflow that would result from broader application of neuromotor assessment among infants. Marker-based, multi-camera, 3D analysis of infant movement has been used to detect both upper [15] and lower extremity movements [16] correlated with GMA assessment of CP. Given that marker-based systems typically require multiple cameras and a laboratory setting; an accurate and reliable markerless computer vision approach that can be operationalized in either the home or clinic setting may make screening more widely accessible. Markerless computer vision technology further preserves the non-intrusive character of the GMA, leveraging, as does the GMA itself, video to assess an infant's movements unhampered by markers or other sensing devices [12].

Computer vision techniques to automate the analysis of infant movements captured on 2D video have been under exploration for over a decade [17,18]. 3D recordings, however, may provide added value through higher spatial resolution, depth information, and higher accuracy and reliability; however, exploration has been limited by high technology cost and computational overhead [19]. Markerless computer vision systems have the ability to implement a kinematic model [20,21] and have been presented as a promising alternative to marker-based systems [22].

Avoiding markers may be particularly appropriate in the case of infants where they may be poorly tolerated and, as a result, introduce movement patterns that are not part of the infant's typical GM repertoire [12].

Some markerless systems have employed a multi-camera approach [23] but a more accessible and practical solution is to use a single camera, which enhances portability and makes it possible to carry out assessment in more confined spaces [24] such as the home or clinic. Use of a commercial RGB-D sensor system that integrates an RGB camera with a depth sensor in the same hardware is a promising approach.

Such an integrated system promises to help fill the current gap in infant movement assessment by providing a low-cost, compact platform that can be implemented repeatedly and longitudinally in the infant's naturalistic environment, where movement repertoires are most likely to characterize the actual behaviors of the infant [19].

RGB-D sensors have been used in upper limb rehabilitation for adult patients post-stroke, as well as for analyzing balance recovery [25]. They have also been used in adult Parkinson's disease patients to evaluate upper limb tasks [26], gait, and postural stability [27,28].

The current study aimed at recording infants' upper body movements with a single, commercial RGB-D sensor; at tracking the 2D trajectories of selected points of interest (PoI) leveraging DeepLabCut [29], a well-established, open-source deep learning algorithm for pose estimation, i.e., generating 2D coordinates for tracked PoI; at obtaining the 3D PoI trajectories by applying a newly developed method; and finally at extracting GM metrics from the PoI 3D trajectories. This study proposes a novel method for assessing infants' GM that features a simplified instrumental setup, suitable for home (or clinic) use.

This article presents in detail first the characteristics of the subjects involved and the experimental setup utilized, then how data were processed distinguishing what was already available from what was newly developed, and finally how GM metrics were obtained from 3D PoI trajectories. In presenting the results obtained, the clinical evaluation of two pediatric physicians with expertise in neurology and physiatry were taken into account.

2. Materials and Methods

The parents of eight infants recruited from the community volunteered to perform video recordings of their babies at their home. Infants, sitting in a baby seat covered with a green cloth to facilitate background exclusion during identification of infant PoI, were positioned on the floor in front of an RGB camera with an integrated depth sensor. The children's natural movements were recorded for a target duration of three minutes at three different timepoints (3, 4, and 5 months from birth).

Infants used the same, washable seat throughout testing across different timepoints for consistency. Light conditions and interactions with humans were controlled to the degree possible in a home environment replicating the most natural conditions and guaranteeing the simplicity of the protocol.

Two expert physicians analyzed the recorded videos at each timepoint and were asked to report if they observed any cause for concern in the development of the infants.

The camera used for the recordings was a commercial RGB-D sensor (Intel RealSense D435, Intel, Santa Clara, CA, USA, combining a pre-calibrated RGB camera with 1280×720 native resolution and frame rate of $\sim$30 fps with a depth sensor generating depth-coded images with 640×480 native resolution and frame rate of $\sim$30 fps). Each pixel of the depth image had an intensity proportional to the distance of the surfaces in the image from the camera. Depth images were generated by the stereo vision of two infrared sensors mounted on the device with the left sensor used as point of view. If a region is seen only by the left sensor, the resulting depth image in that region remains black ("black area"). RGB and depth images were pre-calibrated; however, a residual misalignment between the two remained.

The markerless motion tracking software used in this study was DeepLabCut (Swiss Federal Institute of Technology, Lausanne, Switzerland) [29], an open-source toolkit for pose estimation in which a training set of images is manually labeled and returns the x, y coordinates of the tracked points along with a confidence level, varying from 0 (lowest confidence) to 1 (maximum confidence). DeepLabCut provides a framework for supervised, deep learning to tune an existing, high-performance convolutional neural network (ResNet, Microsoft Research, Redmond, WA, USA) to the features of a specialized dataset to produce a high level of recognition accuracy.

A sequence of steps was implemented to reconstruct the time series of the 3D coordinates of the selected points from the recorded RGB-D videos and to extract the associated GM metrics (Figure 1).

All blocks are explained in detail in the following sections.

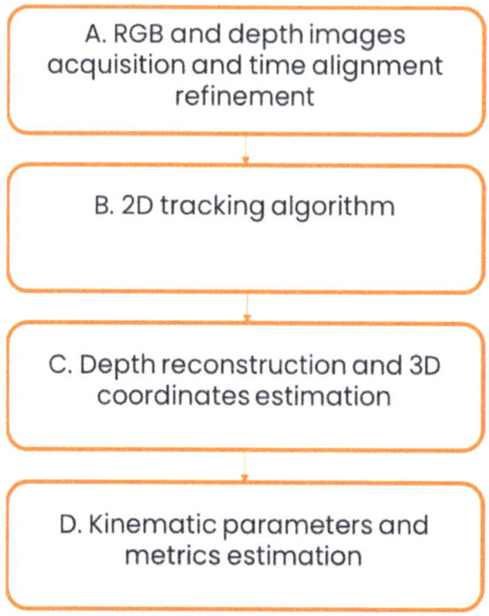

Figure 1. Block diagram of the proposed markerless-based method.

2.1. RGB and Depth Images Acquisition and Time Alignment Refinement

The images acquired with the RGB camera and the depth sensor required refinement of the time alignment with respect to that obtained using the manufacturer's proprietary software, since neither frame rate is exactly constant. The timestamps provided by the acquisition software were used for this purpose. Three alignment scenarios were, each of them requiring a different countermeasure:

1. The timestamp of an RGB image was closer to one or more RGB image timestamps than the closest depth image timestamp. Countermeasure: a gap of the proper number of frames was inserted in the sequence of depth frames;
2. The timestamp of a depth image was closer to one or more depth image timestamps than the closest RGB image timestamp. Countermeasure: a gap of the proper number of frames is inserted in the sequence of RGB frames;
3. The difference between the RGB and depth image timestamp was less than half the duration of the nominal sampling interval (<17 ms). The two frames were considered time aligned.

All gaps generated were then filled by applying a cubic spline interpolation.

2.2. 2D Tracking Algorithm

RGB images were converted into video files using ImageJ (National Institute of Health, Bethesda, MD, USA) [30] and fed to the DeepLabCut (Swiss Federal Institute of Technology, Lausanne, Switzerland) image processing tool. The tracking software was trained to identify six PoIs on the infant's upper body: left and right shoulders (LS and RS), elbows (LE and RE), and wrists (LW and RW). All PoIs were manually labeled on 10% of the video frames (identified by DeepLabCut using a k-means algorithm that selected frames based on pixel characteristic variability) to create a training set.

After the network was trained and validated, DeepLabCut provided the PoI positions in all RGB frames, together with their confidence levels. When a PoI was occluded, its position was provided with a low confidence level. A recognition network was trained individually for each infant video to achieve the greatest possible accuracy prior to the association of RGB and depth coordinates of PoI.

2.3. Depth Reconstruction and 3D Coordinates Estimation

The 3D position of PoIs tracked in the RGB images was obtained by developing a method which exploits the depth sensor recordings. The location in the depth image corresponding to that of a tracked PoI in the concurrent RGB image was determined after addressing the three possible causes of incorrect or undefined PoI 3D positions:

1. the RGB location of a tracked PoI falling over the "black area" in the corresponding depth image, therefore lacking depth information (Figure 2a);
2. PoI occlusions corrupting the estimation of PoI 3D positions. Since the tracking algorithm determines PoI locations exclusively from RGB information, the estimated location of a PoI could fall over a body segment covering the PoI in the RGB image (as when the head covers a shoulder). In such cases the estimate of the relevant depth coordinate would be affected by an error equal to the distance, along the depth direction, between the surfaces of the two body parts. To compensate for this error, the prediction confidence level values provided by the tracking software were used. The depth values obtained for frames with a confidence level lower than 0.6 were not considered (Figure 2b);
3. a residual spatial misalignment between RGB and depth images causing errors in the estimation of the tracked PoI depth coordinate. Such a misalignment is responsible for errors in estimating depth coordinates when a tracked PoI is near a substantial depth discontinuity (Figure 2c). To compensate for the consequences of this error, the following procedure was implemented: the first derivative of the PoI depth coordinate was calculated; when its value was higher than a threshold value set based on the physical limits of the subject motion speed, the relevant depth value was removed.

   All resulting depth coordinate gaps were then filled by applying a cubic spline interpolation.

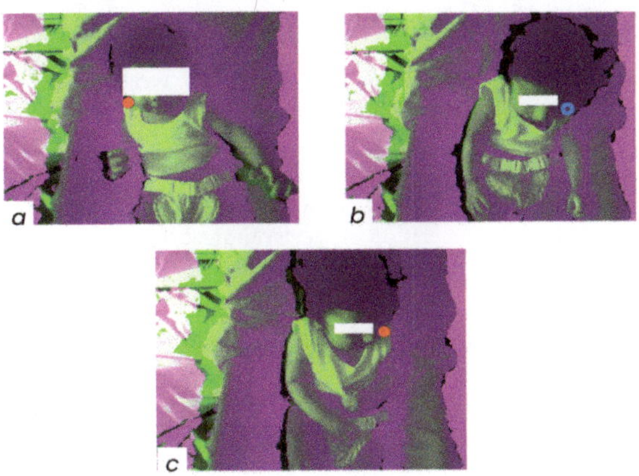

Figure 2. Issues causing undefined 3D PoI positions: (**a**) RS falling on the "black area", (**b**) occlusion of the LS from the head, and (**c**) residual spatial misalignment between RGB and depth images. Subjects are made unidentifiable by using white patches. The coloured circles represent PoIs.

Finally, the identification of PoIs in the RGB images is conditioned by the way the PoI area is seen by the camera. Depending on the RGB frame, a single PoI may be identified in different areas of the infant's body surface (Figure 3).

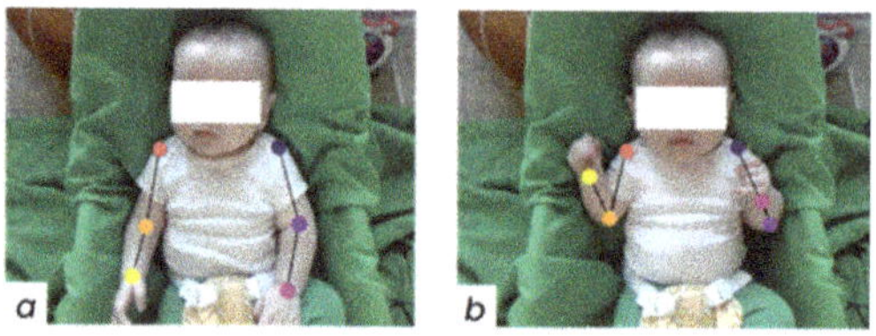

Figure 3. (**a**) RE is identified in the middle between epicondyles and (**b**) RE is positioned on the medial epicondyle. Subjects are made unidentifiable by using white patches. The coloured circles represent PoIs.

The performance of the markerless method described above was evaluated both on a physical model [31] as well as on real infants [32].

2.4. Kinematic Parameters and Metrics Estimation

From PoI 3D trajectories, the following metrics for quantifying GMs were estimated [16]:

- area in which the trajectories of the wrists differed from the moving average for the same trajectories, normalized with respect to the length of the moving average window (two seconds);
- area in which the trajectories of the wrists were outside of the standard deviation of the moving average for the same trajectories, normalized with respect to the samples in which the trajectories were outside the standard deviation (no information regarding the normalization was provided in the reference work);
- periodicity in the wrist trajectories;
- area in which the speed profiles of the wrists differed from the moving average for the velocity profiles, normalized with respect to the length of the moving average window (2 s);
- area in which the speed profiles of the wrists were outside of the standard deviation of the moving average for the velocity profiles, normalized with respect to the length of the moving average window (2 s);
- periodicity in the wrist velocities;
- the skewness of the velocities of the wrists;
- the cross-correlation of accelerations between left and right wrists.

In addition, we estimated the range of motion (ROM) of the elbow angle (EA), defined as the angle between the forearm segment and the upper arm segment.

To limit the influence of extended intervals of lack of upper limb movements to the estimated parameters described above, bouts of activity were introduced. The time intervals during which the infants' wrists were moving were extracted from the rest of the acquisition. Bouts were defined as intervals of time characterized by wrist speed higher than a fixed threshold (5% of the wrist maximum velocity).

The blocks A, C, and D of Figure 1 were implemented in MATLAB R2021b (The MathWorks Inc., Natick, MA, USA).

3. Results

The two expert physicians involved in the study evaluated the RGB videos of the infants at 3, 4, and 5 months from birth to identify any features raising concern that an infant might not be typically developing (TD). Not all videos were commented on, but an overall evaluation of each infant was provided. The two physicians agreed that four infants (S1, S5, S7, and S8) appeared to be TD, and that one infant (S2) did show signs of possible atypical development. The physicians did not agree on the evaluation of the remaining three infants (S3, S4, and S6). Table 1 provides a complete description of the evaluations. The values of the nine GM metrics estimated from the upper body 3D PoI kinematics obtained by applying the proposed markerless method are reported in Figure 4 for each infant at each timepoint for left and right sides, both separately and together. To link the clinical evaluation to the metrics extracted, the range defined by the values found for those infants not suggesting atypical development was grayed in each plot.

Table 1. Report of the evaluation of the RGB videos performed by two expert physicians (A and B); subjects not suggesting atypical development are grayed.

Sub#	*Clinician*	*3 Months*	*4 Months*	*5 Months*	*Overall Evaluation* **No:** *Nothing Here Suggests the Infant Is Not Developing Typically* **Yes:** *I Did Observe Some Features that Raise Concern*
S1	A	-	-	-	No
	B	*lots of midline gaze*	*midline/R gaze but toddler on R*		No
S2	A	-	-	*Slow upper limb movements; no hands to mouth and midline; opens hands; thumbs frequently adducted*	Yes
	B				Yes
S3	A	-	-	-	No
	B				Yes
S4	A	-	-	-	No
	B	*decreased fidgety movements; subtle R hand preference? more fidgety movements on R*	*decreased fidgety movements; subtle R hand preference? more fidgety movements on R*	-	Yes *Lots of midline hand clasping and midline gaze preference at all ages*
S5	A	-	-	-	No
	B	-	-	-	No *Hands at midline; great gen and fidgety movements*
S6	A	-	-	-	No
	B	*non social smile; midline grasp*	*social smile; L fingers in mouth 65% of video;*	*L fingers in mouth entire video; no clear fidgety movements*	Yes
S7	A	-	-	-	No
	B	*Great visual attention; great gen and fidgety movements*	-	-	No *Grabbing toes; sucking on fingers; social*
S8	A	-	-	-	No
	B	-	-	-	No *Appears sleepy; improved visual attention and social engagement; good general movements; fingers or thumb in mouth*

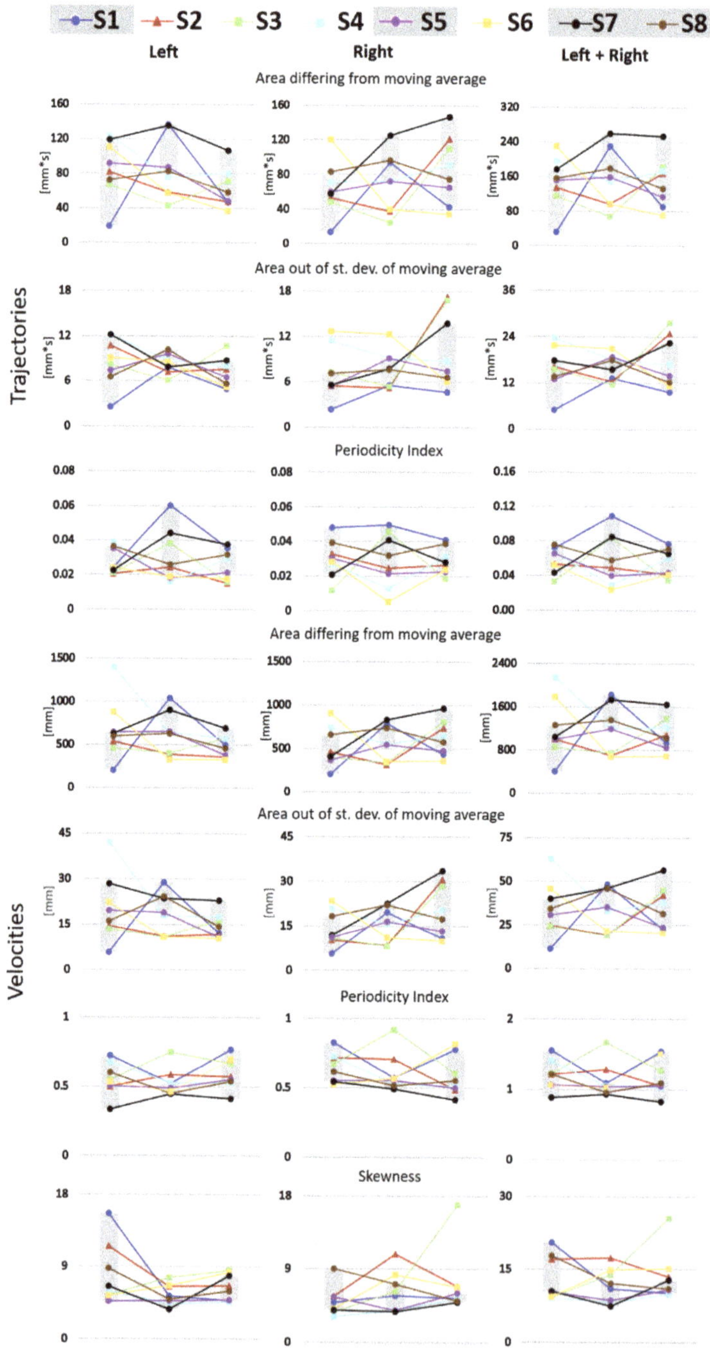

Figure 4. Metrics obtained from wrist trajectories and velocities for each infant at each timepoint (3, 4, and 5 months). Infants not suggesting atypical development to both physicians are identified with circles, infants raising the concern of both physicians are identified with triangles, and infants differently evaluated by the physicians are identified by squares. Infants not suggesting atypical development define the gray interval at each time point.

Figures 5 and 6 show the cross-correlation between left and right wrists accelerations and the range of motion of the elbow angle for each subject at each timepoint (3, 4, and 5 months), respectively.

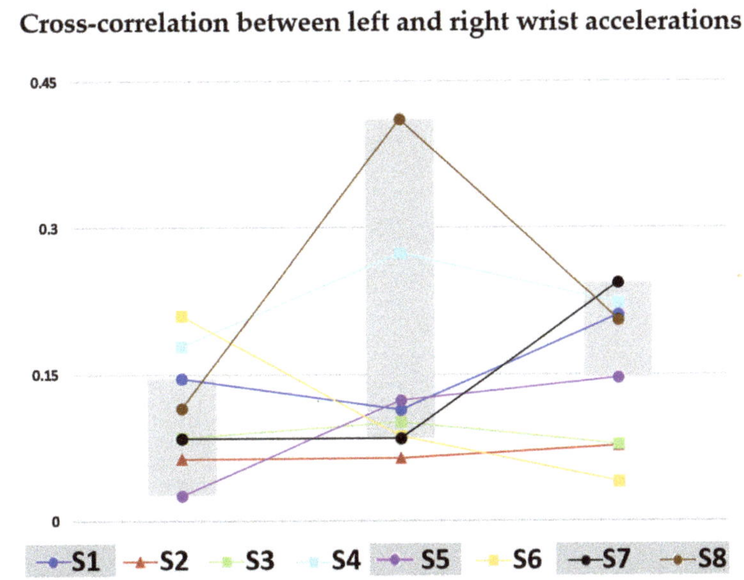

Figure 5. Cross-correlation between the left and right wrist acceleration for each infant at each timepoint (3, 4, and 5 months). Infants not suggesting atypical development to both physicians are identified with circles, infants raising the concern of both physicians are identified with triangles, and infants differently evaluated by the physicians are identified by squares. Infants not suggesting atypical development define the gray interval at each time point.

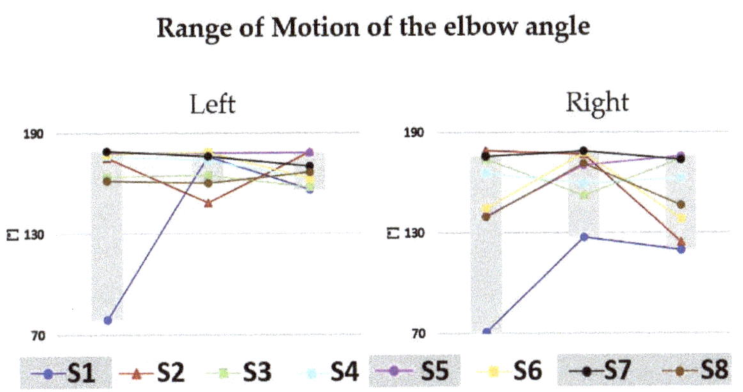

Figure 6. Range of motion of the elbow angle for the left and right side for each infant at each time point (3, 4, and 5 months). Infants not suggesting atypical development to both physicians are identified with circles, infants raising the concern of both physicians are identified with triangles, and infants differently evaluated by the physicians are identified by squares. Infants not suggesting atypical development define the gray interval at each time point.

Figure 7 shows the mean and standard deviation of the bout durations for each subject at each timepoint (3, 4, and 5 months) together with the number of bouts and movement duration, calculated as percentage of the acquisition time. Infants not suggesting atypical development (grayed) appeared to move their arms more than the other infants especially at the 4 and 5 month time points.

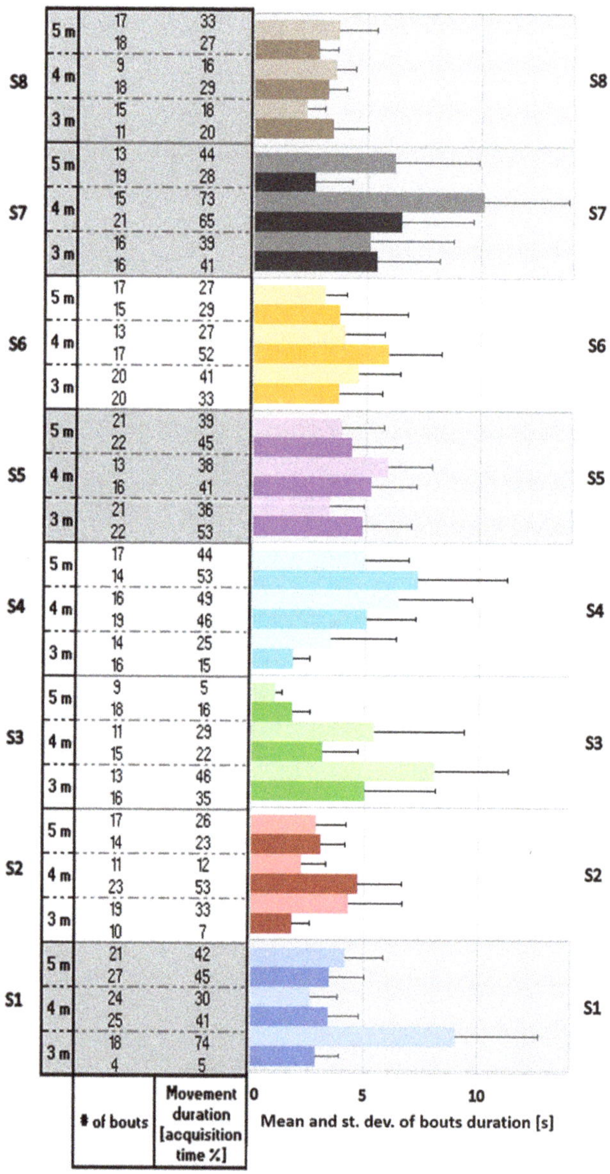

Figure 7. Mean and standard deviation of bouts duration for each infant at each timepoint (3, 4, and 5 months). The left side is the lighter color while the right side is the darker one. The number of bouts and the movement duration, calculated as percentage of the acquisition time, are reported in the table on the left. Subjects not suggesting atypical development are grayed.

Numeric values are reported in Appendix A.

4. Discussion

The recording and the analysis of an infant's upper body movements in a familiar environment has been a difficult and challenging task due to both technological and environmental factors. The technology used in previous studies has shown some limitations, since it is normally developed for and applied to the analysis of adults' or children's movements [33]. Markerless methods for the analysis of human motion have opened new possibilities for movement analysis, although initially it was mainly in two dimensions [20,21,34]. The analysis of infants' movements may benefit substantially from markerless methods given the problems normally encountered in securely and safely positioning markers on their small body segments.

The low cost RGB-D cameras currently available in the consumer electronics market allow extending markerless techniques to 3D movement analysis without increasing the complexity of the experimental setup, a factor that allows leverage of the technique in environments outside the lab and promotes repeated measurement over time. The latter observation is of primary importance when infants' movements are studied. Sensorimotor integration occurs rapidly in the first months of life through a process of activity-dependent neuronal modeling [35]. More frequent, routinized monitoring of infants' movement in the convenient and familiar environment of the home increases the likelihood that infants who display abnormal movement repertoires will be identified promptly and interventions to prevent loss of neural connections and their specific functions instituted [7].

In this work we applied a markerless method to the RGB images recorded from a commercial RGB-D camera and used selected upper body PoIs extracted from the RGB video frames together with the recorded depth information to reconstruct 3D PoI kinematics [31] from which both some novel and already published metrics were calculated. The metrics used in this study were originally proposed to quantify GM [16,36], given the demonstrated power of GM assessment to predict the development of movement disorders very early in infancy [11]. Notably, since the key requirement of our approach to infant screening for neuromotor delay was that measurement be easily carried out in an informal environment such as the home, we did not attempt to replicate the General Movements Assessment in our protocol. For example, our infants were videoed in whatever attire their parents had chosen for the temperature in their homes, they were seated in a standard infant seat, versus lying supine, and videoed from the front using a commercial camera tripod versus from overhead requiring a special, suspended camera apparatus. The shift in infant posture likely caused the trends we calculated for GM parameters to vary from those reported by [36] for infants from three to five months of age.

Due to the small size of our sample, it was not possible to conduct meaningful statistical analyses. Rather, we describe visualized trends across three-, four-, and five-month measurement timepoints. Refer to Figure 4 for plots of parameters 1–7, to Figure 5 for a plot of parameter 8, and to Figure 6 for a plot of range of motion of the elbow angle. Table 2 summarizes the relationship between observed GM patterns and parameters, as well as the expected fluctuation in parameter values from three to five months of age in both TD and those later identified with CP. However, it should be noted that large differences between TD and non-TD infants would not be expected in our sample, as there was no documented injury that would have classified any of our non-TD infants as at-risk. In most of the metrics, variability in the data made it difficult to compare to predictions. However, in two metrics, trends were consistent with literature (described below).

Table 2. Mapping of observed patterns of general movements at 3–5 months of age to kinematic parameters. Adapted from ref. [16].

Metrics			Movements				
#	Class	Description	Aspect	Observed TD Characteristics	Measured TD Characteristics	Observed Non-TD Characteristics	Measured Non-TD Characteristics
1	Trajectories	Area where wrist trajectories differ from the moving average of the same trajectories	Diversity and Variability	Fluid and congruent	No significant change in area	Chaotic	Area smaller than TD, continues to diminish
2	Trajectories	Area where wrist trajectories of are outside the SD of the moving average of the same trajectories	Diversity and Variability	Multi-facetted	Smaller area, less diversity at 3 months (Increases after 5 months)	Poor-repertoire, spastic	Area smaller than TD, continues to diminish
3	Trajectories	Periodicity in the wrist trajectories	Unpredictability and Complexity	Fidgety	Periodicity decreases with age	Poor-repertoire	Periodicity greater than in TD
4	Velocities	Area where the wrist speed profiles differs from their moving average	Diversity and Variability	Fluid and congruent	Area does not change	Chaotic	Area decreases
5	Velocities	Area where the wrist speed profiles are outside the SD of the moving average the speed profiles	Diversity and Variability	Fidgety	Variation in velocity is constant	Cramped	Variation in velocity continuously decreases
6	Velocities	Periodicity in the wrist velocities	Equability of Velocity	Fidgety	Periodicity does not change	Cramped or chaotic	Periodicity does not change
7	Velocities	Skewness of the velocities of the wrists	Velocity Distribution of the Movement	Slow, small in amplitude	Skewness increases with age	Cramped, spastic	Skewness already increased by 3 months relative to TD
8	NA	The cross-correlation of accelerations between left and right wrists	Similarity and Coordination of Movement	Similar, coordinated, synchronous	Cross-correlation increases	Dissimilar, uncoordinated, asynchronous	Cross-correlation does not increase

The area where the wrist trajectory differs from the moving average of that same trajectory is suggested to quantify the diversity and variability of GM with respect to fluidity and congruence (metric #1). No significant change in the metric is expected from three to five months of age in TD infants, while those who are not TD are expected to exhibit smaller areas. Consistent with expectations, at four months, all infants about whom at least one of the clinicians evaluating videos expressed concern (S2, S3, S4, and S6) had smaller metric values relative to typical development (S1, S5, S7, and S8). The metric is expected to continually decrease in non-TD infants during the three-to-five month window. We did not, however, observe this trend as some infants have large swings in values across the timepoints.

The cross-correlation of acceleration between left and right wrists (metric #8) also showed trends consistent with previous work. This metric is associated with the observed characteristics of similarity and coordination of movement. TD infants may be expected to display movements that are similar, coordinated, and synchronous in the three-to-five-month window. Non-TD infants are expected to demonstrate the opposite movement pattern: dissimilar, uncoordinated, and asynchronous. This metric is expected to increase in TD infants between three and five months of age and to not increase during that time period in non-TD children. The metric values were higher at 5 months than 3 months in all infants for whom there was no concern (S1, S5, S7, and S8). However in S8, the cross-correlation at 4 months peaked sharply then regressed at five months but to a point still greater than the three-month cross-correlation. The cross-correlation measured for S6 (split concern) decreased monotonically between three and five months. Infants S2 (agreed concern) and S3 (split concern) virtually flatlined across all time points and logged cross-correlation metrics near the bottom of the cohort, well below the TD range at 5 months. Infant S4 (split concern) did not follow this pattern of decreasing or flatlined change over time. In summary, seven of the eight infants followed a pattern consistent with expectations.

We introduced elbow range of motion (Figure 6), a novel exploratory metric. Most infants' ROM on the left arm fell within a narrow band changing little from three to five months. S1 (no concern) displayed very limited range at three months but increased to resemble the ROM of the cohort generally at four months. S2 (concern) demonstrated a ROM for the left arm among the highest in the cohort at three and five months but presented as the lowest at four months. The range of ROM angles was more diverse on the right side. S1 (no concern) was markedly low at three months and increased at four and five months, though not as much as had been noted on the left side. S2 (concern) lost range markedly at five months. A larger sample will be needed to determine how useful this metric will be for screening of non-TD infants.

Visualization of infants' movement data suggest that the metrics are not independent. Environment influences that impact one impact others as well. Based on inspection of videos, several recommendations can be made for further work in this area. Better control of environmental factors might decrease variability in the data. While the protocol specifications were to have no one in the infant's visual field during testing, it was difficult to enforce this in infants' homes for all of the videos. In one case, a sibling approached the infant from the right causing him to lateralize in that direction. In another case, the infant had one hand in his mouth for a large portion of the video. For infants with this tendency, multiple capture sessions within the same day might be needed. For younger infants, the use of a baby seat was not optimal, as sitting posture was not fully developed. In the protocol, it was decided infants should use the same, washable seat throughout testing across different time points for consistency, sanitation, and to prevent infants from crawling away at older ages. Both the seat and infants' upright position in the seat constrained their movements to an extent not experienced in the standard GM assessment protocol for which the quantification metrics we used were developed. Future studies should consider use of a standardized postural support method for younger seated infants.

In clinical practice, to determine whether an infant is typically developing or not, clinicians base their judgment on a full range of motor characteristics such as hand opening and closing and whether the infant brings his/her hands to the mouth. Characteristics such as these have assessment validity, but lie outside Prechtl et al.'s criteria for which the eight kinematic parameters proposed by [16,36] and colleagues (and applied in this study) were developed. Midline gaze, bringing hands to the midline, visual field preference, visual attention, social smile, and social engagement figure prominently among the criteria applied by our clinicians in applying their clinical discernment to our infant cohort. It would increase the power of 3D markerless movement assessment in infants to quantify observed clinical criteria such as those just enumerated to apply side-by-side with explicitly GM kinematics.

While evaluating videos, clinicians were also sensitive to the infants' state. Characteristics of seeming to be sleepy, distraction from persons inevitably close to the home-based testing area, and, for infants who had not yet developed trunk control, being slumped to the side introduced ambiguity into the association of movement criteria with typical or pathological development. For example, infant S2, whom both clinicians flagged as exhibiting characteristics that caused concern, was documented as being slumped to the right at both the 3- and 4-month testing session. Being slumped to the side and concurrent lack of trunk control reasonably would predispose this infant to move asymmetrically. Notably, the cross-correlation between the left and right wrist accelerations of infant S2 were quite low, falling toward the bottom of the typical range at three months as defined by the range of values calculated for the cohort of infants whom clinicians evaluated as not of concern. Similarly, infant S6, for whom the highest left/right wrist acceleration cross-correlation was logged at three months, came in at the bottom of the not-of-concern cohort at four months but went on to log the lowest cross-correlation at five months of age. Clinician notes reveal that the infant had his finger in his mouth about 65% of the time at four months and for almost the entire duration of the video at five months. Clearly, when the infant's spontaneous movements are restricted, as in the case of infant S2 who may not have been positioned so that both arms could move freely and as in the case of infant S6 whose side was (self-)constrained, the synchroneity, similarity, and coordination of movement on the left and right sides, summarized by the cross-correlation metric, is not representative of the infant's actual movement characteristics. The constraint should be remediated and, ideally, the test repeated.

5. Conclusions

This work has shown the feasibility of estimating GM metrics with a single low-cost RGB-D sensor. The simplicity and portability of the proposed markerless protocol allows its use as a screening tool at home or any familiar environment and further makes it possible to avoid clinical environments which are artificial from a child's perspective, and hence challenging for the assessment of true neurodevelopmental performance.

Compared with previous research, this article aimed to characterize GM without markers attached to the infants' skin, which might interfere with infants' spontaneous movements and consequentially affect their behavioral state. In addition, this markerless system provides 3D coordinates of each PoI, and is significantly advantageous over 2D motion capture when dealing with out of plane rotations and allowing more reliable characterization of GMs. Thanks to depth information provided by the RGB-D sensor, this protocol is able to deal with PoI occlusions that occur when using single-camera motion analysis. Our markerless system was designed especially for a home environment. This focus could be very beneficial for enhancing screening of neurodevelopmental disorders particularly for infants and families in rural and remote areas, a population with reduced health services. Due to the small size of our sample, it was not possible to conduct meaningful statistical analyses. For this reason, future studies will be devoted to validating the proposed protocol on a larger number of infants for testing its use in clinical practice.

Author Contributions: Conceptualization, M.M.S., P.S.L. and U.D.C.; methodology, D.B., H.K., J.W., M.M.S., P.S.L. and U.D.C.; software, D.B., H.K., J.W. and I.G.P.; validation, U.D.C.; formal analysis, D.B., H.K., I.G.P., P.S.L. and U.D.C.; investigation, D.B., H.K., J.W., I.G.P., M.M.S., U.D.C. and O.M.; resources, H.K., J.W. and M.M.S.; data curation, D.B., H.K., J.W., I.G.P. and M.M.S.; writing—original draft preparation, D.B., M.M.S., P.S.L. and U.D.C.; writing—review and editing, D.B., I.G.P., M.M.S., P.S.L., A.C. and U.D.C.; visualization, D.B. and I.G.P.; supervision, M.M.S., P.S.L. and U.D.C.; funding acquisition, P.S.L. and U.D.C. All authors have read and agreed to the published version of the manuscript.

Funding: This research was funded by the National Institute on Disability, Independent Living and Rehabilitation, Administration for Community Living of the U.S. Department of Health and Human Services, grant number 90REGE0004 and by "Bando Fondazione di Sardegna 2022 e 2023—Progetti di ricerca di base dipartimentali" (University fund for research).

Institutional Review Board Statement: The study was conducted in accordance with the Declaration of Helsinki, and approved by the Institutional Review Board of The Catholic University of America (protocol #19-0012, initially approved 7 May 2019) for studies involving humans.

Informed Consent Statement: Informed consent was obtained from the parents of all infants involved in the study. Parents also provided written informed consent for the publication of the data and images presented in this paper.

Data Availability Statement: Matlab code used in this project can be provided upon request to the corresponding author.

Acknowledgments: We thank the late Taeun Chang, pediatric neurologist of Children's National Hospital, Washington, DC, USA, for her expert assessment of the movement patterns of infants presented in this paper.

Conflicts of Interest: The authors declare no conflict of interest. The funder, the National Institute on Disability, Independent Living, and Rehabilitation Research, had no role in the design of the study; in the collection, analyses, or interpretation of data; in the writing of the manuscript; or in the decision to publish the results.

Appendix A

Table A1. Relevant values of parameters of "Left side" and "Right side" described in Figure 4 obtained from wrist trajectories and velocities for each infant at each timepoint (3, 4 and 5 months). "M" is month, "Area 1" is "Area differing from moving average", "Area 2" is "Area out of standard deviation of moving average", "PI" is "Periodicity Index" and "Sk." is "Skewness". "Area 1" was normalized by the length of the moving average window (2 s) while "Area 2" was normalized by the samples in which the signal was outside the standard deviation. Subjects not suggesting a non-normal development are grayed.

		Left Side							Right Side						
		Trajectory			Velocity				Trajectory			Velocity			
Sub	M	Area 1 [mm·s]	Area 2 [mm·s]	PI	Area 1 [mm]	Area 2 [mm]	PI	Sk.	Area 1 [mm·s]	Area 2 [mm·s]	PI	Area 1 [mm]	Area 2 [mm]	PI	Sk.
S1	3	19	2.5	0.024	203	5.6	0.724	15.61	14	2.4	0.048	203	5.6	0.829	4.91
	4	137	7.9	0.060	1043	28.8	0.521	5.31	94	5.5	0.049	787	19.7	0.572	5.75
	5	47	5.0	0.036	500	12.2	0.769	4.77	43	4.7	0.041	439	10.8	0.780	5.41
S2	3	82	10.8	0.021	532	14.6	0.498	11.47	53	5.5	0.033	463	10.1	0.717	5.63
	4	59	7.3	0.025	387	11.0	0.584	6.52	38	5.2	0.024	316	8.3	0.709	10.78
	5	48	7.6	0.015	355	11.6	0.569	6.63	121	17.3	0.026	736	30.7	0.486	6.89
S3	3	67	8.2	0.021	455	13.4	0.548	5.47	49	7.3	0.012	398	10.8	0.675	3.74
	4	43	6.1	0.038	406	10.8	0.749	7.62	25	5.4	0.046	340	8.2	0.922	6.23
	5	72	10.8	0.016	577	16.4	0.666	8.50	110	16.9	0.019	808	28.5	0.604	16.87
S4	3	125	12.2	0.039	1401	42.1	0.677	5.94	72	11.5	0.027	742	20.9	0.726	3.19
	4	76	8.9	0.016	643	17.8	0.524	4.39	72	9.2	0.013	554	15.5	0.527	3.88
	5	81	8.0	0.029	578	17.7	0.526	4.58	92	8.8	0.032	641	20.5	0.508	5.38
S5	3	92	7.5	0.035	641	19.6	0.503	4.69	60	5.4	0.030	365	11.2	0.555	5.57
	4	87	9.7	0.018	653	18.9	0.490	4.79	72	9.2	0.022	546	16.5	0.551	3.95
	5	49	6.5	0.021	377	10.7	0.552	4.86	66	7.5	0.023	475	13.4	0.501	6.07
S6	3	110	9.1	0.024	878	22.1	0.532	5.34	120	12.7	0.028	907	23.5	0.525	4.05
	4	58	8.6	0.019	328	10.8	0.463	6.59	40	12.4	0.005	355	11.0	0.565	8.24
	5	37	5.2	0.018	329	10.4	0.694	8.26	35	6.1	0.024	360	10.0	0.817	6.81
S7	3	119	12.2	0.023	626	28.4	0.334	6.55	58	5.6	0.021	409	11.8	0.543	3.91
	4	135	7.9	0.044	898	23.6	0.444	3.70	125	7.8	0.041	831	22.6	0.489	3.76
	5	106	8.8	0.038	691	22.8	0.411	7.85	146	13.7	0.028	958	33.5	0.414	4.88
S8	3	72	6.5	0.036	600	16.0	0.597	8.76	83	7.2	0.039	658	18.3	0.615	9.01
	4	83	10.2	0.026	631	24.3	0.449	4.95	96	7.7	0.032	735	22.0	0.513	7.09
	5	59	5.7	0.032	456	14.2	0.538	5.97	75	6.6	0.039	572	17.5	0.554	5.07

Table A2. Relevant values of parameters of "Left + Right sides" described in Figure 4 obtained from wrist trajectories and velocities for each infant at each timepoint (3, 4 and 5 months). "M" is month, "Area 1" is "Area differing from moving average", "Area 2" is "Area out of standard deviation of moving average", "PI" is "Periodicity Index" and "Sk." is "Skewness". "Area 1" was normalized by the length of the moving average window (2 s) while "Area 2" was normalized by the samples in which the signal was outside the standard deviation. Subjects not suggesting a non- normal development are grayed.

		Left + Right						
		Trajectory			Velocity			
Sub	**M**	**Area 1 [mm·s]**	**Area 2 [mm·s]**	**PI**	**Area 1 [mm]**	**Area 2 [mm]**	**PI**	**Sk.**
S1	3	33	4.9	0.072	407	11.3	1.553	20.52
	4	230	13.4	0.110	1830	48.5	1.093	11.07
	5	91	9.6	0.077	938	23.0	1.549	10.19
S2	3	135	16.3	0.053	995	24.7	1.215	17.10
	4	97	12.5	0.049	703	19.3	1.293	17.30
	5	169	24.9	0.041	1091	42.3	1.055	13.53
S3	3	116	15.5	0.033	853	24.2	1.224	9.21
	4	68	11.5	0.084	745	19.0	1.670	13.85
	5	182	27.7	0.035	1385	44.9	1.269	25.37
S4	3	197	23.7	0.066	2143	63.0	1.403	9.13
	4	148	18.1	0.029	1197	33.3	1.051	8.27
	5	173	16.9	0.061	1219	38.1	1.034	9.97
S5	3	152	12.9	0.065	1007	30.8	1.058	10.27
	4	160	18.9	0.040	1199	35.4	1.041	8.74
	5	115	14.0	0.044	852	24.1	1.053	10.94
S6	3	231	21.8	0.052	1784	45.6	1.057	9.39
	4	98	21.0	0.024	683	21.8	1.028	14.83
	5	72	11.3	0.042	689	20.4	1.511	15.08
S7	3	176	17.8	0.043	1035	40.1	0.878	10.47
	4	260	15.6	0.085	1729	46.3	0.933	7.46
	5	253	22.5	0.066	1649	56.3	0.825	12.73
S8	3	156	13.7	0.075	1259	34.4	1.212	17.77
	4	179	17.9	0.058	1365	46.3	0.962	12.04
	5	134	12.3	0.071	1028	31.7	1.092	11.05

Table A3. Relevant values of parameters described in Figures 5 and 6 for each infant at each timepoint (3, 4, and 5 months). Subjects not suggesting.

Sub	M	Elbow's Range of Motion [°]		Cross-Correlation between Left and Right Wrists Accelerations
		Left	Right	
S1	3	79	71	0.146
	4	176	127	0.114
	5	156	119	0.211
S2	3	175	179	0.063
	4	148	177	0.064
	5	179	124	0.077
S3	3	164	174	0.085
	4	165	152	0.101
	5	158	174	0.078
S4	3	176	166	0.179
	4	174	159	0.274
	5	163	163	0.223
S5	3	178	140	0.026
	4	178	170	0.123
	5	178	176	0.147
S6	3	177	145	0.210
	4	179	178	0.086
	5	163	138	0.040
S7	3	179	176	0.084
	4	176	179	0.084
	5	170	174	0.244
S8	3	162	139	0.116
	4	160	171	0.412
	5	167	146	0.206

References

1. Metz, C.; Jaster, M.; Walch, E.; Sarpong-Bengelsdorf, A.; Kaindl, A.M.; Schneider, J. Clinical Phenotype of Cerebral Palsy Depends on the Cause: Is It Really Cerebral Palsy? A Retrospective Study. *J. Child Neurol.* **2022**, *37*, 112–118. [CrossRef] [PubMed]
2. Oskoui, M.; Coutinho, F.; Dykeman, J.; Jetté, N.; Pringsheim, T. An update on the prevalence of cerebral palsy: A systematic review and meta-analysis. *Dev. Med. Child Neurol.* **2013**, *55*, 509–519. [CrossRef] [PubMed]
3. El-Tallawy, H.N.; Farghaly, W.M.; Shehata, G.A.; Rageh, T.A.; Metwally, N.A.; Badry, R.; Sayed, M.A.; el Hamed, M.A.; Abd-Elwarth, A.; Kandil, M.R. Cerebral palsy in Al-Quseir City, Egypt: Prevalence, subtypes, and risk factors. *Neuropsychiatr. Dis. Treat.* **2014**, *10*, 1267–1272. [CrossRef]
4. Wang, H.H.; Hwang, Y.S.; Ho, C.H.; Lai, M.C.; Chen, Y.C.; Tsai, W.H. Prevalence and initial diagnosis of cerebral palsy in preterm and term-born children in taiwan: A nationwide, population-based cohort study. *Int. J. Environ. Res. Public Health* **2021**, *18*, 8984. [CrossRef] [PubMed]
5. Christensen, D.; Van Braun, K.; Doernberg, N.S.; Maenner, M.J.; Arneson, C.L.; Durkin, M.S.; Benedict, R.E.; Kirby, R.S.; Wingate, M.S.; Fitzgerald, R. Prevalence of cerebral palsy, co-occurring autism spectrum disorders, and motor functioning-Autism and Developmental Disabilities Monitoring Network, USA, 2008. *Dev. Med. Child Neurol.* **2014**, *56*, 59–65. [CrossRef] [PubMed]
6. Graham, H.K.; Rosenbaum, P.; Paneth, N.; Dan, B.; Lin, J.; Damiano, D.L.; Becher, J.G.; Gaebler-Spira, D.; Colver, A.; Reddihough, D.S.; et al. Cerebral palsy. *Nat. Rev. Dis. Prim.* **2016**, *2*, 16005. [CrossRef]
7. Novak, I.; Morgan, C.; Adde, L.; Blackman, J.; Boyd, R.N.; Brunstrom-Hernandez, J.; Cioni, G.; Damiano, D.; Darrah, J.; Eliasson, A.; et al. Early, Accurate Diagnosis and Early Intervention in Cerebral Palsy: Advances in Diagnosis and Treatment. *JAMA Pediatr.* **2017**, *171*, 897–907. [CrossRef]
8. te Velde, A.; Morgan, C.; Novak, I.; Tantsis, E.; Badawi, N. Early diagnosis and classification of cerebral palsy: An historical perspective and barriers to an early diagnosis. *J. Clin. Med.* **2019**, *8*, 1599. [CrossRef]

9. Prechtl, H.F.R. Qualitative changes of spontaneous movements in fetus and preterm infant are a marker of neurological dysfunction. *Early Hum. Dev.* **1990**, *23*, 151–158. [CrossRef]
10. Heinz, P.; Einspieler, C.; Cioni, G.; Bos, A.F.; Ferrari, F.; Sontheimer, D. An early marker for neurological deficits after perinatal brain lesions. *Lancet* **1997**, *349*, 1361–1363.
11. Einspieler, C.; Prechtl, H.F.R. Prechtl's assessment of general movements: A diagnostic tool for the functional assessment of the young nervous system. *Ment. Retard. Dev. Disabil. Res. Rev.* **2005**, *11*, 61–67. [CrossRef] [PubMed]
12. Silva, N.; Zhang, D.; Kulvicius, T.; Gail, A. The future of General Movement Assessment: The role of computer vision and machine learning—A scoping review. *Res. Dev. Disabil.* **2021**, *110*, 103854. [CrossRef] [PubMed]
13. Hadders-Algra, M.; Boxum, A.G.; Hielkema, T.; Hamer, E.G. Effect of early intervention in infants at very high risk of cerebral palsy: A systematic review. *Dev. Med. Child Neurol.* **2017**, *59*, 246–258. [CrossRef] [PubMed]
14. Hekne, L.; Montgomery, C.; Johansen, K. Early access to physiotherapy for infants with cerebral palsy: A retrospective chart review. *PLoS ONE* **2021**, *16*, e0253846. [CrossRef] [PubMed]
15. Mazzarella, J.; McNally, M.; Richie, D.; Chaudhari, A.M.W.; Buford, J.A.; Pan, X.; Heathcock, J.C. 3d motion capture may detect spatiotemporal changes in pre-reaching upper extremity movements with and without a real-time constraint condition in infants with perinatal stroke and cerebral palsy: A longitudinal case series. *Sensors* **2020**, *20*, 7312. [CrossRef]
16. Disselhorst-Klug, C.; Heinze, F.; Breitbach-Faller, N.; Schmitz-Rode, T.; Rau, G. Introduction of a method for quantitative evaluation of spontaneous motor activity development with age in infants. *Exp. Brain Res.* **2012**, *218*, 305–313. [CrossRef]
17. Adde, L.; Helbostad, J.L.; Jensenius, A.R.; Taraldsen, G.; Grunewaldt, K.H.; StØen, R. Early prediction of cerebral palsy by computer-based video analysis of general movements: A feasibility study. *Dev. Med. Child Neurol.* **2010**, *52*, 773–778. [CrossRef]
18. Ihlen, E.A.F.; Støen, R.; Boswell, L.; de Regnier, R.; Fjørtoft, T.; Gaebler-Spira, D.; Labori, C.; Loennecken, M.C.; Msall, M.E.; Möinichen, U.I.; et al. Machine learning of infant spontaneous movements for the early prediction of cerebral palsy: A multi-site cohort study. *J. Clin. Med.* **2020**, *9*, 5. [CrossRef]
19. Marcroft, C.; Khan, A.; Embleton, N.D.; Trenell, M.; Plötz, T. Movement recognition technology as a method of assessing spontaneous general movements in high risk infants. *Front. Neurol.* **2015**, *6*, 284. [CrossRef]
20. Castelli, A.; Paolini, G.; Cereatti, A.; Della Croce, U. A 2D markerless gait analysis methodology: Validation on healthy subjects. *Comput. Math. Methods Med.* **2015**, *2015*, 186780. [CrossRef]
21. Balta, D.; Salvi, M.; Molinari, F.; Figari, G.; Paolini, G.; Della Croce, U.; Cereatti, A. A two-dimensional clinical gait analysis protocol based on markerless recordings from a single RGB-Depth camera. In Proceedings of the 2020 IEEE International Symposium on Medical Measurements and Applications (MeMeA), Bari, Italy, 1 June–1 July 2020; pp. 1–6. [CrossRef]
22. Chambers, C.; Seethapathi, N.; Saluja, R.; Loeb, H.; Pierce, S.R.; Bogen, D.K.; Prosser, L.; Johnson, M.J.; Kording, K.P. Computer vision to automatically assess infant neuromotor risk. *IEEE Trans. Neural Syst. Rehabil. Eng.* **2020**, *28*, 2431–2442. [CrossRef] [PubMed]
23. Mündermann, L.; Corazza, S.; Andriacchi, T.P. The evolution of methods for the capture of human movement leading to markerless motion capture for biomechanical applications. *J. NeuroEngineering Rehabil.* **2006**, *3*, 6. [CrossRef] [PubMed]
24. Cimolin, V.; Vismara, L.; Ferraris, C.; Amprimo, G.; Pettiti, G.; Lopez, R.; Galli, M.; Cremascoli, R.; Sinagra, S.; Mauro, A.; et al. Computation of Gait Parameters in Post Stroke and Parkinson's Disease: A Comparative Study Using RGB-D Sensors and Optoelectronic Systems. *Sensors* **2022**, *22*, 824. [CrossRef] [PubMed]
25. Bower, K.; Thilarajah, S.; Pua, Y.; Williams, G.; Tan, D.; Mentiplay, B.; Denehy, L.; Clark, R. Dynamic balance and instrumented gait variables are independent predictors of falls following stroke. *J. Neuroeng. Rehabil.* **2019**, *16*, 3. [CrossRef]
26. Kim, W.-S.; Cho, S.; Baek, D.; Bang, H.; Paik, N.-J. Upper Extremity Functional Evaluation by Fugl-Meyer Assessment Scoring Using Depth-Sensing Camera in Hemiplegic Stroke Patients. *PLoS ONE* **2016**, *11*, e0158640. [CrossRef] [PubMed]
27. Albani, G.; Ferraris, C.; Nerino, R.; Chimienti, A.; Pettiti, G.; Parisi, F.; Ferrari, G.; Cau, N.; Cimolin, V.; Azzaro, C.; et al. An Integrated Multi-Sensor Approach for the Remote Monitoring of Parkinson's Disease. *Sensors* **2019**, *19*, 4764. [CrossRef]
28. Pantzar-Castilla, E.; Cereatti, A.; Figari, G.; Valeri, N.; Paolini, G.; Della Croce, U.; Magnuson, A.; Riad, J. Knee joint sagittal plane movement in cerebral palsy: A comparative study of 2-dimensional markerless video and 3-dimensional gait analysis. *Acta Orthop.* **2018**, *89*, 656–661. [CrossRef]
29. Mathis, A.; Mamidanna, P.; Cury, K.M.; Abe, T.; Murthy, V.N.; Mathis, M.W.; Bethge, M. DeepLabCut: Markerless pose estimation of user-defined body parts with deep learning. *Nat. Neurosci.* **2018**, *21*, 1281–1289. [CrossRef]
30. Schindelin, J.; Rueden, C.T.; Hiner, M.C.; Eliceiri, K.W. The ImageJ ecosystem: An open platform for biomedical image analysis. *Mol. Reprod. Dev.* **2015**, *82*, 518–529. [CrossRef]
31. Balta, D.; Kuo, H.; Wang, J.; Porco, I.G.; Schladen, M.; Cereatti, A.; Lum, P.S.; Della, U. Croce Estimating infant upper extremities motion with an RGB-D camera and markerless deep neural network tracking: A validation study. In Proceedings of the 44th Annual International Conference of the IEEE Engineering in Medicine & Biology Society (EMBC), Glasgow, UK, 11–15 July 2022.
32. Balta, D.; Kuo, H.; Wang, J.; Porco, I.G.; Schladen, M.; Cereatti, A.; Lum, P.S.; Croce, U.D. Infant upper body 3D kinematics estimated using a commercial RGB-D sensor and a deep neural network tracking processing tool. In Proceedings of the 17th Edition of IEEE International Symposium on Medical Measurements and Applications (MeMeA), Messina, Italy, 22–24 June 2022.
33. Cappozzo, A.; Della Croce, U.; Leardini, A.; Chiari, L. Human movement analysis using stereophotogrammetry. Part 1: Theoretical background. *Gait Posture* **2005**, *21*, 186–196. [CrossRef]

34. Surer, E.; Cereatti, A.; Grosso, E.; Della Croce, U. A markerless estimation of the ankle-foot complex 2D kinematics during stance. *Gait Posture* **2011**, *33*, 532–537. [CrossRef] [PubMed]
35. McIntyre, S.; Morgan, C.; Walker, K.; Novak, I. Cerebral palsy-Don't delay. *Dev. Disabil. Res. Rev.* **2011**, *17*, 114–129. [CrossRef] [PubMed]
36. Meinecke, L.; Breitbach-Faller, N.; Bartz, C.; Damen, R.; Rau, G.; Disselhorst-Klug, C. Movement analysis in the early detection of newborns at risk for developing spasticity due to infantile cerebral palsy. *Hum. Mov. Sci.* **2006**, *25*, 125–144. [CrossRef] [PubMed]

sensors

MDPI

Article

Multi-Modal Deep Learning for Assessing Surgeon Technical Skill

Kevin Kasa [1], David Burns [1,2,3], Mitchell G. Goldenberg [4], Omar Selim [5], Cari Whyne [1,2,3] and Michael Hardisty [1,3,*]

1 Orthopaedic Biomechanics Lab, Holland Bone and Joint Program, Sunnybrook Research Institute, Toronto, ON M4N 3M5, Canada; kevin.kasa@ryerson.ca (K.K.); d.burns@utoronto.ca (D.B.); cwhyne@sri.utoronto.ca (C.W.)
2 Institute of Biomedical Engineering, University of Toronto, Toronto, ON M5S 1A1, Canada
3 Division of Orthopaedic Surgery, Department of Surgery, University of Toronto, Toronto, ON M5S 1A1, Canada
4 Division of Urology, Department of Surgery, University of Toronto, Toronto, ON M5S 1A1, Canada; mitchell.goldenberg@med.usc.edu
5 Department of Surgery, Royal Victoria Regional Health Center, Barrie, ON L4M 6M2, Canada; omar.selim@medportal.ca
* Correspondence: m.hardisty@utoronto.ca

Abstract: This paper introduces a new dataset of a surgical knot-tying task, and a multi-modal deep learning model that achieves comparable performance to expert human raters on this skill assessment task. Seventy-two surgical trainees and faculty were recruited for the knot-tying task, and were recorded using video, kinematic, and image data. Three expert human raters conducted the skills assessment using the Objective Structured Assessment of Technical Skill (OSATS) Global Rating Scale (GRS). We also designed and developed three deep learning models: a ResNet-based image model, a ResNet-LSTM kinematic model, and a multi-modal model leveraging the image and time-series kinematic data. All three models demonstrate performance comparable to the expert human raters on most GRS domains. The multi-modal model demonstrates the best overall performance, as measured using the mean squared error (MSE) and intraclass correlation coefficient (ICC). This work is significant since it demonstrates that multi-modal deep learning has the potential to replicate human raters on a challenging human-performed knot-tying task. The study demonstrates an algorithm with state-of-the-art performance in surgical skill assessment. As objective assessment of technical skill continues to be a growing, but resource-heavy, element of surgical education, this study is an important step towards automated surgical skill assessment, ultimately leading to reduced burden on training faculty and institutes.

Keywords: deep learning; surgical skills assessment; machine learning; computer vision; surgical education; biomedical engineering; multi-modal; human activity recognition

Citation: Kasa, K.; Burns, D.; Goldenberg, M.G.; Selim, O.; Whyne, C.; Hardisty, M. Multi-Modal Deep Learning for Assessing Surgeon Technical Skill. *Sensors* **2022**, *22*, 7328. https://doi.org/10.3390/s22197328

Academic Editor: Giovanni Saggio

Received: 18 August 2022
Accepted: 23 September 2022
Published: 27 September 2022

1. Introduction

There has been a gradual evolution in surgical education towards objective assessment of competence as a requirement for trainee advancement and an increased reliance on simulation-based training [1]. This paradigm responds to mounting pressures to shorten the surgical trainee workweek, and improve operating room efficiency and safety at teaching institutions. However, competency-based medical education (CBME) can increase the burden on supervising surgical faculty and increase program reliance on the objectivity and validity of their CBME assessments [2].

Machine learning techniques, along with increased data-collection abilities across a variety of settings may offer the ability to tackle these challenges by automating some surgical skills assessments, potentially improving their objectivity and reducing the burden of CBME on training faculty and institutes. Deep learning in particular is well suited for tackling technical skills assessment due to its robustness to noise and flexibility to learn an optimal feature set representative of task performance from large, unstructured, and multi-modal data sources. Further, new innovations allow for the collection of large-scale multi-modal data in previously unwelcoming environments, such as the operating room [3].

However, existing work on surgical skills assessment has yet to fully exploit deep learning networks and large-scale data availability to automate skills assessment. Instead, previous research relies on classical machine learning algorithms [4], only classify high-level categories of performance [5], and rely on small datasets [6].

In this study, we investigate a unique multi-modal model to automate surgical skills assessment across multiple categories and evaluate its performance on a novel dataset. Specifically, our main contributions are as follows:

- Development of a multi-modal deep learning model that combines data from both images of the final surgical product and kinematic data of the procedure. We demonstrate that this model can assess surgical performance with comparable performance to the expert human raters on several assessment domains. This is significant since existing approaches are limited in scope and predominately focus on predicting solely high-level categories.
- Ablation studies comparing the image-based, kinematic-based, and combined multi-modal networks. We show that the multi-modal network demonstrates the best overall performance.
- A new dataset of seventy-two surgical trainees and surgeons collected during a University of Toronto Department of Surgery Prep Camp and Orthopaedics Bootcamp. This consists of image, video, and kinematic data of the simulated surgical task, as well as skills assessment evaluations performed by three expert raters. This large dataset will present new and challenging opportunities for data-driven approaches to surgical skills assessment and gesture recognition tasks. (The dataset can be downloaded here: https://osf.io/rg35w/).

In the following section we provide a brief synopsis of previous works related to surgical skills assessment and activity recognition. In Section 2 we describe the details of our data-collection, processing, and deep learning model development. We present the experimental results in Section 3, with a discussion of the results, comparisons with existing studies, and motivations behind the methodologies presented in Section 4. Finally, we summarize our main findings and discuss the broad impact of this work in Section 5.

Related Work

Successful CBME is dependent on domain-specific assessment and feedback for trainees, as is currently provided by faculty members. Previous research in automating surgical skills assessment has shown promising results in effectively assessing global performance. For example, several recent studies [4,7–9] use machine learning techniques to classify surgical performance into "novice" or "expert" categories from kinematic time-series data. Other studies employ standard assessment frameworks, such as the Objective Structured Assessment of Technical Skills (OSATS) [10], to assess skill on various domains. However, many of these studies only classify performance in each domain into high-level categories (beginner, intermediate, advanced) [5,11,12]. Some studies do predict OSATS scores in a regression framework [11,13], however, the score prediction is only a small part of their work, and limited performance metrics are presented.

Instead of directly quantifying surgical performance, previous studies also focus on capturing proxies indicative of surgical performance, such as detecting surgical instruments [14], tracking instruments [15], or identifying events such as incisions [5]. Our work directly predicts the OSATS scores across five domains in a continuous regression framework. This is advantageous as it provides specific fine-grained assessment akin to that performed by a real faculty member, and eliminates ambiguities caused by broader discrete categories. Further, we present numerous performance metrics to understand the model's performance, including direct comparisons with three expert human raters.

Machine learning algorithms have been applied to surgical skills assessment by previous studies [4,7,9,12,16]. Classical machine learning, combining engineered features with learned classifiers, as well as deep learning models have shown promising results for both skills assessment works [17], as well as human activity recognition tasks (HAR) [18–23]. More recent work has focused on deep learning networks because of their ability to better exploit rich data sources (e.g., images, videos, motion tracking), which has led to improvements in performance. The deep learning models applied generally use fully convolutional or convolutional-recurrent networks; leveraging one-dimensional convolutional layers as feature extractors and recurrent layers to capture temporal dependencies. This investigation expands upon the convolutional-recurrent networks [18,19] by applying a much deeper ResNet-18 based architecture, combined with a multi-modal approach. To our knowledge, no other works have reported leveraging deeper ResNet-LSTM based models to analyze kinematic data for surgical skills assessment tasks. We discuss this approach in more detail in Sections 2 and 3.

Further, no existing studies use multi-modality approaches in surgical skills assessment. Multiple data sources (i.e., images of the final product, kinematic data of the procedure) can capture different information necessary for good performance across multiple domains of surgical skills assessment. Similarly to some image-based approaches [24], we employ a late-fusion approach. Some HAR studies investigate concatenating extracted features from different gestures for classical machine learning algorithms, and report that which features were extracted was more important than the fusion technique [23]. Unlike the studies in [23,24], we investigate fusing features extracted from disparate modalities (kinematic time-series + images) and not a single modality (images), and fuse learned features extracted from the raw data by the neural networks, instead of fusing handcrafted features.

Previous investigations applying machine learning to surgical skills assessment have relied on small custom datasets, or the open-source JIGSAWS dataset [6]. The JIGSAWS dataset consists of video and kinematic data captured using a DaVinci Robotic system [25] from eight subjects (four beginner, two intermediate, two expert). These small datasets have presented a large limitation for data-driven methods such as machine learning. Many previous studies focus solely on data acquired using robotic systems or virtual simulators [5,11,17], and not on human-performed surgical tasks. In contrast, the dataset presented in this work larger and encompasses greater participant skill levels, containing 360 total samples from 72 participants across ten surgical divisions, with experience levels ranging from first year residents to staff surgeons. This challenging real-world dataset will enable new opportunities for research into automated surgical skills assessment. The dataset is described further in Section 2.

2. Materials and Methods

This project sought to develop and validate deep learning models for automated surgical skill assessment, specifically for the assessment of technical skill for a simulated knot-tying task. To facilitate this, 72 participants performed a knot-tying task, which were subsequently rated by human experts. Video and kinematic data of the task was recorded, as well as a photograph of the final product. In this study, the anonymized video recording was used for assessment by the human raters; the machine learning models used only image and kinematic data.

2.1. Surgical Task

Seventy-two surgical trainees and surgeons were recruited for participation in this study during the 2018 University of Toronto Department of Surgery Prep Camp and Orthopaedics Bootcamp suturing modules. Participants performed a simulated vessel ligature task using one-handed knot-tying with 0-silk ties on polypropylene tubing. Each participant performed the task five times consecutively, with each performance as a separate task. No feedback was provided to participants between executions of the task. The overall goal of this task is to determine if the trainees can correctly tie off, or occlude, the simulated blood vessel using the silk suture.

2.2. Data Collection

The vessel ligature tasks were recorded using three modalities, which are visualized in Figure 1:

- High resolution digital photograph of the final product
- Anonymized video recording of the operative field
- 3D kinematic motion tracking of the hands using a Leap Sensor

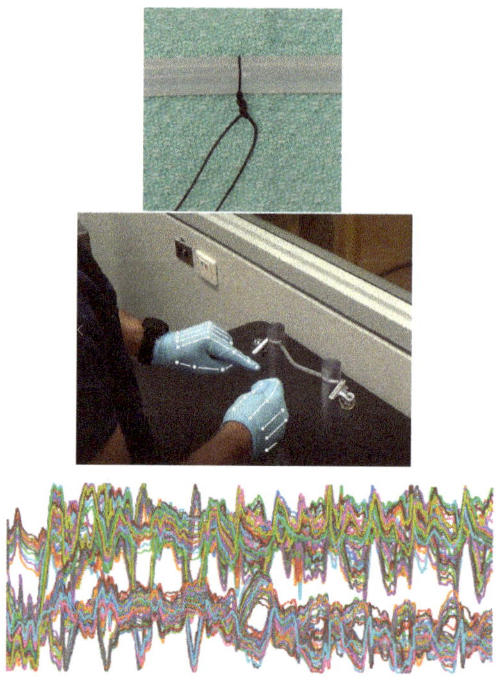

Figure 1. The trials were recorded using three modalities. The top is an image of the final product, the middle is a screen capture of the video data with a visualization of the joints tracked by the Leap sensor. The bottom is an example of the kinematic time series data, representing the temporal 3-dimensional movement of the hand joints during the knot tying task.

2.3. Task Ratings

Three blinded independent raters conducted the technical skills assessment from the recorded video and photograph of the final product. The raters were senior surgical residents (PGY4 and above) with expertise in the assessed skill. Performance at the simulated surgical task was assessed by each rater using the Objective Structured Assessment of Technical Skill (OSATS) Global Rating Scale (GRS) [10] on the following four domains:

1. Respect for Tissue
2. Time and Motion
3. Quality of Final Product
4. Overall Performance

Each domain was scored on a 5-point scale (1–5). All raters were oriented to the OSATS GRS and domain specific anchors using example performances and suggested ratings. An example of the rating scale used by the human raters can be seen in Table 1.

It was also important to ensure that the dataset was collected from a diverse and representative set of participants, including diversity in aspects such as surgical division, and prior experience level. The plurality of participants were from the division of orthopaedics, with participants from nine other surgical divisions included. Most participants were Post-Graduate Year 1 (PGY1) trainees, with experience levels ranging up to Fellows and Staff surgeons. A summary of the experience level and surgical division of the participants can be seen in Figure 2.

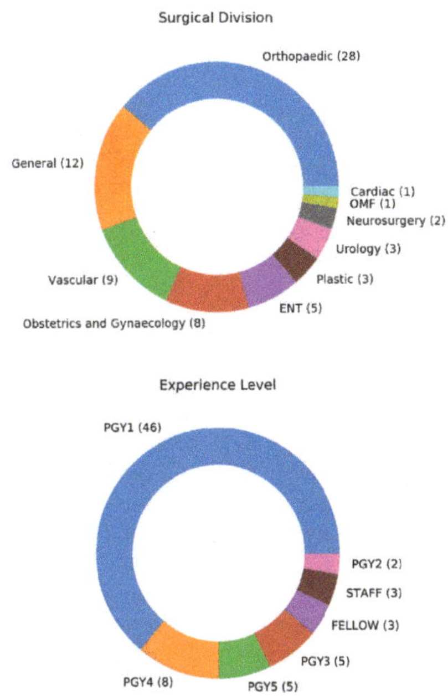

Figure 2. Participants came from 10 surgical divisions, with experiences ranging from PGY1 to Fellow.

Table 1. Rating scale used when evaluating surgical skill on the GRS Domains.

Domain	Rating Scale
Respect for Tissue	1—Very poor: Frequent or excessive pulling or sawing of tissue 3—Competent: Careful handling of tissue with occasional sawing or pulling 5—Clearly superior: Consistent atraumatic handling of tissue
Time and Motion	1—Very poor: Many unnecessary movements 3—Competent: Efficient time/motion but some unnecessary moves 5—Clearly superior: Clear economy of movement and maximum efficiency
Quality of Final Product	1—Very poor 3—Competent 5—Clearly superior
Overall Performance	1—Very poor 3—Competent 5—Clearly superior

The sequence of tasks was randomized so that the raters were not consecutively exposed to tasks performed by the same individual. Further, the randomization was seeded separately for each rater, providing each rater with a different random order of tasks to assess. Forty random samples were also selected to be rated a second time by each rater for test-retest reliability assessment.

2.4. Data Pre-Processing

The three-dimensional position data of each joint in the phalanges from both hands was extracted from the Leap Motion Sensor's kinematic data capture. This 120 channel timeseries data was used as input into the deep learning models. The kinematic models require a fixed-length input, and the trials were not uniform in length. The Seglearn library [26] was used to truncate or zero-pad each data sample to a length of 4223 samples, which represents the 90th percentile of the sample lengths. This means that most samples were padded instead of truncated, so that as much information as possible was preserved. With a sampling rate of 110 Hz, this 4223-timestamp sequence is approximately 36 s long.

The Python implementation of OpenCV was used to pre-process the image data. The images were first temporarily masked to a binary image, isolating the black suture from the background. A dilating operation was applied to this image to enlarge the knot center. The OpenCV blob detector was then used to detect the suture knot, and a 512×512 bounding box was drawn around the center. The cropped image was then unmasked back to full RGB color. The kinematic and image data were also normalized between [0,1]. This is a standard deep learning procedure to speed computation time and avoid local minima in model optimization.

2.5. Data Augmentation

Although our dataset is not small relative to other relevant datasets, deep learning almost always benefits from larger quantities of data. Thus, the entire dataset was randomly oversampled to increase the number of training examples. Additionally, the trials with ratings that were greater or less than one standard deviation from the mean were further oversampled by a factor of three. By more evenly balancing the score distribution, the network can better learn to predict these minority classes.

However, increasing the size of the dataset without introducing any variation may lead to degraded performance, as the network may rely on memorizing specific features of the training data and fail to generalize to unseen data. Data augmentation may be used to alter the input instances, thus artificially increasing the variety of training data and the network's ability to generalize. To minimize the model overfitting to the training data, the oversampled data was also augmented prior to input into the networks. The images were augmented with random 90-degree rotations and reflections about the x- or y-axis, largely to help mirror the varying knot orientation in the real data. The kinematic data was augmented based on recommendations in previous literature [27]: random rotations, reflections, and injection of Gaussian noise.

2.6. Machine Learning Models

We developed and analyzed three deep learning models. The first uses the RGB image data of the simulated vessel and ligature as input and the Quality rating as output. The second model uses the hand kinematic data as input and predicted the three other domains (Respect for Tissue, Time and Motion, and Overall Performance). The final is a composite model containing both RGB and kinematic modalities and output all four GRS rating domains. The video data was not used by the model.

The models were trained in a supervised regression learning framework, with the mean scores of the three expert raters as the ground truth. We trained the models to minimize a mean-squared error loss, however the number of output targets varied between the models since some predicted only one OSATS domain and others multiple.

$$L = \frac{1}{N}\sum_{i=1}^{N}\sum_{j=1}^{K}(y_i^k - \hat{y}_j^k) \tag{1}$$

Here, N is the number of samples in the training batch, and K is the number of output targets. For example, the image-only model has a $K = 1$ since it predicts only the Quality score, whereas the multi-modal has $K = 4$ since all four domains are predicted.

Deep residual models (ResNets) are particularly powerful in training deeper neural networks with increased capacity to learn and model complicated relationships, achieving state-of-the-art performance on many image recognition tasks [28]. These improvements largely stem from the use of "skip connections", or residual blocks, between layers which allow for deeper networks without suffering from vanishing gradient problems. This ability to effectively train very deep networks is the major advantage of the ResNet architecture. Although ResNet's are often employed in image related tasks, they can also be implemented using one-dimensional convolutions for time-series data.

The image model is depicted in the bottom branch of Figure 3, and consists of a ResNet-50 backbone with pre-trained weights from the ImageNet dataset. Prior to input, the images were resampled to 1024 × 1024, further cropped 30% tighter, and normalized based on the ImageNet metrics. Following best-practises, the pre-trained networks were initially frozen for the first 200 epochs, and only the final dense layer was trained. This is to avoid the large gradient magnitudes from the new randomly initialized dense layer destroying the pre-trained weights [29]. Subsequently, the learning rate was reduced and the top layers of the ResNet model were fine-tuned for another 200 epochs. This freezing/fine-tuning method was followed for all subsequent pre-trained models and experiments.

Previous works demonstrate that convolutional-recurrent neural networks can been used to successfully perform human activity recognition from kinematic data [18,19]. In our work, the network was tasked with scoring surgical skill across multiple domains from a relatively high-dimensional dataset (120 channels). To ensure the network had the capacity to perform these tasks, a one-dimensional ResNet-18 model was used as a feature extractor on the kinematic data. The extracted features were then inputted into two bi-directional LSTM layers to model the temporal nature of the data. Finally, three dense layers were used to score the 'Overall Performance', 'Respect for Tissue', and 'Time and Motion' from the learned features. This model was trained for 200 epochs, and the architecture can be seen in the top branch of Figure 3.

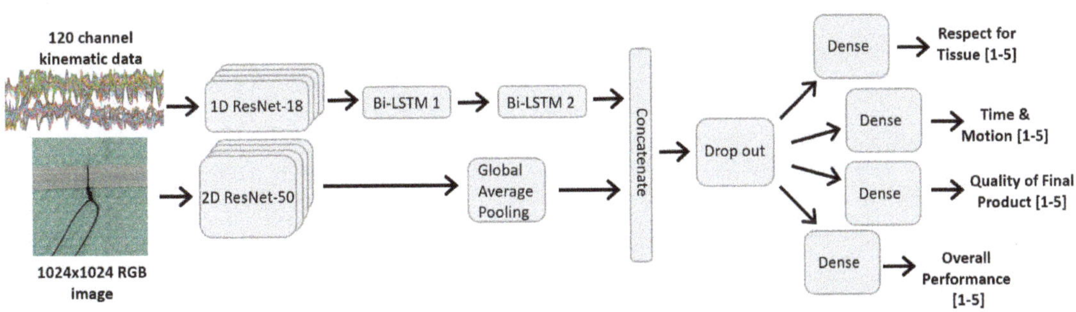

Figure 3. Images were analyzed using a ResNet-based network, and the kinematic data was analyzed using a 1D ResNet-18 as a 'feature extractor', followed by 2 bidirectional LSTM layers. The combined multi-modal network is concurrently trained on both the image and kinematic data as input, and predicts all four GRS domains.

The previous two models are combined so that all four GRS domains can be scored. The time series and image networks are trained concurrently, and the extracted feature sets are concatenated. These are then inputted into fully-connected layers to perform the final task scoring for each domain, as seen in Figure 3. The 2D ResNet network also leveraged pre-trained ImageNet weights and followed a fine-tuning scheme similar to that described above, where the ResNet layers were initially frozen for 50 epochs and used solely as a feature extractor, followed by fine-tuning the top layers of the ResNet for another 50 epochs.

The dataset was randomly split into 80%/10%/10% training/validation/testing sets. This means there were 58 participants (and 290 trials) in the training set, and 7 participants (35 trials) in the validation and testing sets. Further, the training epochs were tuned heuristically; we trained either until we saw substantial overfitting, or our computing resources were exhausted. Table 2 summarizes the hyper-parameters of the final multi-modal model.

Table 2. Summary of the hyper-parameters used to train the multi-modal network. Hyper-parameters were tuned heuristically.

Hyperparameter	Value
Learning rate	$1 \cdot 10^{-4}$
Optimizer	Adam ($\beta_1 = 0.9, \beta_2 = 0.999$)
Batch size	16
Dropout	0.50
Epochs (frozen backbone)	50
Epochs (fine-tuning backbone)	50
Loss function	Mean Squared Error
Image dimensions	(1024, 1024)
Timeseries length	4223 timestamps

2.7. Statistical Analysis

The collected dataset was analyzed to ensure its reliability and validity prior to being used for training and evaluating the deep learning models. The analysis of the expert human raters also serve as a baseline for understanding the model's best achievable performance. The Intraclass Correlation Coefficient (ICC) and Standard Error of Measure (SEM) were used to analyze the human and AI ratings for agreement and consistency. To assess the interrater reliability on the entire collected dataset, the ICC (2,3), ICC(2,1), and SEM scores were calculated for each of the GRS domains [30]. The ICC (2,3) model is selected since our raters are chosen as representative of a larger population, and the mean of the three raters is used as the ground-truth. The ICC (2,1) was also used to assess the human raters on their test-retest consistency, using the randomly repeated trials that were rated twice. Our hypothesis was that the human raters show moderate to good agreement on the GRS domains and good consistency in their ratings.

In addition to measuring the average human rater reliability on the entire dataset, we also looked at the ICC score of the raters on the held-out testing subset of the data. Since the AI models were evaluated on this test set, finding the human rater's reliability on this subset alone can allow for a more direct comparison with the network performance.

The experience levels of the participants and their ratings were also investigated to help establish construct validity. A one-way ANOVA was performed between the beginner (PGY1 & PGY2, n = 48), intermediate (PGY3, PGY4, & PGY5, n = 18), and expert (Staff & Fellow, n = 6) level participants. A Tukey–Kramer post hoc test was then done to determine which groups were different from each other. These tests were all done using the participants performance on the "Overall Performance" GRS domain.

Several tests were done to evaluate the model's performance. The point difference between the model's predictions and the human ratings with the ground truth was evaluated using the mean squared error (MSE). The goodness of fit of the model was evaluated using R^2. Finally, the agreement amongst each (human or AI) rater and the ground truth was determined using the ICC (2,1) score. This means that the ICC between the AI ratings and the ground truth was determined, as well as the ICC between each human rater and the ground truth. This allows us to consider how our model performs as a single generalized rater [30] in terms of its agreement with the ground-truth data, as well as compare the AI agreement with that exhibited by the humans. Our hypothesis was that the AI would demonstrate comparable point errors (MSE) and agreement (ICC) with the ground truth data as the human raters.

Although previous research seeking to directly predict GRS scores is sparse, existing studies report performance using the mean Spearman Correlation Coefficient ρ across the predicted vs. true GRS scores [13]. For consistency in the reported metrics, we also evaluate the Spearman Coefficient on the multi-modal model.

Finally, some studies that directly predict the GRS domain scores report their performance in terms of accuracy [11]. For a comparable metric, we also find the accuracy of our multi-modal model. Since our predictions are continuous and accuracy deals with discrete data, we first round the ground-truth and model predictions; for example, a score of 2.7 will get rounded to 3.0, which is necessary to compute the accuracy metric. Our model is designed to predict continuous scores so this is not a perfect metric, but serves to gain a general comparison with previous studies.

3. Results

3.1. Dataset Analysis

The human raters showed ICC scores corresponding to moderate agreement on the four GRS domains, when measured on the entire collected dataset, as summarized in Table 3.

Table 3. The expert human raters demonstrate moderate to good agreement on their evaluations when as measured using the mean. The AI model was trained & evaluated on the mean value of the ratings.

GRS Domain	ICC (2,3)	SEM (2,3)	ICC (2,1)	SEM (2,1)
Respect for Tissue	0.71	0.45	0.47	0.62
Time and Motion	0.70	0.47	0.44	0.64
Quality of Final Product	0.83	0.40	0.63	0.61
Overall Performance	0.73	0.39	0.47	0.55

There was some variance in the test-retest performance of the human raters, with ICC scores ranging from 0.49 to 0.88, and SEM ranging from 0.37 to 0.58. Overall, Rater 1 demonstrated better consistency amongst their ratings than Rater 2 or 3. Although some raters performed better than others, overall, they all showed moderate to good consistency, and the results are summarized in Table 4.

Table 4. Test-retest performance of the human raters on the forty repeated trials. Although the raters performance varies, they all show moderate to good consistency.

GRS Domains	Rater 1		Rater 2		Rater 3	
	ICC	SEM	ICC	SEM	ICC	SEM
Respect for Tissue	0.84	0.43	0.49	0.55	0.55	0.54
Time and Motion	0.83	0.46	0.57	0.58	0.62	0.48
Quality of Final Product	0.88	0.40	0.79	0.47	0.69	0.43
Overall Performance	0.85	0.37	0.60	0.49	0.58	0.48

On the held-out test set, the human raters showed good to excellent agreement, as seen in Table 5. Greater agreement was seen on this smaller subset of the overall data likely because there are fewer samples for the human raters to disagree on.

The one-way ANOVA returned a p-value of 0.0038, suggesting there was a significant performance difference amongst the surgeon experience groups. The Tukey analysis resulted in a significant difference between the Beginner (n = 48, mean = 2.31) and Intermediate (n = 36, mean = 2.79) groups (p = 0.003), and no significance between the Expert group (n = 6, mean = 2.50) and either of the two groups. The lack of significance in the

Expert group may be due to the relatively small sample size compared to the other two. The results of the ANOVA are depicted in Figure 4.

Table 5. Human raters show good to excellent agreement on the held-out test set. Determining agreement on the same test set the AI model is evaluated on can help provide a better baseline for expected performance.

GRS Domain	ICC (2,3)	SEM (2,3)	ICC (2,1)	SEM (2,1)
Respect for Tissue	0.78	0.44	0.54	0.63
Time and Motion	0.81	0.41	0.58	0.61
Quality of Final Product	0.93	0.30	0.82	0.49
Overall Performance	0.86	0.30	0.68	0.30

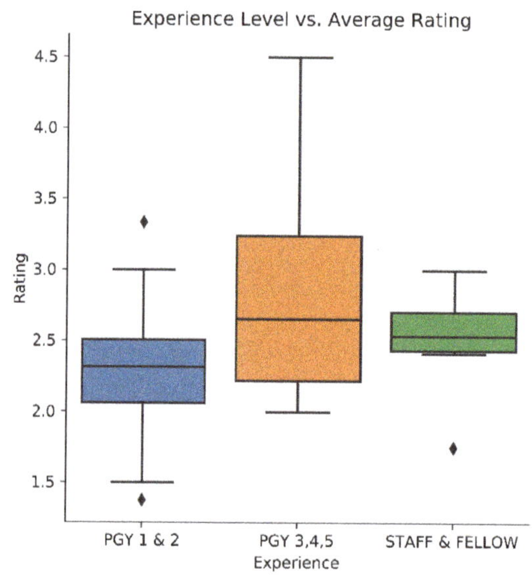

Figure 4. Participant experience and rating on the 'Overall Performance' domain. A significant difference was found between the Beginner and Intermediate groups.

3.2. Deep Learning Model Performance

The kinematic, image, and multi-modal models were all trained and evaluated independently of each other on the same reserved testing set. The model performance was evaluated by how well it can predict the mean OSATS GRS ratings provided by the raters, as well as the intrarater reliability between the model predictions and the expert raters.

Table 6 highlights the performance relative to the ground-truth. For a direct comparison with the human performance, the same metrics are presented for each individual rater's score compared to their mean scores, for the test-set trials. These metrics serve as an understanding for how close the model predictions are to the dataset's ground truth. The model's predictions do appear close to the ground-truth, with lower point errors than two of three human raters, and with the multi-modal model exhibiting the lowest point error overall.

The error between the ground-truth and the model predictions, as well as human ratings, is also seen in Figure 5. The improvements of the multi-modal model were particularly noted on the Overall Performance domain.

Table 7 summarizes the agreement between the AI model and the ground truth scores (i.e., mean of the human ratings). For comparison, we also considered the ICC scores between each individual rater and their mean score. The AI model demonstrated ICC scores ranging from 0.3 to 0.90, with the human raters ranging from 0.60 to 0.92. The multi-modal model demonstrated better agreement based on the ICC and SEM than the kinematic or image-only models on all domains except for Respect for Tissue. The multi-modal model also demonstrated better agreement with the ground truth than 2 of the 3 human raters on the Overall Performance and Quality of Final Product domains, however its performance was poorer on the remaining two domains.

Table 6. Performance metrics, including mean squared Error (MSE), of the AI predictions and human ratings, compared to the ground truth (mean of human scores).

Model	Metric	Respect for Tissue	Time and Motion	Quality of Final Product	Overall Performance
Image Model	MSE	-	-	**0.146**	-
	RMSE	-	-	0.392	-
	MAE	-	-	0.293	-
	R2	-	-	0.778	-
Kinematic Model	MSE	**0.336**	0.420	-	0.373
	RMSE	0.579	0.648	-	0.610
	MAE	0.523	0.456	-	0.431
	R2	**0.337**	0.244	-	0.453
Multi-modal Model	MSE	0.480	**0.356**	0.186	**0.194**
	RMSE	0.693	0.597	0.431	0.440
	MAE	0.545	0.459	0.331	0.315
	R2	0.136	**0.476**	**0.838**	**0.618**
Rater 1	MSE	0.464	**0.348**	0.531	0.505
	RMSE	0.681	0.590	0.729	0.710
	MAE	0.528	0.474	0.449	0.407
Rater 2	MSE	0.546	0.553	0.545	0.466
	RMSE	0.739	0.744	0.738	0.683
	MAE	0.586	0.483	0.425	0.436
Rater 3	MSE	**0.288**	0.363	0.193	0.290
	RMSE	0.537	0.602	0.439	0.539
	MAE	0.409	0.426	0.291	0.336

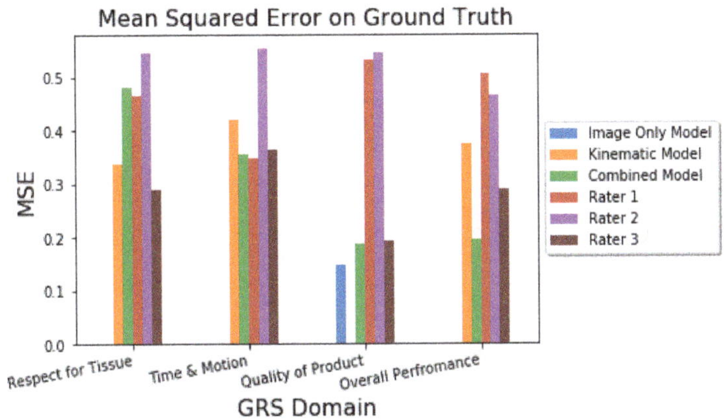

Figure 5. Graphical comparison of the MSE on the GRS Domains—lower MSE is better.

The Spearman Correlation Coefficient, ρ, of our multi-modal model is reported in Table 8. This represents the correlation between the model's predictions and the ground truth.

The discretized scores are used to evaluate the model's accuracy, and are summarized in Table 9. As mentioned, accuracy is not a perfect metric for our continuous data predictions, however it is indicative of the difference between the predictions and ground-truth on the datasets.

Overall, the multi-modal model demonstrated comparable results to the humans on most of the GRS domains. The AI had a lower point error on the ground truth scores than the human raters on three of the four GRS domains, as exhibited by the lower MSE. The ICC metrics suggest that in general, the human raters were in better agreement with the ground-truth scores. The multi-modal model demonstrated the best performance, with higher ICC on some domains (e.g., Quality of Final Product) than two of the three raters.

Table 7. Intraclass Correlation Coefficient (ICC) and Standard Error of Measurement (SEM) scores between the ground truth and the AI models & human raters.

Model	Metric	Respect for Tissue	Time and Motion	Quality of Final Product	Overall Performance
Image Model	ICC(2,1)	-	-	0.888	-
	SEM(2,1)	-	-	**0.257**	-
Kinematic Model	ICC(2,1)	**0.477**	***0.621***	-	0.534
	SEM(2,1)	**0.464**	0.441	-	0.416
Multi-modal Model	ICC(2,1)	0.301	0.591	**0.904**	***0.746***
	SEM(2,1)	0.499	**0.428**	0.309	**0.305**
Rater 1	ICC(2,1)	0.717	0.779	0.823	0.616
	SEM(2,1)	0.476	0.414	0.512	0.502
Rater 2	ICC(2,1)	0.606	0.627	0.758	0.508
	SEM(2,1)	0.516	0.524	0.521	0.689
Rater 3	ICC(2,1)	**0.797**	**0.797**	**0.924**	**0.789**
	SEM(2,1)	**0.377**	**0.423**	**0.308**	0.379

Table 8. Spearman Correlation Coefficient between the multi-modal AI predictions and the ground truth. Best performing model on the JIGSAWS dataset included as reference [13].

GRS Domain	ρ	
	Multi-Modal Model (Ours)	**FCN [13]**
Respect for Tissue	0.18	-
Time and Motion	0.73	-
Quality of Final Product	0.95	-
Overall Performance	0.82	-
Mean	**0.67**	0.65

Table 9. Accuracy of the multi-modal model, determined by first rounding the continuous ground-truth and predicted scores. Best performing model on the JIGSAWS dataset included as reference [11].

GRS Domain	Accuracy	
	Multi-Modal Model (Ours)	**Embedding Analysis [11]**
Time and Motion	**0.54**	0.32
Quality of Final Product	**0.76**	0.51
Overall Performance	**0.76**	0.41

4. Discussion

This paper presented a new dataset consisting of multi-modal recordings (image, video, & kinematic) of a simulated surgical knot-tying task, with skill assessment conducted by expert human raters based on the OSATS GRS framework. A thorough statistical analysis was conducted to ensure the validity of the dataset. Three deep-learning models were trained and evaluated on this dataset: a ResNet-50 image model, a unique "ResLSTM" kinematic model, and a combined multi-modal model.

All three models were able to successfully perform the skills assessment, with the multi-modal model performing the best overall. In comparison to previous studies conducted on the JIGSAWS dataset [6], which contains video and kinematic data from eight surgeons performing three surgical actions (knot tying, needle passing, and suturing) using the DaVinci Robotic System [25], our multi-modal model achieves better performance. For comparison, previous literature report a mean Spearman Correlation of $\rho = 0.65$ on the knot-tying task in the JIGSAWS dataset [13], as seen in Table 8. This means that on average, our multi-modal model demonstrates better correlation between its predictions and the ground-truth on our dataset, than reported on similar datasets in previous literature. Further, Khalid et al. [11] present a study that directly predicts the GRS scores in a regression fashion, using the video data of the JIGSAWS dataset. As seen in Table 9, they report a mean accuracy of 0.32 for Time and Motion, 0.51 for Quality of Final Product, and 0.41 for Overall Performance.

This is particularly encouraging as assessing surgical skill from human performed knot-tying is seemingly more challenging than evaluating a robotically operated dataset. This result means that our model can be used in a wider range of environments and facilities, where robotic surgery systems are not available for surgical trainees or faculty. Further, while some studies attempt to indirectly compute performance metrics for surgical skill [14,15], our model directly predicts performance on the GRS domains and provides the most pertinent assessment of surgical skill to trainees.

The AI performance was comparable to the human rater on three out of the four GRS domains. Further experiments are required to determine why the model consistently struggles on the Respect for Tissue domain. A possible explanation is that since the Leap Sensor is only tracking the subjects' hands, important information on the handling of the "tissue" (or polypropylene tubing) is not captured using this modality. Respect for Tissue was better assessed on video which was available to raters but not used by the model. Future analysis will investigate leveraging the video modality within the multimodal model to improve performance on this domain.

The image-only model was trained solely on the Quality of Final Product domain, since it is not likely that the images alone contain enough relevant information for this model to perform well on the other categories (e.g., Time and Motion). Smaller models were investigated for this task, such as 5- and 7-layer convolutional neural networks, however these all exhibited poor performance in the rating task and were abandoned. This suggests that the ResNet's increased capacity to extract important and meaningful features from the image data is important in assessing surgical skill. We also explored using a pre-trained MobileNet as the imaging backbone, however found the performance to be poorer than the ResNet-50. The ResNet-50 presents a good balance between performance (better ImageNet performance than VGG [28]), and reasonable computing requirements. Future studies may investigate the use of alternate backbone networks, including models such as Vision Transformers [31].

Similarly, shallow recurrent neural networks exhibited poor performance on the kinematic data and were also discarded. Learning to score various categories of surgical skill is a complex task and these models likely did not have the capacity to extract the necessary features from the kinematic data. This justifies the development of a deeper, more powerful "ResLSTM" model; the one-dimensional ResNet-18 backbone and bi-directional LSTM layers exhibited far better performance on our dataset than shallower networks. This outperforms a LSTM-only network for two likely reasons: (1) the ResNet extracts meaningful features from the raw sensor data, and (2) the convolution operators reduce the length of the time-series sequences, which are easier for the LSTM layers to learn than longer sequences.

Leveraging transfer learning was also important to increasing the image model's performance. Training a ResNet-50 model without weights pre-trained on ImageNet leads to an RMSE of 0.523 (0.274) for the quality of final product score, compared to the RMSE score of 0.392 (0.146) exhibited with pre-training. Although the ImageNet dataset does not contain examples of surgical sutures, the low-level features learned on the large-scale generic dataset are helpful starting points when transitioning to a domain-specific task. Our results further suggest the need for even larger datasets that can be used for pre-training the kinematic portion of the model. The image only model performed better than the kinematic model, likely in part due to the availability of ImageNet pre-trained weights for the image feature extractor.

Combining both the kinematic and image modalities allows for a single model to rate all four surgical skill assessment categories. Further, training a single model on both modalities led to an increase in performance across all the categories, except for Respect for Tissue. It is unclear why this model sees a degradation in performance in this category compared to the kinematic-only model; further experiments are required to discern this. Notably, the Overall Performance category saw a large increase in MSE and R^2 scores. Training on both kinematic and image data allows for the combined model to learn a more optimal feature set that is better representative of the task performances.

This study is limited in that the AI was trained and evaluated on data collected from a single training center. It remains to be studied how the model performance is affected by increased participant diversity, e.g., trainees from different institutes or countries. Future studies can investigate how the model generalizes to new participants. Further, while the OSATS was used in this study to evaluate the knot tying performance, improved assessment tools, such as a modified OSATS score which incorporates additional domains [32], may be more suitable in future studies as more complex tasks are considered in more physiologically challenging environments.

5. Conclusions

This study demonstrated a multi-modal deep learning model for surgical skill assessment with performance comparable to expert raters. This investigation highlights the importance of multi-modal data sources (image, kinematic, video) in surgical skill assessment. Automation of surgical skill assessment has the potential to transform surgical education, making training more effective, equitable, and efficient; trainees can receive quicker and more frequent feedback, while surgical faculty will have less of a burden to evaluate, allowing for greater focus on educational and clinical tasks. Further, with the addition of data collection systems to the operating room, skill assessment technology has the potential to lead to greater surgeon skill and improved patient outcomes.

Author Contributions: Conceptualization, K.K., D.B., M.H., O.S. and C.W.; methodology, K.K, D.B. and M.H.; software, K.K.; validation, K.K., D.B. and M.H.; formal analysis, K.K.; investigation, K.K., D.B., M.G.G. and O.S.; resources, C.W. and M.H.; data curation, K.K., D.B., M.G.G. and O.S.; writing—original draft preparation, K.K.; writing—review and editing, K.K., D.B., M.G.G., O.S., M.H. and C.W.; visualization, K.K.; supervision, M.H. and C.W.; project administration, M.H. and C.W.; funding acquisition, M.H. and C.W. All authors have read and agreed to the published version of the manuscript.

Funding: This research was funded by the Wyss Medical Foundation and Feldberg Chair for Spinal Research.

Institutional Review Board Statement: This study was approved by the Sunnybrook Health Sciences Center Research Ethics Board (REB) on August 1, 2018 (REB protocol # 248-2018), and the Mount Sinai Hospital REB on June 22, 2018 (REB protocol # 18-0149-E).

Informed Consent Statement: Informed consent was obtained from all subjects involved in the study

Data Availability Statement: The dataset and code will be made publicly available and shared once posted online.

Acknowledgments: The authors would like to acknowledge Lisa Satterthwaite, Oleg Safir, and the staff at the Mount Sinai Hospital Surgical Skills Centre for their support of this work.

Conflicts of Interest: The authors declare no conflict of interest.

References

1. Reznick, R.K.; MacRae, H. Teaching surgical skills–changes in the wind. *N. Engl. J. Med.* **2006**, *355*, 2664–2669. [CrossRef] [PubMed]
2. Sonnadara, R.R.; Mui, C.; McQueen, S.; Mironova, P.; Nousiainen, M.; Safir, O.; Kraemer, W.; Ferguson, P.; Alman, B.; Reznick, R. Reflections on Competency-Based Education and Training for Surgical Residents. *J. Surg. Educ.* **2014**, *71*, 151–158. [CrossRef] [PubMed]
3. Boet, S.; Etherington, C.; Lam, S.; Lê, M.; Proulx, L.; Britton, M.; Kenna, J.; Przybylak-Brouillard, A.; Grimshaw, J.; Grantcharov, T.; et al. Implementation of the Operating Room Black Box Research Program at the Ottawa Hospital Through Patient, Clinical, and Organizational Engagement: Case Study. *J. Med. Internet Res.* **2021**, *23*, e15443. [CrossRef] [PubMed]
4. Poursartip, B.; LeBel, M.E.; McCracken, L.C.; Escoto, A.; Patel, R.V.; Naish, M.D.; Trejos, A.L. Energy-Based Metrics for Arthroscopic Skills Assessment. *Sensors* **2017**, *17*, 1808. [CrossRef] [PubMed]
5. Yanik, E.; Intes, X.; Kruger, U.; Yan, P.; Diller, D.; Voorst, B.; Makled, B.; Norfleet, J.; De, S. Deep neural networks for the assessment of surgical skills: A systematic review. *J. Def. Model. Simul. Appl. Methodol. Technol.* **2021**, *19*, 159–171. [CrossRef]
6. Gao, Y.; Vedula, S.S.; Reiley, C.E.; Ahmidi, N.; Varadarajan, B.; Lin, H.C.; Tao, L.; Zappella, L.; Béjar, B.; Yuh, D.D.; et al. JHU-ISI Gesture and Skill Assessment Working Set (JIGSAWS): A Surgical Activity Dataset for Human Motion Modeling. In Proceedings of the Modeling and Monitoring of Computer Assisted Interventions (M2CAI)—MICCAI Workshop, Boston, MA, USA, 14–18 September 2014.
7. Fard, M.J.; Ameri, S.; Ellis, R.D.; Chinnam, R.B.; Pandya, A.K.; Klein, M.D. Automated robot-assisted surgical skill evaluation: Predictive analytics approach. *Int. J. Med Robot. Comput. Assist. Surg.* **2018**, *14*, e1850. [CrossRef]
8. Law, H.; Ghani, K.; Deng, J. Surgeon Technical Skill Assessment Using Computer Vision Based Analysis. In Proceedings of the 2nd Machine Learning for Healthcare Conference, Boston, MA, USA, 18–19 August 2017. Available online: https://proceedings.mlr.press/v68/law17a.html (accessed on 17 August 2022).
9. Watson, R.A. Use of a machine learning algorithm to classify expertise: Analysis of hand motion patterns during a simulated surgical task. *Acad. Med.* **2014**, *89*, 1163–1167. [CrossRef]
10. Martin, J.A.; Regehr, G.; Reznick, R.; Macrae, H.; Murnaghan, J.; Hutchison, C.; Brown, M. Objective structured assessment of technical skill (OSATS) for surgical residents. *Br. J. Surg.* **1997**, *84*, 273–278.
11. Khalid, S.; Goldenberg, M.G.; Grantcharov, T.P.; Taati, B.; Rudzicz, F. Evaluation of Deep Learning Models for Identifying Surgical Actions and Measuring Performance. *JAMA Netw. Open* **2020**, *3*, e201664. [CrossRef]
12. Aneeq, Z.; Yachna, S.; Vinay, B. Video and accelerometer-based motion analysis for automated surgical skills assessment. *Int. J. Comput. Assist. Radiol. Surg.* **2018**, *13*, 443–455.
13. Ismail Fawaz, H.; Forestier, G.; Weber, J.; Idoumghar, L.; Muller, P.A. Accurate and interpretable evaluation of surgical skills from kinematic data using fully convolutional neural networks. *Int. J. Comput. Assist. Radiol. Surg.* **2019**, *14*, 1611–1617. [CrossRef] [PubMed]
14. O'Driscoll, O.; Hisey, R.; Camire, D.; Erb, J.; Howes, D.; Fichtinger, G.; Ungi, T. Object detection to compute performance metrics for skill assessment in central venous catheterization. In *SPIE 11598, Proceedings of the Medical Imaging 2021: Image-Guided Procedures, Robotic Interventions, and Modeling, Online, 15–19 February 2021*; Linte, C.A., Siewerdsen, J.H., Eds.; International Society for Optics and Photonics, SPIE: Bellingham, WA, USA, 2021; Volume 11598, pp. 315–322. [CrossRef]
15. O'Driscoll, O.; Hisey, R.; Holden, M.; Camire, D.; Erb, J.; Howes, D.; Ungi, T.; Fichtinger, G. Feasibility of object detection for skill assessment in central venous catheterization. In *SPIE 12034, Proceedings of the Medical Imaging 2022: Image-Guided Procedures, Robotic Interventions, and Modeling, San Diego, CA, USA, 20–23 February 2022*; Linte, C.A., Siewerdsen, J.H., Eds.; International Society for Optics and Photonics, SPIE: Bellingham, WA, USA, 2022; Volume 12034, pp. 358–365. [CrossRef]
16. Zia, A.; Essa, I. Automated surgical skill assessment in RMIS training. *Int. J. Comput. Assist. Radiol. Surg.* **2018**, *13*, 731–739. [CrossRef] [PubMed]
17. Ismail Fawaz, H.; Forestier, G.; Weber, J.; Idoumghar, L.; Muller, P.A. Evaluating Surgical Skills from Kinematic Data Using Convolutional Neural Networks. In Proceedings of the Medical Image Computing and Computer Assisted Intervention—MICCAI 2018, Granada, Spain, 16–20 September 2018; Frangi, A.F., Schnabel, J.A., Davatzikos, C., Alberola-López, C., Fichtinger, G., Eds.; Springer International Publishing: Cham, Switzerland, 2018; pp. 214–221.
18. Ordonez, F.J.; Roggen, D. Deep Convolutional and LSTM Recurrent Neural Networks for Multimodal Wearable Activity Recognition. *Sensors* **2016**, *16*, 115. [CrossRef] [PubMed]

19. Burns, D.M.; Leung, N.; Hardisty, M.; Whyne, C.M.; Henry, P.; McLachlin, S. Shoulder Physiotherapy Exercise Recognition: Machine Learning the Inertial Signals from a Smartwatch. *Physiol. Meas.* **2018**, *39*, 075007. [CrossRef]
20. Hammerla, N.Y.; Halloran, S.; Ploetz, T. Deep, Convolutional, and Recurrent Models for Human Activity Recognition using Wearables. *arXiv* **2016**, arXiv:1604.08880.
21. Rueda, F.M.; Grzeszick, R.; Fink, G.A.; Feldhorst, S.; ten Hompel, M. Convolutional Neural Networks for Human Activity Recognition Using Body-Worn Sensors. *Informatics* **2018**, *5*, 26. [CrossRef]
22. Huang, J.; Lin, S.; Wang, N.; Dai, G.; Xie, Y.; Zhou, J. TSE-CNN: A Two-Stage End-to-End CNN for Human Activity Recognition. *IEEE J. Biomed. Health Inform.* **2020**, *24*, 292–299. [CrossRef]
23. Cheng, Y.; Ji, X.; Li, X.; Zhang, T.; Malebary, S.J.; Qu, X.; Xu, W. Identifying Child Users via Touchscreen Interactions. *ACM Trans. Sens. Netw. (TOSN)* **2020**, *16*, 1–25. [CrossRef]
24. Seeland, M.; Mäder, P. Multi-view classification with convolutional neural networks. *PLoS ONE* **2021**, *16*, e0245230. [CrossRef]
25. DiMaio, S.; Hanuschik, M.; Kreaden, U. The da Vinci Surgical System. In *Surgical Robotics*; Rosen, J., Hannaford, B., Satava, R., Eds.; Springer: Boston, MA, USA, 2011. [CrossRef]
26. Burns, D.M.; Whyne, C.M. Seglearn: A Python Package for Learning Sequences and Time Series. *J. Mach. Learn. Res.* **2018**, *19*, 3238–3244.
27. Itzkovich, D.; Sharon, Y.; Jarc, A.; Refaely, Y.; Nisky, I. Using Augmentation to Improve the Robustness to Rotation of Deep Learning Segmentation in Robotic-Assisted Surgical Data. In Proceedings of the 2019 International Conference on Robotics and Automation (ICRA), Montreal, QC, Canada, 20–24 May 2019; pp. 5068–5075. [CrossRef]
28. He, K.; Zhang, X.; Ren, S.; Sun, J. Deep Residual Learning for Image Recognition. In Proceedings of the IEEE Conference on Computer Vision and Pattern Recognition (CVPR), Las Vegas, NV, USA, 27–30 June 2016; pp. 770–778. [CrossRef]
29. Varno, F.; Soleimani, B.H.; Saghayi, M.; Di-Jorio, L.; Matwin, S. Efficient Neural Task Adaptation by Maximum Entropy Initialization. *arXiv* **2019**, arXiv:1905.10698.
30. Koo, T.; Li, M. A Guideline of Selecting and Reporting Intraclass Correlation Coefficients for Reliability Research. *J. Chiropr. Med.* **2016**, *15*, 155–163. [CrossRef] [PubMed]
31. Dosovitskiy, A.; Beyer, L.; Kolesnikov, A.; Weissenborn, D.; Zhai, X.; Unterthiner, T.; Dehghani, M.; Minderer, M.; Heigold, G.; Gelly, S.; et al. An Image is Worth 16 ×16 Words: Transformers for Image Recognition at Scale. *arXiv* **2020**, arXiv:2010.11929.
32. Hopmans, C.J.; den Hoed, P.T.; van der Laan, L.; van der Harst, E.; van der Elst, M.; Mannaerts, G.H.H.; Dawson, I.; Timman, R.; Wijnhoven, B.P.; Ijzermans, J.N.M. Assessment of surgery residents' operative skills in the operating theater using a modified Objective Structured Assessment of Technical Skills (OSATS): A prospective multicenter study. *Surgery* **2014**, *156*, 1078–1088. [CrossRef] [PubMed]

sensors

MDPI

Article

Improved Fully Convolutional Siamese Networks for Visual Object Tracking Based on Response Behaviour Analysis

Xianyun Huang [1], Songxiao Cao [2,*], Chenguang Dong [1], Tao Song [2] and Zhipeng Xu [2]

1 Scientific Research Post, Suzhou Institute of Metrology, Suzhou 215128, China
2 College of Metrology and Measurement Engineering, China Jiliang University, Hangzhou 310018, China
* Correspondence: caosongxiao@cjlu.edu.cn

Abstract: Siamese networks have recently attracted significant attention in the visual tracking community due to their balanced accuracy and speed. However, as a result of the non-update of the appearance model and the changing appearance of the target, the problem of tracking drift is a regular occurrence, particularly in background clutter scenarios. As a means of addressing this problem, this paper proposes an improved fully convolutional Siamese tracker that is based on response behaviour analysis (SiamFC-RBA). Firstly, the response map of the SiamFC is normalised to an 8-bit grey image, and the isohypse contours that represent the candidate target region are generated through thresholding. Secondly, the dynamic behaviour of the contours is analysed in order to check if there are distractors approaching the tracked target. Finally, a peak switching strategy is used as a means of determining the real tracking position of all candidates. Extensive experiments conducted on visual tracking benchmarks, including OTB100, GOT-10k and LaSOT, demonstrated that the proposed tracker outperformed the compared trackers such as DaSiamRPN, SiamRPN, SiamFC, CSK, CFNet and Staple and achieved state-of-the-art performance. In addition, the response behaviour analysis module was embedded into DiMP, with the experimental results showing the performance of the tracker to be improved through the use of the proposed architecture.

Keywords: visual tracking; Siamese tracker; tracking drift; background clutter

Citation: Huang, X.; Cao, S.; Dong, C.; Song, T.; Xu, Z. Improved Fully Convolutional Siamese Networks for Visual Object Tracking Based on Response Behaviour Analysis. *Sensors* **2022**, 22, 6550. https://doi.org/10.3390/s22176550

Academic Editor: Petros Daras

Received: 27 June 2022
Accepted: 25 August 2022
Published: 30 August 2022

Publisher's Note: MDPI stays neutral with regard to jurisdictional claims in published maps and institutional affiliations.

1. Introduction

Visual object tracking has become increasingly important in many application fields, including surveillance, robotics and human–computer interfaces. However, the challenges of reliable tracking due to cluttered backgrounds, occlusion and different illuminations still remain.

Inspired by artificial neural networks [1–5] and deep learning [6], breakthroughs in many areas such as deep learning-based methods have attracted growing interest in the visual object tracking field. According to the network architecture, there are four categories of deep learning trackers: convolutional neural network- or CNN-based trackers, recurrent neural network- or RNN-based trackers, generative adversarial network- or GAN-based trackers and Siamese neural network- or SNN-based trackers [6].

(1) CNN was the first deep learning model to be used in the visual object tracking field due to its powerful representation of a target. Wang [7] proposed a tracking algorithm that used fully convolutional networks pre-trained on image classification tasks, and this performed better than the majority of other trackers regarding both precision and success rate at that time. Nam [8] pre-trained a CNN using a large set of videos with tracking ground truths for obtaining a generic target representation. CNN-based trackers have inherent limitations, including computational complexities and the requirement of large-scale supervised training data.

(2) RNN-based trackers are excellent for dealing with temporal information of video frames, including object movement or motion. Yang [9] embedded a long short-term memory (LSTM) network into a recurrent filter learning network as a means of achieving state-of-the-art tracking. Ma [10] exploited a pyramid multi-directional recurrent network to memorise target appearance. However, RNN-based trackers are generally difficult to train and have a considerable number of parameters that require tuning, and the number of these trackers is limited.
(3) GAN-based trackers can generate desired training positive images in the feature space for tackling the issue of sample imbalance [11]. Guo [12] proposed a task-guided generative adversarial network (TGGAN) to learn the general appearance distribution that a target may undergo through a sequence. As RNN trackers, it is also difficult to train and evaluate GAN-based trackers, so their number is also limited.
(4) Recently, Siamese networks (SNN), which follow a tracking using a similarity comparison strategy, have received significant attention from the visual tracking community due to their favourable performance [13–17]. SNN-based trackers formulate the visual object tracking problem by learning a general similarity map through cross-correlation between the feature representations learned for the target template and the search region. Due to the satisfactory balance between performance and efficiency, SNN-based trackers have become the most widely used and researched trackers in recent years.

Although these tracking approaches can obtain balanced accuracy and speed, some problems must be urgently addressed, the most important of which is the object locating strategy or motion model. With traditional Siamese trackers, the new position of the target is always the location with the highest score in the response map for every input image frame. This strategy can potentially result in tracking drift if distractors exist near to the real target, particularly if one of them has a higher response score than the real target. In order to address this problem, an improved SiamFC tracker based on response map analysis is proposed. Extensive experiments on visual tracking benchmarks including OTB100, GOT-10k and LaSOT demonstrated that the proposed tracker improves the performance in terms of both tracking accuracy and robustness.

The main contributions of this work are as follows:

- A new distractor detecting method is proposed that analyses the response map without training. Following an experimental comparison, it is proven that the proposed response behaviour analysis module can be embedded into other response map- or score map-based trackers as a means of improving tracking performance, making this a common strategy for many other trackers.
- The behaviour of real targets and distractors can be observed and recognised through the analysis of the dynamic pattern of the contours in the response map. This method enables a simple, effective and dynamic analysis of the movement trend of the target and the surrounding distractors over a period of time to be performed for the prediction of the potential impact the distractors have on the target object.
- The performance of the classic SiamFC can be significantly improved through the adoption of the response analysis model during the tracking process. This shows that for certain problems with classical visual target tracking algorithms such as SiamFC, tracking performance can be improved more substantially through the use of well-designed but simple strategies, which do not necessarily require the reconstruction of complex network structures or long training periods.

This paper is organised in the following way. A basic introduction and work relating to Siamese trackers are introduced in Section 2. Section 3 outlines the proposed response analysis method that includes the response map contour, distractor approaching analysis and peak switching strategy. In Section 4, the proposed method is compared to DaSiamRPN [15], SiamFC [17], SiamRPN [18], CFNet [19], CSK [20] and Staple [21] using the OTB100, GOT-10k and LaSOT benchmarks. In addition, the experimental results and analyses are also provided. Finally, Section 5 presents conclusions and suggests future research directions.

2. Related Work

The Siamese network consists of two subnetworks with identical network architectures and shared weights. It was initially proposed by Bromley et al. [22] for signature verification, and the pioneering work of the use of the Siamese network in the visual object tracking field is SINT [23], which simply searches for the candidate that is most similar to the exemplar that is provided in the starting frame.

2.1. SNN-Based Trackers

Bertinetto et al. proposed a fully convolutional Siamese network (SiamFC) [17] for the estimation of the feature similarity between two frames. SiamFC adopts the Siamese network as a feature extractor, introducing the correlation layer for combining response maps, and the position of the target is determined by locating the maximum value of the response map.

Following the proposal of the classic SiamFC, many further works have been proposed on its basis, including CFNet, DCFNet, RASNet, SiamRPN, CHASE and COMET. CFNet [19] interprets the correlation filters as a differentiable layer in a Siamese tracking framework, thereby achieving end-to-end representation learning. However, the performance improvement is limited in comparison to SiamFC. In order to improve the tracking performance when faced with challenges such as partial occlusion and deformation, FlowTrack [24] exploits motion information in the Siamese architecture as a means of improving the feature representation and tracking accuracy. RASNet [25] was proposed by Wang et al. and embedded diverse attention mechanisms into the Siamese network for adapting the tracking model to the current target. For more accurately estimating the target bounding boxes, Li et al. integrated the regional proposal network (RPN) into the Siamese network and proposed the SiamRPN tracker [18]. The results demonstrated superior tracking performance in comparison to classical trackers with the presence of RPN. Following the proposal of the SiamRPN tracker, many researchers have attempted to improve tracker performance. One typical tracker is DaSiamRPN [15], which utilises a distractor-aware module for performing incremental learning of background distractors. SiamRPN++ [13] made further improvements based on DaSiamRPN, using a spatial-aware sampling strategy and training a ResNet-driven Siamese tracker with a significant performance gain. CHASE [26] was proposed by Marvasti-Zadeh et al., and it is a novel cell-level differentiable architecture search mechanism with early stopping for automating the network design of the tracking module. It has the objective of adapting backbone features to the objective of Siamese tracking networks during offline training. In order to address the problem of tracking an unknown small target from aerial videos at medium to high altitudes, the researchers also proposed a context-aware IoU-guided tracker (COMET) [27] to exploit a multitask two-stream network and an offline reference proposal generation strategy. Several trackers have recently been introduced using transformers, including TransT [28] and ToMP [29]. They have gained significant attention in the visual tracking community. Similar to Siamese-based trackers, these transformer trackers take a pair of image patches as the inputs of the backbone network and employ a feature fusion network consisting of multiple self- and cross-attention modules.

2.2. Discriminative Object Representation and Improvement Solutions

The object representation model plays a crucial role in all visual tracking algorithms. A good representation model can help a tracker distinguish between real targets and distractors. A disadvantage of Siamese trackers is poor performance when distractors are close to the true target, as the Siamese network does not have the strategy of discovering distractors during tracking and is only concerned with the highest score of the response map in tracking without any focus on the background clutter situation. Many solutions have been proposed by scholars for solving this problem. They can be classified into the following five categories: (1) Learning distractor-aware. Zhu et al. [15] discovered that the imbalanced distribution of training data makes the learned features less discriminative, proposing the DaSiamRPN algorithm. This method introduced a new sampling strategy and made the model focus on semantic distractors. Similarly, target-aware deep tracking (TADT) [30] chose the target-aware features based on activations to represent the targets. As both trackers utilised pre-trained deep features, and due to the fact that the targets of interest can be arbitrary objects in visual tracking, the problem of being less effective in modelling arbitrary targets to distinguish them from the background still exists.

(2) Combing confidence map. R-FCSN [31] adaptively weighted each region response as a means of forming a joint confidence map. This confidence map placed greater emphasis on reliable regions and eliminated the clutter that is caused by drifting regions. LTSN [32] used a multi-confidence map strategy as a means of improving the adaptiveness of appearance changes and background distractors. The advantage of these algorithms is that they require no training and are fast, but the disadvantage is that they are too simple and do not consider the motion information of the target.

(3) Mining hard samples. Siam R-CNN [33] proposed an embedding network for extracting an embedding vector for every ground truth bounding box that represents the appearance of the object. In this way, the tracker discovered hard examples for re-detection conditioned on the reference object through the retrieval of objects from other videos. DaSiamRPN also used hard sample mining technology to improve object representation. Mining and training hard samples represent an incredibly useful method that leads to the improvement of the performance of distinguishing similar objects, but finding and training hard samples are generally quite difficult.

(4) Integrating background appearance. DiMP [34] proposed an end-to-end architecture based on a target model prediction network, which is derived from a discriminative learning loss, and integrated background appearance as a means of achieving state-of-the-art performance.

(5) Using classification components. ATOM [35] designed special dedicated target estimation and classification components, combining them to create a novel tracking architecture. Both DiMP and ATOM utilised a similar state update strategy based on the comparison of the two maximum peaks of the response map. With this strategy, when some distractors were near to the target, the response map scores were below a certain threshold, resulting in the tracking state being labelled as 'uncertain'. The position had the highest score returned as the new tracking position, which was not reasonable as this could result in tracking drift as the position with the highest score has a greater probability of being a distractor.

Different to the aforementioned Siamese-based trackers where the problems of background clutter and distractors were addressed through training with different network structures or different data samples, this paper proposes a distractor analysis method for tracking without retraining the network based on a Siamese tracker. The proposed method can handle the tracking drift problem, particularly in background clutter scenarios.

The proposed method process is as follows. Firstly, the response map of the SiamFC is normalised to an 8-bit grey image, and the isohypse contours that represent the candidate target region are generated through thresholding. Secondly, the dynamic behaviour of the contours is analysed to ascertain whether there are distractors approaching the tracked target. Finally, a peak switching strategy is used for determining the real tracking position of all the candidates. In addition, a new Siamese network does not need to be constructed for this method, and there is only a need to modify the tracking update process. Following the use of this response analysis method, classic SiamFC tracking performance can be improved to state-of-the-art level. An overview of this proposed method can be seen in Figure 1.

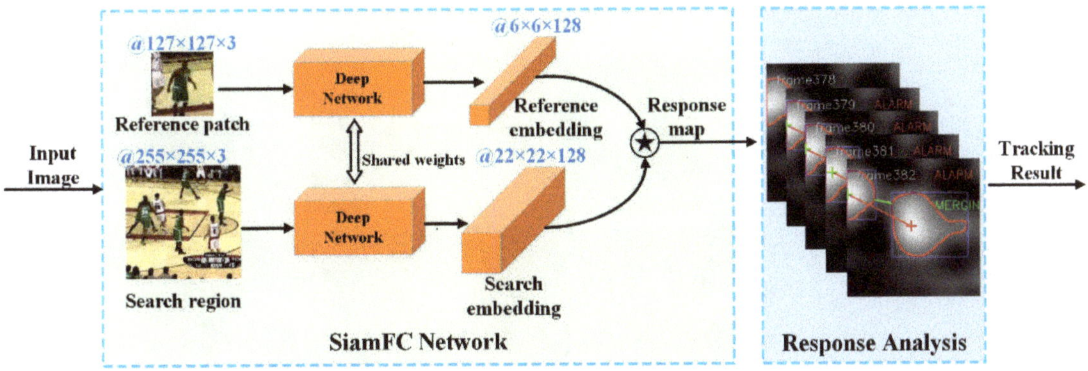

Figure 1. Overview of the proposed tracking method.

3. Proposed Method

In Figure 1, an overview of the visual object tracking method proposed in this paper is presented. In this section, the details of the algorithm will be introduced, including a brief introduction to Siamese trackers, details of response behaviour analysis and the pseudo-code for the proposed method.

3.1. Siamese Trackers

With a typical Siamese network, a pair of images (x, z), where x and z are the target template patch and search patch, is used for training. The images are sent into a deep network as a means of obtaining two feature maps:

$$g_\rho(p_t, p_s) = f_\rho(p_t) * f_\rho(z) + b \tag{1}$$

where $f_\rho(p_t)$ is a deep convolution network, ρ is a learnable parameter, b is a scalar offset value, $*$ denotes the cross-correlation operation and $g_\rho(p_t, p_s)$ represents the response map, denoting the similarity between p_t and p_s. The training goal is to enable the maximum value of the response map to correspond to the target position.

During the testing stage, similarities between the target template patch and the search patch are presented by a single channel response map, and the estimated location of the target can be predicted as follows:

$$q = \text{argmax} f(p_t) * f(p_s) \tag{2}$$

where q is the central position of the target.

A more detailed explanation of Siamese trackers can be found in [17].

3.2. Improved SiamFC Tracker Based on Response Behaviour Analysis

With traditional Siamese trackers, the new position of the target is predicted using the location with the highest score of the response map for every input image frame. For the frame without background cluttering, the response map is a single model, but if some distractors exist that are similar to the template patch in the searching region, the response map has a general tendency to be multi-model. In certain cases, the distractor has a higher score than the true object. If the tracking strategy involves always changing to the position with the highest score in every frame, the tracking will drift to other background distractors, as can be seen in Figure 2.

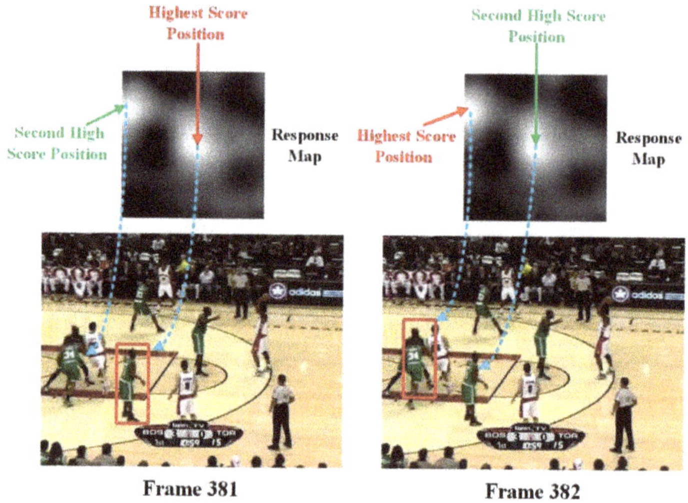

Figure 2. Tracking drift from frame 381 to frame 382.

Detailed analysis of the response map is essential for improving tracking performance and addressing this problem. It was found that changing the response map from frame to frame exhibited interesting behaviour that could be used to analyse whether distractors are approaching. Based on this motivation, this paper proposes an improved SiamFC tracker based on response behaviour analysis. An overview of the proposed method can be seen in Figure 3.

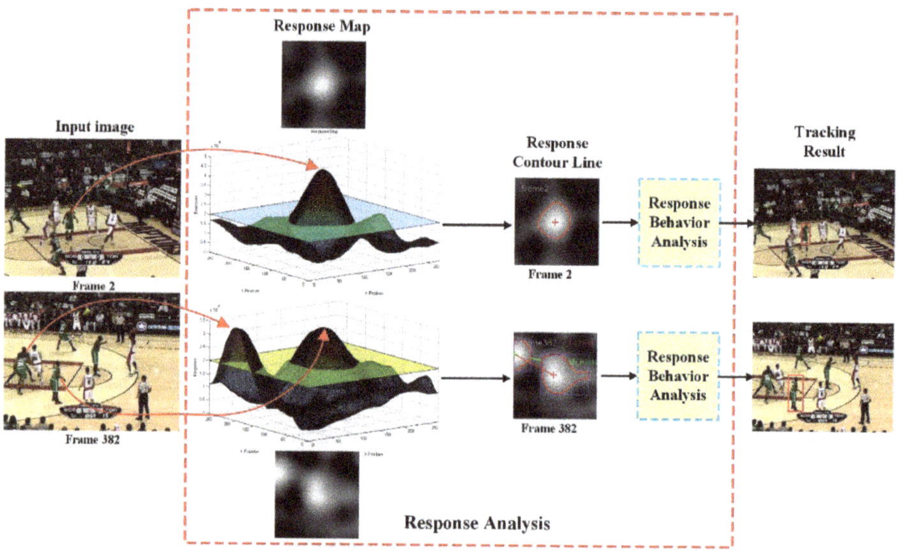

Figure 3. Overview of the proposed response analysis.

3.2.1. Response Isohypse Contour

A response map can be normalised to an 8-bit grey-level image where a higher value represents a higher score of the original response map. In this normalised response map, a binarisation operation with a certain threshold is equivalent to drawing a contour plane of the original response map. The blob regions of the binarised image can then be used for analysing the behaviour of the response map frame by frame. An overview of the response isohypse contour method can be seen in Figure 4.

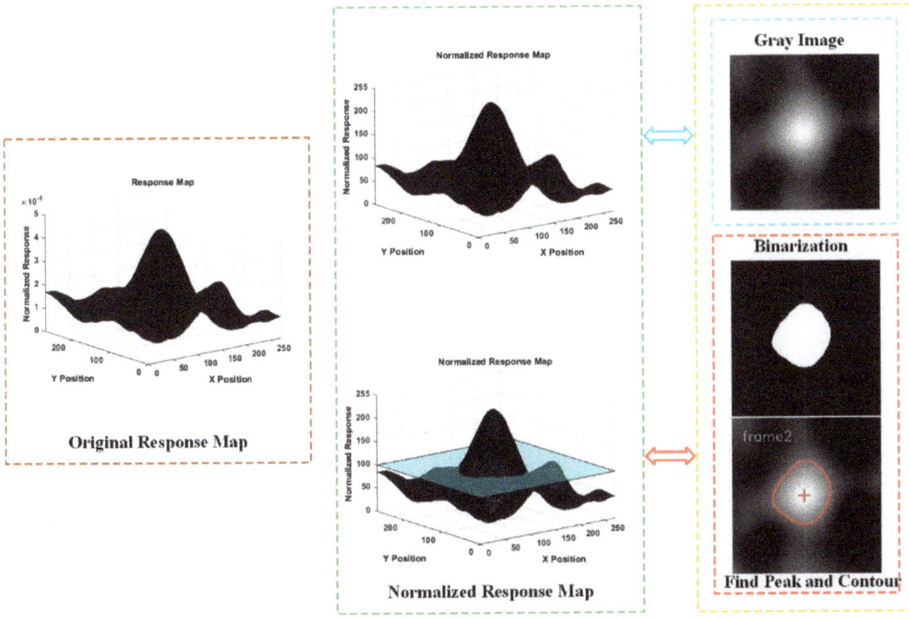

Figure 4. Illustrations of response isohypse contour.

3.2.2. Distractor Approaching Analysis

An obvious phenomenon can be witnessed when an analysis of the response map is performed. When there is no distractor with a similar appearance around the target, the response map is unimodal, but when an obvious distractor is nearby, it will generally be multi-modal. By transferring the response map to the response isohypse contour map, the situation where there is only one contour in the middle of the map represents tracking without background cluttering, and when the map has more than one contour, this indicates the presence of distractors around the target. In addition, if one contour gradually becomes closer to the centre contour in each frame, this indicates an approaching distractor to the true target. The process of distractor approaching analysis can be seen in Figure 5. On this basis, the approaching distractor can be analysed through the following three steps:

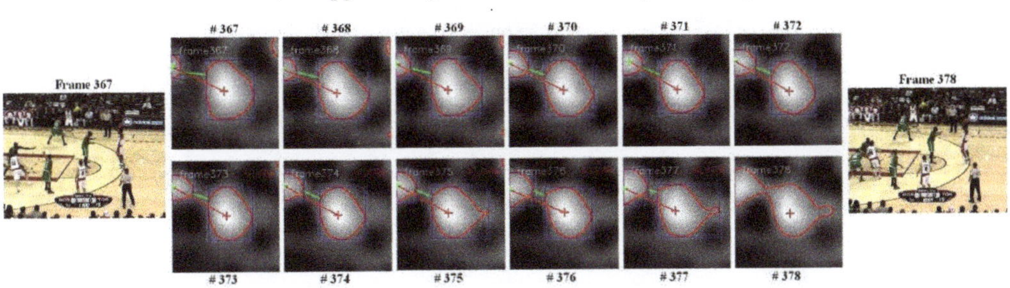

Figure 5. The process of distractor approaching analysis.

Step one: Contour number judgement. If only one contour exists and is located in the central part of the response map, this represents good tracking conditions. Otherwise, if there is more than one contour, distractor approaching analysis is utilised.

Step two: Calculate the minimum distance between contours.

$$d_{k,i,j} = \sqrt{\left(x_{k,i}^{c1} - x_{k,j}^{c2}\right)^2 + \left(y_{k,i}^{c1} - y_{k,j}^{c2}\right)^2} \tag{3}$$

$$d_k = \min\left\{d_{k,i,j}\right\} \tag{4}$$

where k is the kth frame of the tracking sequence; $c1$ represents the contour with the highest response score and $c2$ represents the contour with the second highest response score; $i \in (0, N_{c1})$ $j \in (0, N_{c2})$, where N_{c1} and N_{c2} are the total points of contours $c1$ and $c2$; $d_{k,i,j}$ represents the distance between the ith point of contour $c1$ and the jth point of contour $c2$ in the kth frame; and d_k is the minimum distance between contours $c1$ and $c2$.

Step three: Analyse the trend of the distance change.

$$MD_k = \frac{\sum_{k-M \leq n \leq k}(d_n - d_{n-1})}{M} \tag{5}$$

where M means there are M frames that are used for analysing the approaching trend, and MD_k is the mean different distance in the M frames before the kth frame. If MD_k is less than a certain threshold T_{md}, this indicates that some distractors are approaching.

3.2.3. Object Centre Switching Strategy

Most Siamese trackers choose the location with the highest response score as the target position in each frame, although this strategy can result in tracking drift in certain situations. Consider this situation: if in the (k − 1)th frame, a response peak with a score of 255 (after being normalised) is located in the central part of the response map and another peak with a score of 254 is located at the edge of the response map, while the score of the central peak in the kth frame changes to 254 and the score of the edge peak becomes 255, then the target position will change to the edge peak, ultimately resulting in tracking drift. This is obviously not an ideal tracking strategy when there are distractors nearby.

In order to address this problem, most existing Siamese trackers employ the strategy of restricting the possible region of the response peak, focusing only on a small region that is close to the position in the previous frame. However, if the peak of the response is at the edge of the response map, it will be abandoned. In certain cases, this strategy can improve performance, but the information of the location of distractors will be lost, and this is useful for further analysis. Unlike traditional Siamese trackers, the proposed method utilises a new strategy that is based on peak angle judgement, as seen in Figure 6.

$$\theta = \arctan\left(\frac{Dh}{Dp}\right) \tag{6}$$

where Dh is the difference in height between two peaks, and Dp is the distance between two peaks. With the proposed strategy, the object centre can only be changed to the edge peak if the angle θ is above a certain threshold.

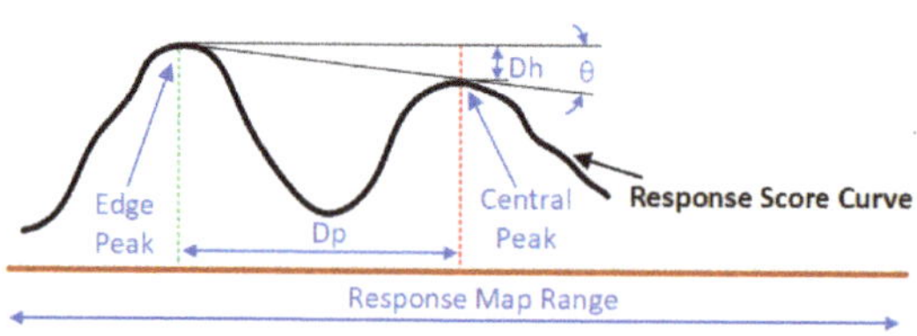

Figure 6. Object centre switching strategy.

Figure 7 shows that the peak angle θ changes throughout the entire sequence. In these plots, angles larger than 0 demonstrate that there is more than one peak in the response map and the central peak is higher than the edge peak. At the same time, when the angle is less than 0, this means the edge peak will have a higher score, and the rest of the points with an angle equal to 0 indicate only one peak in the response map, which means that no distractors can be found near to the true target. The figure demonstrates that the angle changes from a positive number to a negative number at approximately the 380th and 690th frames, and these frames are distractors that are moving close to the true target.

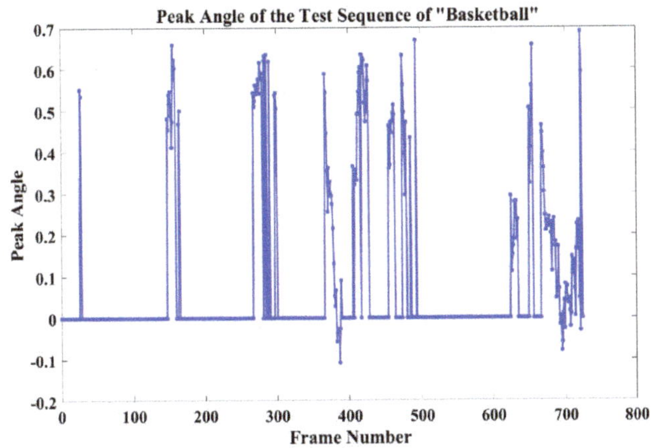

Figure 7. Peak angle plots of sequence 'Basketball' of OTB100.

3.2.4. Pseudo-Code of the Proposed Method

The possibility of distractors approaching and their positions can be calculated by using the above-introduced response behaviour analysis, the pseudo-code can be seen in Algorithm 1.

Algorithm 1: Proposed tracking method

Input: $I = \{i_n\}_{n=1}^{N}c$ (N is the total number of sequences)
for $i = 1, \ldots N$ **do**
  $R_i \leftarrow g_i(x, z)$ # Response map
  $PO_i \leftarrow \text{argmax} f(x) * f(z)$ # Target position offset
  $NR_i \leftarrow \frac{R_i}{\max(R_i)} \times 255$ #Normalisation
  $C_{i,j} \leftarrow NR_i$ # Find j isohypse contours
  if $j > 2$
    $d_{k,i,j} \leftarrow \sqrt{\left(x_{k,i}^{c1} - x_{k,j}^{c2}\right)^2 + \left(y_{k,i}^{c1} - y_{k,j}^{c2}\right)^2}$ # (Equation (3))
    $MD_k \leftarrow \frac{\sum_{k-M \le n \le k}(d_n - d_{n-1})}{M}$ # (Equation (5))
    if $MD_k < T_{md}$
      $flag_{distracotr} \leftarrow true$
   end if
   $\theta \leftarrow \arctan\left(\frac{Dh}{Dp}\right)$ # (Equation (6))
   if $\theta > T_{pa}$
    $PO_n = PO_i$
   end if
   else
    $PO_n = PO_{i-1}$ # use the position of previous frame
   end else
else
  $PO_n = PO_i$
end for

4. Experiments and Discussion

The approach in this study was implemented in Python using PyTorch on a PC with Intel i7, 32G RAM, NVIDIA GeForce RTX 3060. In this section, detailed results are provided. All tracking results are provided by official implementations in order to ensure a fair comparison.

4.1. Datasets

As a means of verifying the efficiency of the proposed method, it was evaluated using the well-known OTB100, GOT-10k and LaSOT tracking benchmarks. OTB100 [36] consists of 100 videos of 22 object categories with 11 tracking attributes. These attributes include abrupt motion, background clutter, blur and deformation. The average resolution of OTB100 is 356 × 530, while the length ranges between 71 and 3872 frames. GOT-10k [37] consists of 10,000 videos from the semantic hierarchy of WordNet [38]. This is divided into training, validation and test splits. The training split contains 9340 sequences with 480 object categories, while the test split contains 420 videos with 83 object categories, each sequence having an average length of 127 frames. LaSOT [39] is a high-quality benchmark that applies to large-scale single-object tracking. LaSOT consists of 1400 sequences with a total of over 3.5 million frames.

4.2. Evaluation Metrics

OTB100 evaluation is based on two metrics: precision plot and success plot.

The precision plot is based on the central location error, which is defined as the average Euclidean distance between the predicted centres of the target object and the ground truth centres in a frame. This is generated by plotting the distance precision over a range of thresholds. Distance precision is defined as the percentage of frames in which the target object is located within a centre location error of 20 pixels.

However, the precision plot does not reflect the size or scale accuracy of the target, so the IoU (Intersection over Union) is utilised for the measurement of prediction error. Given the estimated bounding box p and the ground truth bounding box g, IoU is defined as $(p \cap g)/(p \cup g)$. Therefore, the success rate is the percentage of frames in which the IoU is below a certain threshold, and the success plot is generated by varying the overlap threshold from 0 to 1.

For the GOT-10k dataset, the average overlap (AO) and success rate (SR) are utilised as the metrics. AO is measured by calculating the average of overlaps between all ground truth and predicted bounding boxes. SR is measured by calculating the percentage of successfully tracked frames where overlaps exceed a certain threshold. In the evaluation, AO is exploited for the overall performance ranking.

4.3. Implementation Details

Training: The parameters that were used in the training stage were the same as SiamFC. The ILSVRC15 dataset was used, and the training was performed over 50 epochs, each consisting of 50,000 sampled pairs. The gradients for each iteration were estimated using mini-batches of size 8, and the learning rate was annealed geometrically at each epoch from 10^{-2} to 10_{-5}.

Tracking: Unlike with classic SiamFC, the proposed response analysis was added to the tracking pipeline as a means of optimising tracking accuracy.

The peak angle threshold T_{pa} that was used in the peak switching strategy was determined through experiments (as can be seen in Figure 8) and was set to 0.15. The T_{md} in distractor approaching analysis was set to 15, and the frame number M was set to 10.

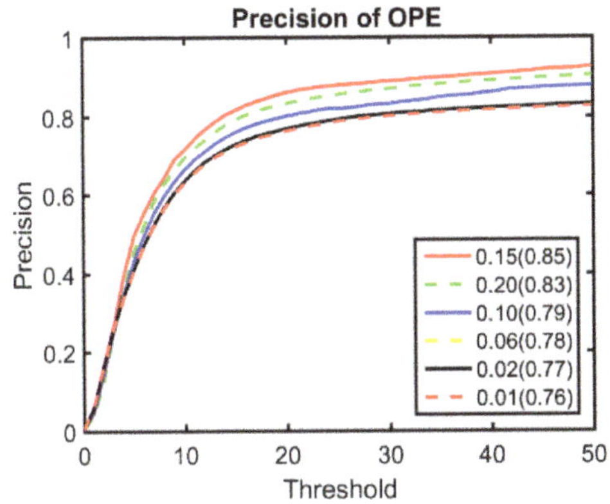

Figure 8. Experimental results of different angle thresholds of the OTB100 benchmark.

4.4. Performance Evaluation

For evaluating the performance of the proposed method, which is known as SiamFC-RBA, the tracker was compared against six different trackers: SiamFC, SiamRPN, DaSiamRPN, CFNet, CSK and Staple. SiamFC is the classic Siamese tracker, and SiamRPN is an advanced Siamese tracker that exhibits state-of-the-art performance.

In addition, the response behaviour analysis module was embedded into DiMP, named DiMP-RBA, as a means of testing the effectiveness of the proposed response behaviour analysis module in the majority of response map-based trackers.

The precision and success plots of OTB100 can be seen in Figure 9. The results demonstrate that the two DiMP-based trackers performed better than the others, and the DiMP-RBA that uses the proposed response behaviour analysis method was approximately 0.2% more precise than the original DiMP. The precision of SiamFC-RBA was 10% greater than that of the classic SiamFC and approximately 0.2% higher than that of SiamRPN, but it was 0.3% lower than the state-of-the-art tracker DaSiamRPN. Although the result of the tracker in this study is almost at the same level as DaSiamRPN, as the training structure was not rebuilt, the same training result as classic SiamFC was used and the update strategy was modified during the tracking process, the performance is still quite impressive. The same phenomenon can be observed in the success plot result. It can also be seen that the proposed method, DaSiamRPN and SiamRPN performed far better than the four other trackers (SiamFC, CFNet, CSK and Staple).

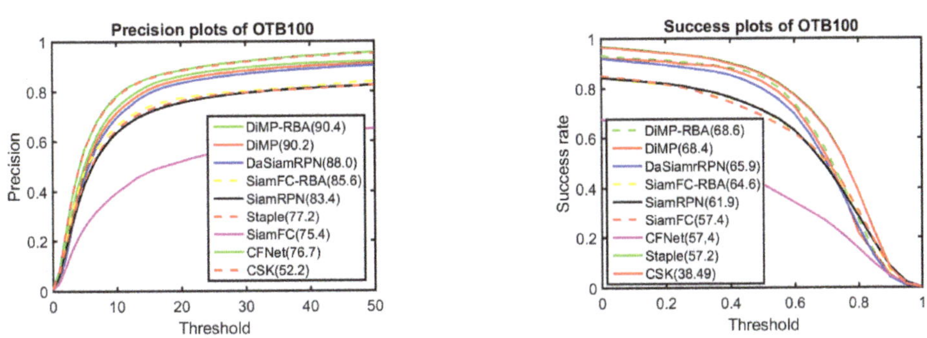

(**a**) (**b**)

Figure 9. Comparison between the proposed method and baseline trackers on the OTB100 benchmark: (**a**) precision plots; (**b**) success plots.

The comparison results for GOT-10k are shown in Figure 10. The performances of the two DiMP-based trackers were far better than those of the other trackers, and the DiMP-RBA which utilises the proposed response behaviour analysis method had approximately 0.9% better precision than the original DiMP. The overall scores of the tracker in this study and SiamRPN are almost identical (0.517) and far better than those of the four other trackers. Although the proposed tracker and SiamRPN have similar scores, they exhibit different performance patterns. When the overlap threshold was below 40, the tracker in this study exhibited better performance than SiamRPN, whereas SiamRPN was better in the opposite situation.

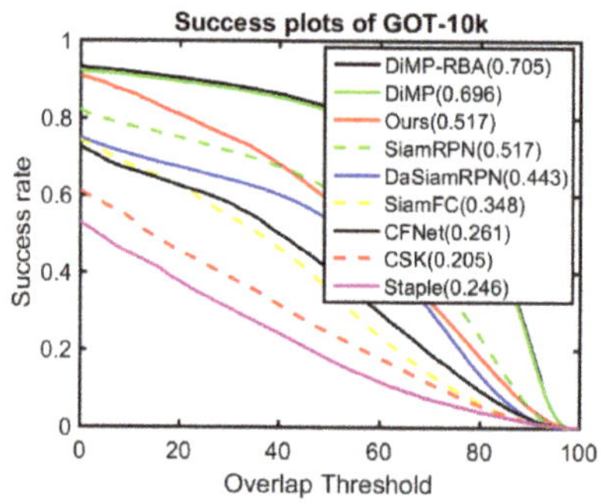

Figure 10. Comparison between the proposed method and baseline trackers on the GOT-10k benchmark.

The comparison results of LaSOT can be seen in Figure 11. The overall trend is the same as for the other two benchmarks. The performance of DiMP-RBA was improved by approximately 1% following the use of response behaviour analysis, and SiamFC-RBA demonstrated both a higher precision and success rate than the original SiamFC and the remaining trackers.

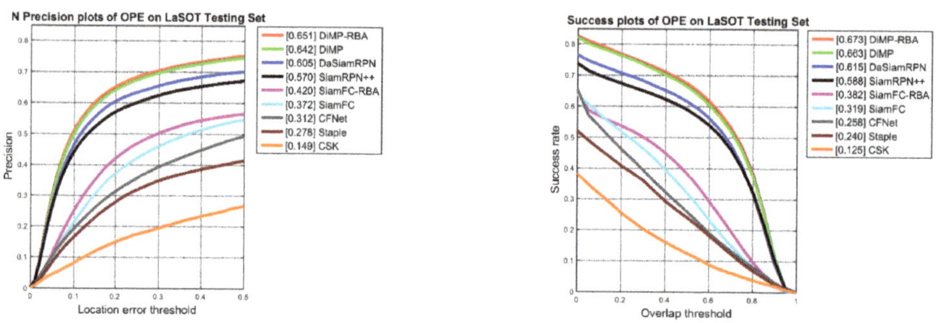

(**a**) (**b**)

Figure 11. Comparison between the proposed method and baseline trackers on the LaSOT benchmark: (**a**) precision plots; (**b**) success plots.

The experiment results are shown in Table 1, and the qualitative results for some typical challenging scenarios are shown in Figure 12. The performance of the proposed method was the same as that of DaSiamRPN. The tracker in this study was better with the GOT-10k benchmark than DaSiamRPN, while the opposite is true with the OTB100 benchmark. As has previously been mentioned, as the training structure was not rebuilt, meaning that the same training result as the classic SiamFC was used and the update strategy was just modified during the tracking process, the performance of the proposed method showed a different method for addressing the problem of tracking drift in background clutter scenarios. This proves that this type of strategy can also be used to achieve the state-of-the-art level.

Table 1. Comparison between the proposed method and baseline trackers on the OTB100, GOT-10k and LaSOT benchmarks. Red and blue indicate the two trackers that use the proposed response behaviour analysis method.

Trackers	OTB100		GOT-10k		LaSOT		FPS
	Precision	Success	AO	SR	Precision	Success	
DiMP-BRA	0.904	68.6	0.705	0.819	0.651	0.673	15.1
DiMP	0.902	68.4	0.696	0.816	0.642	0.663	15.2
DaSiamRPN	0.88	65.9	0.444	0.53	0.605	0.615	134.4
SiamFC-RBA	0.85	62.6	0.517	0.584	0.420	0.382	42.7
SiamRPN	0.83	61.9	0.517	0.615	0.570	0.588	3.17
SiamFC	0.77	57.4	0.348	0.353	0.372	0.319	43.8
CFNet	0.76	57.4	0.261	0.243	0.312	0.258	2541
Staple	0.77	38.49	0.246	0.248	0.278	0.240	28.7
CSK	0.52	57.2	0.205	0.174	0.149	0.125	133.3

CFNet; blue—SiamFC; pink—SiamRPN; red—Staple; green—SiamFC-RBA

Figure 12. Qualitative results for some typical challenging scenarios.

In addition to the benchmark dataset evaluation, the algorithm was implemented using the online real-time video stream of the surveillance camera that is installed in our laboratory. In this scenario, two men wearing similar clothes walked into the lab and crossed paths several times. The distractor approaching process can be seen in the corresponding response map in Figure 13, with the tracker still performing well in most cases. Due to the data of the new scenario not being included in the benchmark and the situation with real-time online application, the performance of the method that was used in this study was not compared to other trackers.

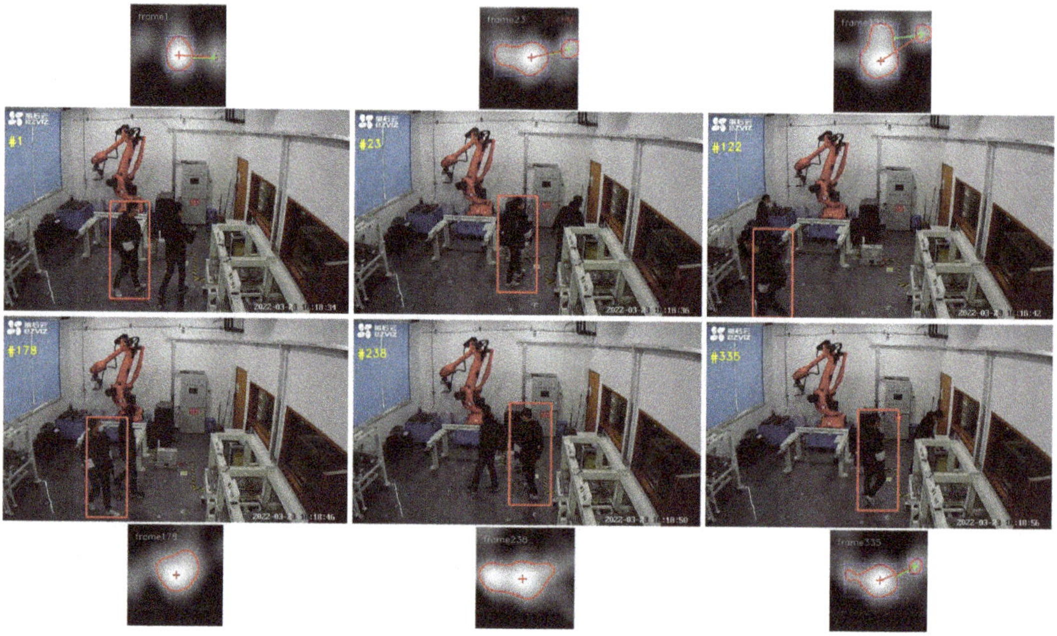

Figure 13. Human tracking with similar objects online in real time.

5. Conclusions

This paper proposes an improved SiamFC tracker that is based on response map analysis as a means of addressing the problem of tracking drift in background clutter scenarios. The key point of this method is that it can be used for judging whether there are distractors near the real target by analysing the behaviour of the response map and by updating the target positioning strategy on the basis of this information. Extensive experiments on visual tracking benchmarks including OTB100, GOT-10k and LaSOT found that by using the proposed method, in comparison to the original SiamFC, the precision performance of SiamFC-RBA increased by approximately 8%, 16% and 5%, respectively, while also outperforming SiamRPN, CSK, CFNet and Staple. The response behaviour analysis module was also embedded into DiMP, which is known as DiMP-RBA, for testing the effectiveness of the proposed response behaviour analysis module in most response map-based trackers. The experimental results found that DiMP-RBA outperformed the original DiMP by 0.2%, 0.9% and 0.9%, respectively, in the three benchmarks. Although the DiMP improvement was relatively small, this proved that the proposed response behaviour analysis module can be embedded into other response map- or score map-based trackers as a means of improving tracking performance.

Author Contributions: Conceptualisation, X.H. and S.C.; methodology, S.C.; software, T.S.; validation, X.H., C.D. and S.C.; formal analysis, Z.X.; investigation, T.S.; resources, X.H.; data curation, T.S.; writing—original draft preparation, X.H.; writing—review and editing, X.H. and S.C.; visualisation, X.H.; supervision, S.C.; project administration, S.C.; funding acquisition, S.C. All authors have read and agreed to the published version of the manuscript.

Funding: This research was funded by the China Scholarship Council, number 202108330169, and the Science and Technology Plan Project of Jiangsu Provincial Market Supervision Administration, number KJ21125091.

Institutional Review Board Statement: Not applicable.

Informed Consent Statement: Not applicable.

Data Availability Statement: Not applicable.

Conflicts of Interest: The authors declare that there are no conflict of interest.

Source Code: The code for this paper is available at: https://github.com/cramkl/SiamFC-RBA, accessed on 27 August 2022.

Abbreviations

Symbols and abbreviations	Full meaning
p_t	Target template patch
p_s	Search patch
ρ	Learnable parameter of Siamese trackers
b	Scalar offset value
q	Central position of the target
$g_\rho(p_t, p_s)$	Response map denoting the similarity between p_t and p_s
$x_{k,i}^{c1}$	x coordinate of the ith point in the $c1$ contour of the kth frame
$d_{k,i,j}$	Distance between the ith point of contour $c1$ and the jth point of contour $c2$ in the kth frame
MD_k	Mean distance before the kth frame
R_i	Response map of the ith frame
NR_i	Normalised response map of the ith frame
$C_{i,j}$	The jth contour of the ith frame
T_{md}	Distance threshold
θ	Angle between the two highest peaks
AO	Average overlap
SR	Success rate
IoU	Interaction over Union
FPS	Frame per second

References

1. Duer, S.; Bernatowicz, D.; Wrzesień, P.; Duer, R. The diagnostic system with an artificial neural network for identifying states in multi-valued logic of a device wind power. In Proceedings of the International Conference: Beyond Databases, Architectures and Structures, Poznan, Poland, 18–20 September 2018; pp. 442–454.
2. Majewski, M.; Kacalak, W. Smart control of lifting devices using patterns and antipatterns. In Proceedings of the Computer Science Online Conference, Prague, Czech Republic, 26–29 April 2017; pp. 486–493.
3. Duer, S.; Zajkowski, K.; Płocha, I.; Duer, R. Training of an artificial neural network in the diagnostic system of a technical object. *Neural Comput. Appl.* **2013**, *22*, 1581–1590. [CrossRef]
4. Duer, S.; Zajkowski, K. Taking decisions in the expert intelligent system to support maintenance of a technical object on the basis information from an artificial neural network. *Neural Comput. Appl.* **2013**, *23*, 2185–2197. [CrossRef]
5. Kacalak, W.; Majewski, M. New intelligent interactive automated systems for design of machine elements and assemblies. In Proceedings of the International Conference on Neural Information Processing, Doha, Qatar, 12–15 November 2012; pp. 115–122.
6. Marvasti-Zadeh, S.M.; Cheng, L.; Ghanei-Yakhdan, H.; Kasaei, S. Deep Learning for Visual Tracking: A Comprehensive Survey. *IEEE Trans. Intell. Transp. Syst.* **2022**, *23*, 3943–3968. [CrossRef]
7. Wang, L.; Ouyang, W.; Wang, X.; Lu, H. Visual tracking with fully convolutional networks. In Proceedings of the IEEE International Conference on Computer Vision, Santiago, Chile, 11–18 December 2015; pp. 3119–3127.
8. Nam, H.; Han, B. Learning multi-domain convolutional neural networks for visual tracking. In Proceedings of the IEEE Conference on Computer Vision and Pattern Recognition, Las Vegas, NV, USA, 27–30 June 2016; pp. 4293–4302.
9. Yang, T.; Chan, A.B. Recurrent filter learning for visual tracking. In Proceedings of the IEEE International Conference on Computer Vision Workshops, Venice, Italy, 22–29 October 2017; pp. 2010–2019.
10. Ma, D.; Bu, W.; Wu, X. Multi-Scale Recurrent Tracking via Pyramid Recurrent Network and Optical Flow. In Proceedings of the BMVC, Newcastle, UK, 3–6 September 2018; p. 242.
11. Song, Y.; Ma, C.; Wu, X.; Gong, L.; Bao, L.; Zuo, W.; Shen, C.; Lau, R.W.; Yang, M.-H. Vital: Visual tracking via adversarial learning. In Proceedings of the IEEE Conference on Computer Vision and Pattern Recognition, Salt Lake City, UT, USA, 18–23 June 2018; pp. 8990–8999.
12. Guo, J.; Xu, T.; Jiang, S.; Shen, Z. Generating reliable online adaptive templates for visual tracking. In Proceedings of the 2018 25th IEEE International Conference on Image Processing (ICIP), Athens, Greece, 7–10 October 2018; pp. 226–230.
13. Li, B.; Wu, W.; Wang, Q.; Zhang, F.; Xing, J.; Yan, J. SiamRPN++: Evolution of Siamese Visual Tracking With Very Deep Networks. In Proceedings of the 2019 IEEE/CVF Conference on Computer Vision and Pattern Recognition (CVPR), Long Beach, CA, USA, 16–20 June 2019; pp. 4277–4286.

14. Chen, Z.; Zhong, B.; Li, G.; Zhang, S.; Ji, R. Siamese Box Adaptive Network for Visual Tracking. In Proceedings of the 2020 IEEE/CVF Conference on Computer Vision and Pattern Recognition (CVPR), Seattle, WA, USA, 13–19 June 2020; pp. 6667–6676.
15. Zhu, Z.; Wang, Q.; Li, B.; Wu, W.; Yan, J.; Hu, W. Distractor-aware Siamese Networks for Visual Object Tracking. In Proceedings of the European Conference on Computer Vision, Marseille, France, 12–18 October 2008.
16. Guo, Q.; Wei, F.; Zhou, C.; Rui, H.; Song, W. Learning Dynamic Siamese Network for Visual Object Tracking. In Proceedings of the 2017 IEEE International Conference on Computer Vision (ICCV), Venice, Italy, 22–29 October 2017.
17. Bertinetto, L.; Valmadre, J.; Henriques, J.F.; Vedaldi, A.; Torr, P. Fully-Convolutional Siamese Networks for Object Tracking. In Proceedings of the European Conference on Computer Vision, Amsterdam, The Netherlands, 11–14 October 2016.
18. Bo, L.; Yan, J.; Wei, W.; Zheng, Z.; Hu, X. High Performance Visual Tracking with Siamese Region Proposal Network. In Proceedings of the 2018 IEEE/CVF Conference on Computer Vision and Pattern Recognition (CVPR), Salt Lake City, UT, USA, 18–23 June 2018.
19. Valmadre, J.; Bertinetto, L.; Henriques, J.F.; Vedaldi, A.; Torr, P. End-to-End Representation Learning for Correlation Filter Based Tracking. In Proceedings of the IEEE Conference on Computer Vision and Pattern Recognition, Honolulu, HI, USA, 21–26 July 2017.
20. Henriques, J.F.; Caseiro, R.; Martins, P.; Batista, J. Exploiting the circulant structure of tracking-by-detection with kernels. In Proceedings of the European Conference on Computer Vision, Florence, Italy, 7–13 October 2012; Springer: Berlin/Heidelberg, Germany, 2012; pp. 702–715.
21. Bertinetto, L.; Valmadre, J.; Golodetz, S.; Miksik, O.; Torr, P.H. Staple: Complementary learners for real-time tracking. In Proceedings of the IEEE Conference on Computer Vision and Pattern Recognition, Las Vegas, NV, USA, 27–30 June 2016; pp. 1401–1409.
22. Bromley, J.; Guyon, I.; LeCun, Y.; Säckinger, E.; Shah, R. Signature verification using a "siamese" time delay neural network. *Adv. Neural Inf. Processing Syst.* **1993**, *6*, 737–744. [CrossRef]
23. Tao, R.; Gavves, E.; Smeulders, A.W. Siamese instance search for tracking. In Proceedings of the IEEE Conference on Computer Vision and Pattern Recognition, Las Vegas, NV, USA, 27–30 June 2016; pp. 1420–1429.
24. Zhu, Z.; Wu, W.; Zou, W.; Yan, J. End-to-end flow correlation tracking with spatial-temporal attention. In Proceedings of the IEEE Conference on Computer Vision and Pattern Recognition, Salt Lake City, UT, USA, 18–23 June 2018; pp. 548–557.
25. Wang, Q.; Teng, Z.; Xing, J.; Gao, J.; Hu, W.; Maybank, S. Learning attentions: Residual attentional siamese network for high performance online visual tracking. In Proceedings of the IEEE Conference on Computer Vision and Pattern Recognition, Salt Lake City, UT, USA, 18–23 June 2018; pp. 4854–4863.
26. Marvasti-Zadeh, S.M.; Khaghani, J.; Cheng, L.; Ghanei-Yakhdan, H.; Kasaei, S. CHASE: Robust Visual Tracking via Cell-Level Differentiable Neural Architecture Search. In Proceedings of the BMVC, Online, 22–25 November 2021.
27. Marvasti-Zadeh, S.M.; Khaghani, J.; Ghanei-Yakhdan, H.; Kasaei, S.; Cheng, L. COMET: Context-aware IoU-guided network for small object tracking. In Proceedings of the Asian Conference on Computer Vision, Kyoto, Japan, 30 November–4 December 2020.
28. Chen, X.; Yan, B.; Zhu, J.; Wang, D.; Yang, X.; Lu, H. Transformer tracking. In Proceedings of the IEEE/CVF Conference on Computer Vision and Pattern Recognition, Nashville, TN, USA, 20–25 June 2021; pp. 8126–8135.
29. Mayer, C.; Danelljan, M.; Bhat, G.; Paul, M.; Paudel, D.P.; Yu, F.; Van Gool, L. Transforming model prediction for tracking. In Proceedings of the IEEE/CVF Conference on Computer Vision and Pattern Recognition, New Orleans, LA, USA, 19–24 June 2022; pp. 8731–8740.
30. Li, X.; Ma, C.; Wu, B.; He, Z.; Yang, M.-H. Target-aware deep tracking. In Proceedings of the IEEE/CVF Conference on Computer Vision and Pattern Recognition, Long Beach, CA, USA, 15–20 June 2019; pp. 1369–1378.
31. Yang, L.; Jiang, P.; Wang, F.; Wang, X. Region-based fully convolutional siamese networks for robust real-time visual tracking. In Proceedings of the 2017 IEEE International Conference on Image Processing (ICIP), Beijing, China, 17–20 September 2017; pp. 2567–2571.
32. Dai, K.; Wang, Y.; Yan, X. Long-term object tracking based on siamese network. In Proceedings of the 2017 IEEE International Conference on Image Processing (ICIP), Beijing, China, 17–20 September 2017; pp. 3640–3644.
33. Voigtlaender, P.; Luiten, J.; Torr, P.H.; Leibe, B. Siam r-cnn: Visual tracking by re-detection. In Proceedings of the IEEE/CVF Conference on Computer Vision and Pattern Recognition, Seattle, WA, USA, 13–19 June 2020; pp. 6578–6588.
34. Bhat, G.; Danelljan, M.; Gool, L.V.; Timofte, R. Learning Discriminative Model Prediction for Tracking. In Proceedings of the International Conference on Computer Vision, Seoul, Korea, 27 October–2 November 2019.
35. Danelljan, M.; Bhat, G.; Khan, F.S.; Felsberg, M. Atom: Accurate tracking by overlap maximization. In Proceedings of the IEEE/CVF Conference on Computer Vision and Pattern Recognition, Long Beach, CA, USA, 15–20 June 2019; pp. 4660–4669.
36. Wu, Y.; Lim, J.; Yang, M. Object Tracking Benchmark. *IEEE Trans. Pattern Anal. Mach. Intell.* **2015**, *37*, 1834–1848. [CrossRef] [PubMed]
37. Huang, L.; Zhao, X.; Huang, K. Got-10k: A large high-diversity benchmark for generic object tracking in the wild. *IEEE Trans. Pattern Anal. Mach. Intell.* **2019**, *43*, 1562–1577. [CrossRef] [PubMed]
38. Miller, G.A. WordNet: A lexical database for English. *Commun. ACM* **1995**, *38*, 39–41. [CrossRef]
39. Fan, H.; Bai, H.; Lin, L.; Yang, F.; Chu, P.; Deng, G.; Yu, S.; Huang, M.; Liu, J.; Harshit; et al. LaSOT: A High-quality Large-scale Single Object Tracking Benchmark. *Int. J. Comput. Vis.* **2021**, *129*, 439–461. [CrossRef]

sensors

MDPI

Article

Interactive Application of Data Glove Based on Emotion Recognition and Judgment System

Wenqian Lin [1,*], Chao Li [2] and Yunjian Zhang [3]

1 School of Media and Design, Hangzhou Dianzi University, Hangzhou 310018, China
2 College of Computer Science and Technology, Zhejiang University, Hangzhou 310027, China
3 College of Control Science and Technology, Zhejiang University, Hangzhou 310027, China
* Correspondence: jiangnanshui253@126.com

Abstract: In this paper, the interactive application of data gloves based on emotion recognition and judgment system is investigated. A system of emotion recognition and judgment is established based on the set of optimal features of physiological signals, and then a data glove with multi-channel data transmission based on the recognition of hand posture and emotion is constructed. Finally, the system of virtual hand control and a manipulator driven by emotion is built. Five subjects were selected for the test of the above systems. The test results show that the virtual hand and manipulator can be simultaneously controlled by the data glove. In the case that the subjects do not make any hand gesture change, the system can directly control the gesture of the virtual hand by reading the physiological signal of the subject, at which point the gesture control and emotion control can be carried out at the same time. In the test of the manipulator driven by emotion, only the results driven by two emotional trends achieve the desired purpose.

Keywords: human-computer interactive; data glove; virtual hand; emotion driven; test

Citation: Lin, W.; Li, C.; Zhang, Y. Interactive Application of Data Glove Based on Emotion Recognition and Judgment System. *Sensors* **2022**, *22*, 6327. https://doi.org/10.3390/s22176327

Academic Editor: Stefano Berretti

Received: 14 July 2022
Accepted: 22 August 2022
Published: 23 August 2022

1. Introduction

Though virtual reality (VR) is a way for human beings to interact with computers and complex data, its main purpose is to allow users to enter the virtual environment, wherein they can have the same experience and feeling as in real life. VR involves many fields and advanced technologies.

VR systems can be divided by different aspects. In terms of system functionality, the essential function of a VR system is environment simulation, so it can be applied to many fields such as military, medicine, and so on. At present, there are three kinds of VR systems: (1) systems used for simulation exercise or training in military field, (2) systems for planning and designing places and environment in the field of architecture, and (3) entertainment equipment and high-immersion systems in the entertainment field. In terms of interaction mode and user immersion mode, VR systems can be divided into non-interactive experience, human-virtual environment interactive experience, and group-virtual environment interactive experience. In terms of data input channels, VR can be divided into platform data, model data, perception data, and control data. In terms of interaction mode and interaction equipment, VR can be divided into four types: scene display, force/touch interaction, tracking and positioning, and walking interaction. The scene display type includes a helmet such as the popular VR glasses, desktops, projections, handhelds, and free stereoscopic displays. The force/touch interaction type includes the data glove with transmission functions, joysticks with force feedback, etc. The tracking and positioning type includes source and non-source tracking and positioning systems. The walking interaction type includes pedal walking and ground walking. In the design of VR system, attention should be paid to the elements of multi-perception, immersion, interaction, and imagination space.

Hand gesture recognition is an interactive type of VR system that relies on sensor technologies such as the electromyographic (EMG) and inertial measurement unit (IMU). There have been numerous studies on hand gesture recognition based on EMG and IMU. For example, Kundu et al. [1] presented a hand gesture based control of an omnidirectional wheelchair using IMU and myoelectric units as wearable sensors, and recognized and classified seven common gestures using a shape-based feature extraction and a Dendrogram Support Vector Machine (DSVM) classifier. Classification involved recognizing the activity pattern based on periodic shape of trajectories of the triaxial wrist tilt angle and EMG-RMS from the two selected muscles. Classification accuracy of 94% was achieved by DSVM classifier on 'k' fold cross validation data of 5 users. Zhang et al. [2] computed a deep learning technique known as the long short-term memory (LSTM) algorithm to build a model to classify hand gestures by training and testing the collected IMU, EMG, and finger and palm pressure data. The experimental results showed an outstanding performance of the LSTM algorithm. Song et al. [3] proposed a force myography (FMG), EMG, and IMU-based multi-sensor fusion model for hand motion classification, and evaluated the feasibility by motion classification accuracy and qualitative of subjects' questionnaires. They showed that the offline classification accuracy of adopting combined FMG-EMG-IMU was 81.0% for the 12 motions, which was obviously higher than single sensing modality; that is, only EMG, FMG, and IMU were 69.6, 63.2, and 47.8%, respectively. Jiang et al. [4] presented the design and validation of a real-time gesture recognition wristband based on surface EMG and IMU sensing fusion, which can recognize 8 air gestures and 4 surface gestures with 2 distinct force levels. The results showed that classification accuracies for the initial experiment were 92.6% and 88.8% for air and surface gestures, respectively, and there were no changes in accuracy results during testing 1 h and 1 day later. Yang et al. [5] applied the multivariate variational mode decomposition to extract the spatial-temporal features from the multiple channels to the EMG signals and used the separable, convolutional neural network for modeling by proposing an extensible two-stage machine learning lightweight framework for multi-gesture task recognition. The experimental results for a 52 hand gestures recognition task showed that the average accuracy on each stage is about 90%. Alfaro and Trejos [6] presented a user-independent gesture classification method combing EMG data and IMU data. They obtained average classification accuracies in the range of 67.5–84.6%, with the Adaptive Least-Squares Support Vector Machine model obtaining accuracies as high as 92.9%. Wu et al. [7] proposed a wearable system for recognizing American Sign Language (ASL) by fusing information from an inertial sensor and surface EMG sensors. Four popular classification algorithms were evaluated for 80 commonly used ASL signs on four subjects. The results showed 96.16% and 85.24% average accuracies for intra-subject and intra-subject cross session evaluation, respectively, with the selected feature subset and a support vector machine classifier. Shin et al. [8] studied a myoelectric interface that controls a robotic manipulator via neuromuscular electrical signals generated when humans make hand gestures. They proposed a system that recognizes dynamic hand motions and configuration of a hand over time. The results showed that the average real-time classification accuracy of the myoelectric interface was over 95.6%. Shahzad et al. [9] studied the effects of surface EMG signal variation on the performance of a hand motion classifier due to arm position variation, and explored the effect of static position and dynamic movement strategies for classifier training. A wearable system was made position aware (POS) using IMU for different arm movement gestures. The results showed the effectiveness of the dynamic training approach and sensor fusion techniques to improve the performance of existing stand-alone surface EMG-based prosthetic control systems. Ordóñez Flores et al. [10] proposed a new methodology and showed its particular application to the recognition of five hand gestures based on 8 channels of electromyography using a Myo armband device placed on the forearm. Romero et al. [11] presented the application of hand gestures and arm movements to control a dual rotor testbench. Chico et al. [12] employed a hand gesture recognition system and the inertial measurement unit integrated in the Myo armband sensor as a human-machine interface to control the position and orientation of a virtual six-degree-of-freedom (DoF) UR5 robot.

Hand gesture recognition mainly includes two methods. One is gesture recognition based on data gloves (i.e., the motion characteristics such as the bending degree), angle, and displacement of each key joint of the hand are obtained through the motion sensor and are then inversed to the system database as much as possible. The other is image-based gesture recognition (i.e., the image data of the hand are collected through camera), wherein the background segmentation and motion modeling are carried out through image recognition, and the hand motion is ultimately restored in the computer. The above two methods have their own advantages and disadvantages. Data gloves need subjects to wear external equipment, which may affect the user interaction experience and have delay in data processing, but they have strong anti-interference to data acquisition, more accurate data acquisition, and are not easily affected by the external environment. The image recognition method is more convenient, and the user's operation is more natural, but it has certain requirements for the environment and is easy to be disturbed by environmental factors. In this paper, the data glove is selected as the interactive device because its data acquisition is more accurate and the sensor used in this paper must contact the user's hand to obtain the physiological signal. In addition, data gloves are easy to implement modification measures, such as adding additional sensors, and have more advantages and pertinence than other interactive devices in hand movement.

Some achievements have been made in the research and development of data gloves, such as 5DT data gloves, cyberglove force feedback data gloves, measurand high-precision data gloves, X-IST music simulation data gloves, etc. Tarchanidis et al. [13] presented a data glove equipped with a force sensor with a resolution of 0.38 N and a sensitivity of 0.05 V/N. Kamel et al. [14] implement data glove from motion animation to signature verification and showed a high accuracy in finding the similarities between genuine samples as well as those differentiated between genuine-forgery trials. Yoon et al. [15] presented a data glove with adaptive mixture-of-experts model and showed the excellent performance and adaptability through tests. Kim et al. [16] used a data glove to present a sign language recognition system and indicated that the system was useful when employed to smartphones in some situations. Chen et al. [17] presented a data glove with highly stretchable conductive fiber strain sensor, which could recognize various gestures by detecting the finger motion. Fang [18] proposed a data glove to recognize and capture the gestures of 3-D arm motion, and the test results verified its effectiveness. Lin et al. [19] presented a data glove with characteristics of low cost, high reliability, and easy wearability. Wang et al. [20] presented a data glove with the feedback force control of a safe, lightweight, yet powerful and stable passive force feedback. Li [21] developed a data glove to monitor the hand posture and operated the division between sensor and base signal to decrease the test error induced by instability of light sources. Wu et al. [22] presented a data glove for catching finger joint angles and tested its effectiveness. Sarwat et al. [23] used a data glove to construct an automated assessment system for in-home rehabilitation, helping poststroke patients with a high level of recovery. Takigawa et al. [24] developed a controlled functional electrical stimulation to realize multiple grasping postures with data glove.

The previous research on data gloves has mostly focused on improving the accuracy of motion recognition and pressure simulation of force feedback. However, the study on the data glove which can capture the user's behavior and obtain the user's emotion through physiological signal sensor is rare. Therefore, in this paper, a kind of data glove with functions of emotion recognition and interaction between human and computer, or a human and hardware device according to the user's emotion, is presented. The data glove can be used in medicine, health, military training, academic research, and other fields.

2. Classification of Emotion Trends

In order to obtain user's emotion through physiological signal sensor, a system of emotion recognition is needed, while emotion recognition is based on the emotion evaluation [25]. Here the valence-arousal (V-A) model is used for the emotion classification. In the V-A model, as shown in Figure 1, V and A indicate the degree of emotional pleasure and emotional arousal, respectively. Four poles of the emotion classification model are extracted and used to represent tired, tense, happy, and depressed, respectively. The emotion classification system based on the V-A model is extended to a plane, and four quadrants of the plane stand for high-arousal and positive-valence (quadrant I: HAPV), high-arousal and negative-valence (quadrant II: HANV), low-arousal and negative-valence (quadrant III: LANV), and low-arousal and positive-valence (quadrant IV: LAPV), respectively.

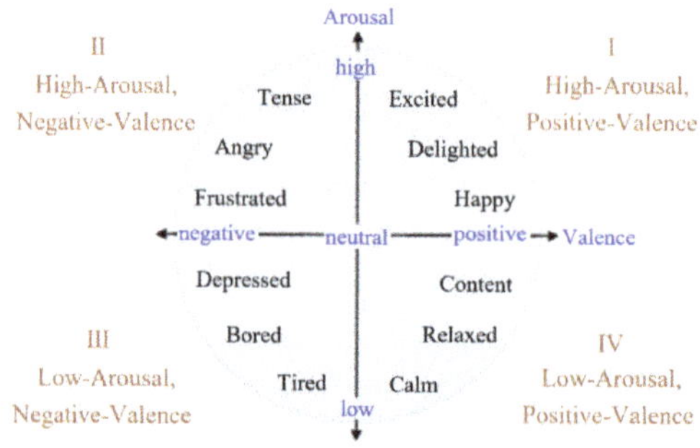

Figure 1. Valence-arousal model.

3. System of Emotion Recognition and Judgment

3.1. Data Analysis of Physiological Signal (PS)

In the present study, skin electricity and pulse wave are taken as PS. The former is easily disturbed by other signals, so the noise interference should be removed before advancing. In order to facilitate computer analysis and processing, the discrete wavelet transform is used to decompose the signal into different frequency bands through low-pass and high-pass filtering. The unit of the frequency used for the filter is Hertz. The wdencmp function in MATLAB 9.0 R2016a is used to denoise the skin electrical signal, and all segments of skin electrical signal were normalized within the range of 0 to 100.

As shown in Figure 2, the signal of pulse wave is composed of main wave, dicrotic anterior wave, dicrotic notch and dicrotic wave. In the figure, the key feature points include: (1) c (peak systolic pressure), (2) e (starting point of left ventricular diastole), (3) g (maximum pressure point of anti tide wave), (4) d (point of aortic dilation depressurization), (5) f (origin of anti tide wave), and (6) b1 (point of aortic valve opening). The key amplitude includes: (1) main wave h1, (2) dicrotic anterior wave h2, (3) dicrotic notch h3, and (4) dicrotic wave h4. The key time includes: (1) the time from the starting point of waveform period to the peak c point of main wave t1, (2) the time from the starting point of waveform cycle to the lowest point of dicrotic notch t2, and (3) duration of one waveform period t. The pulse wave is smoothed and filtered using Butterworth low-pass filter and the relevant parameters of pulse wave are normalized after filtering.

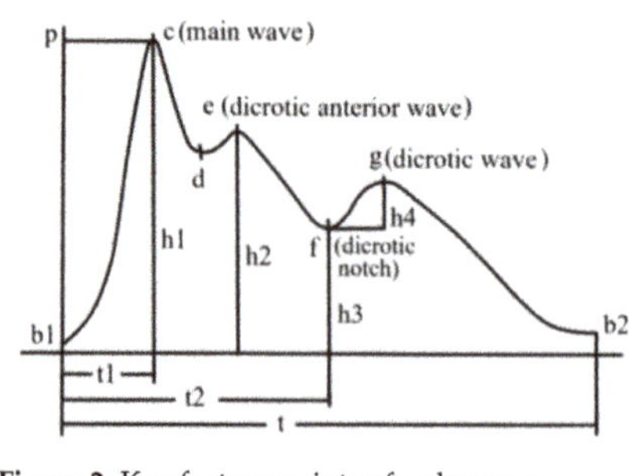

Figure 2. Key feature points of pulse wave.

3.2. Extraction of Optimal Feature of PS

Features of PS are divided into a time domain, a frequency domain, and a feature related with physiological processes [26]. The direct fusion of original signal features will result in too much computation. As such, the dimensionality reduction of original signal feature is performed using the method of principal component analysis (PCA) to make the classifier more efficient and accurate in emotion recognition. Principal components are obtained using PCA, and then the weight threshold of each feature of PS on the principal component is taken as the criterion for selecting feature. Finally, some original features that play a major role can be determined as optimal feature subset. After obtaining optimal feature subset, the Pearson correlation coefficient (PCC) is used to judge the relationship between the emotional interval and these features. The PCC is calculated for features of four emotion trends and can be used to draw the significance P of the features. Based on P and correlation coefficient, the normalized threshold of optimal features correlated with emotional trends is determined. These optimal features include BpNN50 (percentage of main pulse wave interval >50 ms), the "range" of skin electrical signal (the mean value of first order difference for skin electrical signal), and 1dmean (mean value of first order difference of skin electrical signal).

3.3. Establishment of the System of Emotion Judgment

The range of skin electrical signal has a high positive correlation between the two completely opposite emotional trends (i.e., HVLA and LVHA). As such, the skin electrical waveform corresponding to the emotional trend is studied. The results show that it is necessary to add a directional judgment to the range of skin electrical signal. Based on the set of optimal signal feature from the Pearson correlation coefficient, the system of emotion recognition and judgment can be built according to the process as shown in Figure 3.

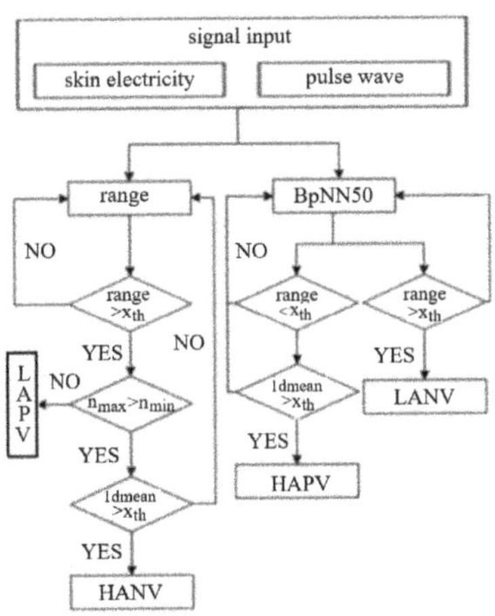

Figure 3. Process of emotion judgment model.

4. Design and Connection of Data Glove

The design framework of data glove with emotion recognition function is shown in Figure 4 where the data glove equipment consists of data acquisition and the controller. The data are acquired from the finger movement and physiological signal. The data glove customized based on DN-01 data module is taken as an example as shown in Figure 5, wherein the data module is an attitude acquisition board, which is used to collect the information of hand motion such as finger motion parameters, angular velocity of hand rotation, hand rotation acceleration, and angle change. The controller processes and integrates the collected information of hand motion, and then packages the processed data and sends it to the host computer for processing through Bluetooth or USB to serial port. The interface of the attitude acquisition board is connected with the sensor, and the acquisition board and the controller are connected by a flat cable as shown in Figure 6.

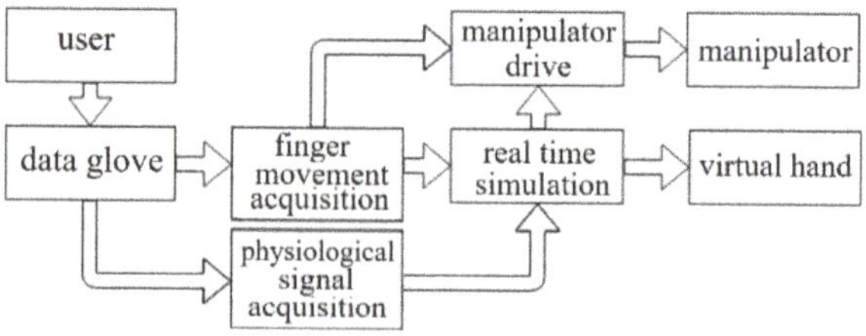

Figure 4. Design framework of data glove with emotion recognition function.

Figure 5. DN-1 composition of data glove module.

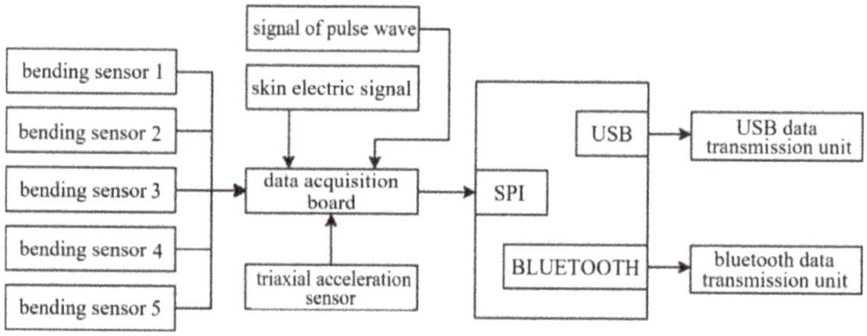

Figure 6. Data glove hardware structure framework.

The data acquisition module in the data glove mainly collects two kinds of data: one is gesture data, and the other is physiological signal data. The prototype design of the data glove is shown in Figure 7, where the sensor of gesture data is Flex2.2 bending sensor which can capture the bending degree of five fingers and the motion posture of the palm, including acceleration, angular velocity, and angle. The length of the bending sensor is 7.7 cm, the non-bending resistance is 9000 Ω, the 90-degree bending resistance is 14,000 Ω, and the 180-degree bending resistance is 22,000 Ω.

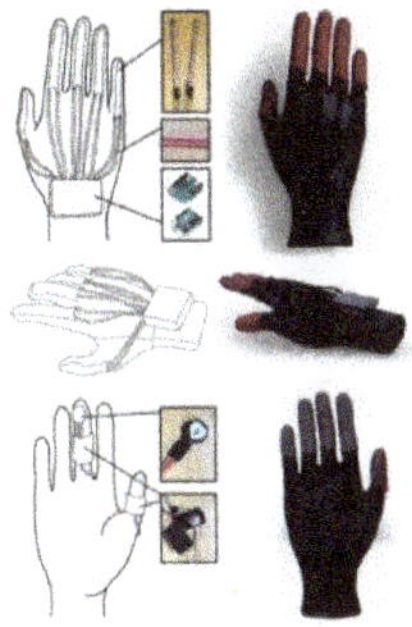

Figure 7. Prototype design of data glove based on emotion recognition of physiological signal.

The skin electrical signal is acquired using a Grove-GSR skin electrical kit, as shown in Figure 8 (left). Two finger sleeves containing electrodes were put on the middle part of the middle finger and the thumb of the left hand, and the frequency of signal sampling was 20 Hz. A pulse sensor, as shown in Figure 8 (right), was used to acquire the signals of the pulse wave and heart rate. The pulse sensor was fixed on the tip of the middle finger of the left hand with a bandage, and the frequency of signal sampling was 100 Hz.

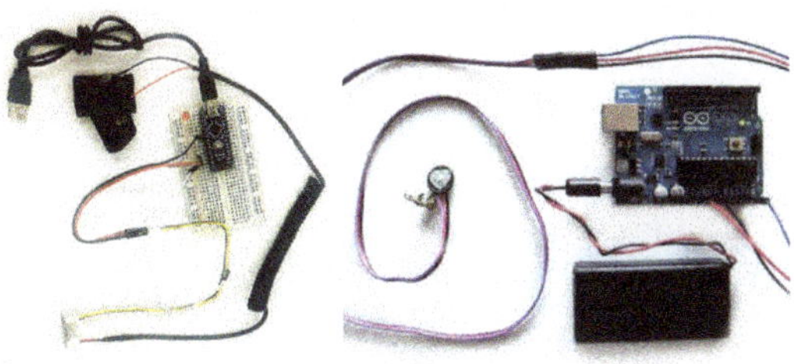

Figure 8. GSR skin electrical kit (**left**) and pulse sensor (**right**).

Gestures are reflected by the bending of fingers, and the output format of finger bending is: 0xaa a1 a2 a3 a4 a5 0xbb, where 0xaa and 0xbb are the head and tail of the frame, and a1, a2, a3, a4, and a5 represent the bending data of five fingers from thumb to little thumb, respectively. The data x read on the interface (a1–a5) is the quantization of the voltage value on the bending sensor:

$$x = \frac{V_x \times 3.3}{4096} \tag{1}$$

where V_x is voltage at sensor. Based on

$$V_x = 3.3 \times \frac{R}{(R+20)} \tag{2}$$

the resistance value R can be obtained. The value of R is proportional to the bending degree of the bending sensor—i.e., the values of a1, a2, a3, a4, and a5 are inversely proportional to the degree of finger bending.

The finger bending data can be analyzed through the following functions: (1) the resume function is used to determine whether the data related to the finger part is received correctly (i.e., the correctness of frame header and tail), (2) finger_calculate function is used to calculate the bending data of the finger, (3) judge function is used to determine whether the finger is bent within a reasonable range (i.e., filtering out wrong motion information), and (4) calculate function is used to process data related to finger bending including bending data of a single finger, storing and recording the data, calculating the offset of bending data, etc.

The acquisition sensors of physiological signal are the skin electrical sensor Grov-GSR skin electric kit (fixed on the middle part of the middle finger and the thumb of the glove) and the pulse sensor (fixed on the tip of the middle finger of the glove). The sensors are connected to the data acquisition module with a wire, and then the acquisition module is connected to the PC end and external hardware equipment through the Bluetooth interface. Unity3D receives the data information transmitted by the data acquisition board through the IO interface.

5. Test of Virtual Gesture Change Driven by Emotion

5.1. System Design

The data glove is used to control the hand gesture of the virtual hand and manipulator, and then the gesture of the virtual hand is changed through the awakened emotion of the subjects and compared with the gesture of the manipulator. The technological process of the system is shown in Figure 9, where DN-1 data glove is adopted and two kinds of sensors are added to DN-1 data glove.

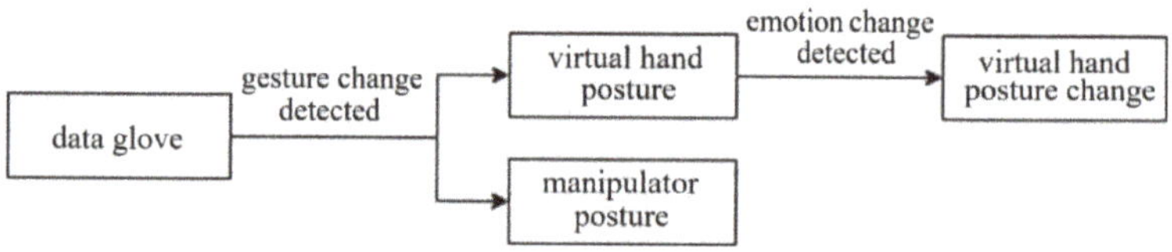

Figure 9. The technological process of the system that virtual gesture change driven by emotion.

The virtual hand model comes from network shared resources. The fingers of index finger, middle finger, ring finger, and little thumb have 5 movable joint points, respectively, and the thumb has four movable joint points. There are 24 movable joint points for changing the gesture of the hand, as shown in Figure 10.

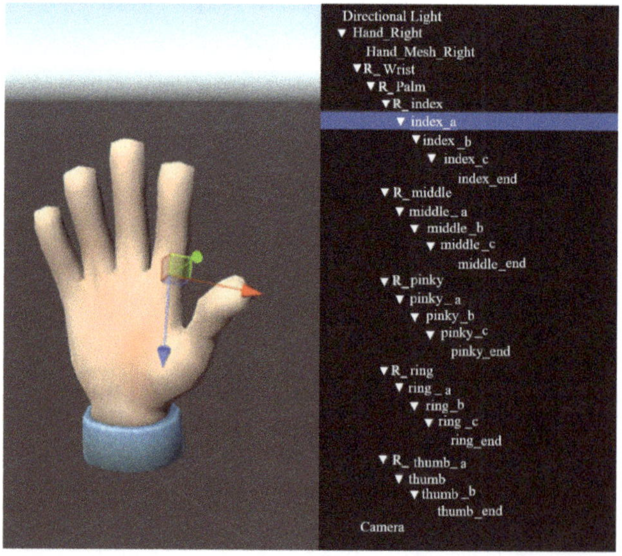

Figure 10. Virtual hand model.

Each joint of the hand is taken as a changeable unit, and two sets of attitude change systems are loaded for the virtual hand model in Unity3D. One is the gesture system which is related to the data of finger gesture transmitted from the data glove; that is, the virtual hand changes the gesture according to the related data of finger gesture, which is consistent with the subject's hand gesture. Another set of action templates is the gesture animation file designed in advance; that is, the corresponding gesture action of virtual hand is activated after the activation conditions are met.

5.2. Virtual Hand Control Driven by Emotion

Five subjects, aged between 24 and 30, participated in the test. Music materials were used to awaken the subjects' emotions in the test. The DN-1 customizable five fingers mechanical claw with the most basic functions is used as external hardware equipment, and the connection of test equipment is shown in Figure 11.

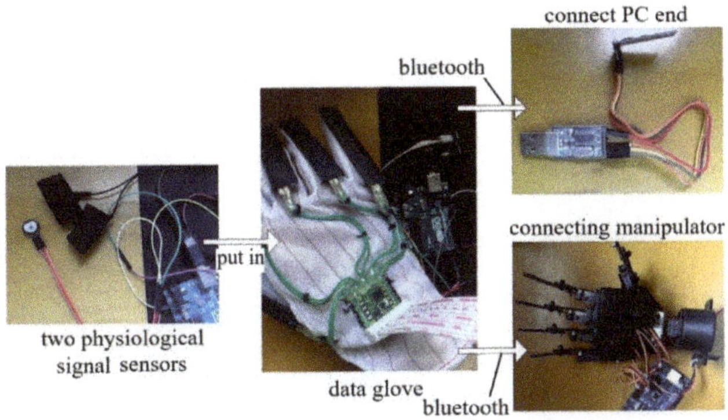

Figure 11. Connection of test equipment that virtual gesture change driven by emotion.

The principle and test process are as follows: (1) playing the music with style of terror, sadness, grandeur, and freshness to awake subjects' emotion; (2) the physiological signals (skin electricity and pulse wave) caused by the subjects' emotion are collected by sensors placed in the data glove; (3) four emotional trends HANV, LANV, HAPV, and LAPV corresponding to terror, sadness, grandeur, and freshness are detected using the system of emotion recognition and judgment as described in Section 3 based on the physiological signals; (4) the emotion changes of the subjects are detected by the system which is built based on the relationship between the emotion and gesture of the virtual hand; and (5) the system drives the virtual hand to make four animation gestures of "1", "2", "3", and "4" corresponding to HANV, LANV, HAPV, and LAPV as shown in Figure 12.

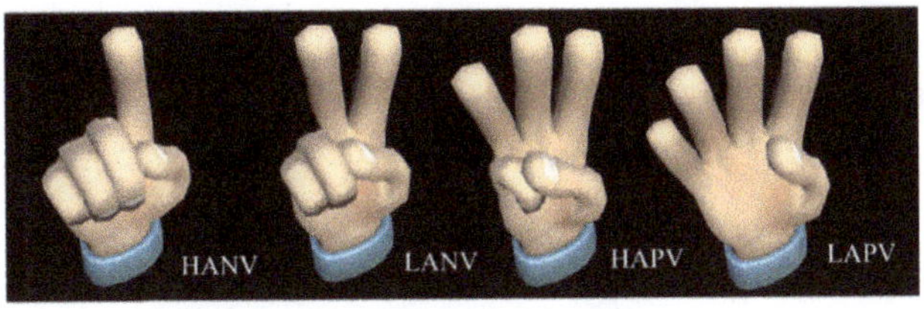

Figure 12. Four hand animation gestures.

The virtual gesture changes of subject 3 as driven by emotion are shown in Figure 13, where we can see the corresponding relationship between physiological signal and virtual gesture change.

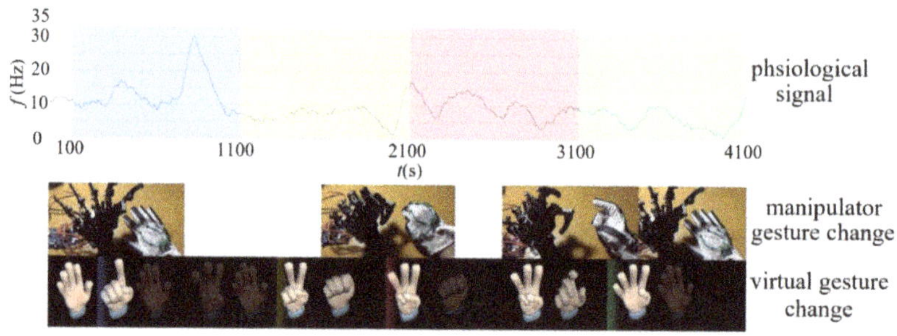

Figure 13. Gesture changes of subject 3 driven by emotion.

A gesture data acquisition module is also placed in the data glove as described in Section 4. The system can drive the virtual hand and the manipulator to make the gesture consistent with the gesture of the subject. In the process described above, when the virtual hand makes the gestures of "1", "2", "3", and "4", the subjects also make the same gesture, which drives the manipulator also to make the same gesture as shown in Figure 13. Therefore, there is a time deviation between the gestures of the virtual hand and the manipulator as show in Table 1.

Table 1. The time deviation between virtual hand change driven by emotion and manipulator change.

Subjects	1	2	3	4
1	−8.44%	+13.42%	−7.12%	−5.03%
2	−14.62%	X	+3.94%	−21.01%
3	−16.38%	−18.93%	+11.27%	−15.50%
4	−15.11%	X	−9.21%	+10.48%
5	−15.57%	−17.81%	−7.44%	x

In Table 1, the time deviation between virtual hand change driven by emotion and manipulator change is basically less than 20%, showing that the virtual hand and manipulator can be controlled synchronously through the data glove. When the user's data glove does not make any gesture change, the system can directly control the gesture of the virtual hand by reading the physiological signal of the subject, and gesture control and emotion control can be carried out at the same time to achieve the desired purpose.

5.3. Manipulator Control Driven by Emotion

The manipulator with six degrees of freedom, weight of 4.5 kg, and load capacity of 5 kg is directly controlled using data glove as shown in Figure 14, where the DN-1 data glove is adopted and two kinds of sensors are added to DN-1 data glove.

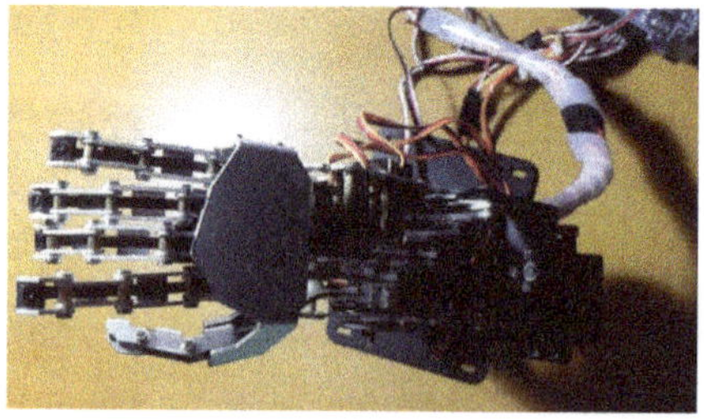

Figure 14. Manipulator in the test.

The control of manipulator is divided into gesture control of finger part and arm part. The control of the finger part can be seen in Section 4, and the arm part controls the movement angle, including the elbow joint (float anglere), the wrist joint (float anglere 1) and the finger root joint on the palm (float anglere 2).

The output angle related data includes acceleration, angular velocity, and angle.

(1) Acceleration:

$$0 \times 55\ 0 \times 51 \text{ AxL AxH AyL AyH AzL AzH TL TH SUM} \tag{3}$$

where 0×55 and 0×51 are the head and tail of the frame; AxL, AyL, and AzL are the low byte of x, y, and z axes; AxH, AyH, and AzH are the high byte of x, y, and z axes; TL and TH are the total data transmission; and SUM is the acceleration output checksum:

$$0 \times 55 + 0 \times 51 + \text{AxH} + \text{AxL} + \text{AyH} + \text{AyL} + \text{AzH} + \text{AzL} + \text{TH} + \text{TL} \tag{4}$$

where the symbols are the same as those in Equation (3).

(2) Angular velocity:

$$0 \times 55\ 0 \times 52 \text{ wxL wxH wyL wyH wzL wzH TL TH SUM} \tag{5}$$

where 0×55 and 0×51 are the head and tail of the frame; wxL, wyL, and wzL are the low byte of x, y, and z axes; wxH, wyH, and wzH are the high byte of x, y, and z axes; TL and TH are the total data transmission; and SUM is the acceleration output checksum:

$$0 \times 55 + 0 \times 52\text{+wxH+wxL+wyH+wyL+wzH+wzL+TH+TL} \tag{6}$$

where the symbols are the same as those in Equation (5).

(3) Angle:

$$0 \times 55\ 0 \times 53 \text{ RollL RollH PitchL PitchH YawL YawH TL TH SUM} \tag{7}$$

where 0×55 and 0×53 are the head and tail of the frame, RollL and RollH are roll angle for x axis, PitchL and PitchH are pitch angle for y axis, YawL and YawH are yaw angle for z axis, TL and TH are total data transmission, and SUM is the acceleration output checksum:

$$0 \times 55 + 0 \times 53 + \text{RollH} + \text{RollL} + \text{PitchH} + \text{PitchL} + \text{YawH} + \text{YawL} + \text{TH} + \text{TL} \tag{8}$$

where the symbols are the same as those in Equation (7).

The angle related data is parsed by the following function:

(1) port_noanglesure function is used to reverse judgment. The data packet is not the angle packet of the hand.
(2) angle_resume function is used to verify whether the data received by the angle package is correct, and to obtain 11-bit data of angle packet (angle_data0, angle_data1, angle_data2, ... , angle_data9, angle_data10).
(3) angledata_calculodegree function is used to convert the received angle data into two-bit angle data in degrees.

The angle drive data are replaced with physiological signal data. The driving conditions of the manipulator steering are determined by the system of emotion recognition and judgment as shown in Section 3. The steering settings are up (HANV), left (LANV), right (HAPV), and down (LAPV), respectively. Each piece of music lasts 30 s $\pm$ 1 s. The manipulator steering change driven by emotion is shown in Figure 15. We can see that the manipulator completes the steering action only driven by the emotions of LANV and HAPV, and there was no response under the emotion conditions of HANV and LAPV, showing that, although the scheme of manipulator driven by emotion is feasible, it needs to be further improved in the recognition rate of emotion and the response speed of the manipulator.

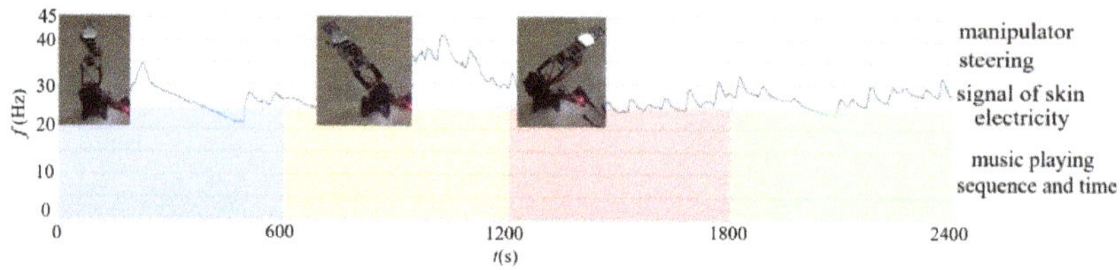

Figure 15. Manipulator steering change driven by emotion.

6. Conclusions

In this paper, the interactive application of data glove based on emotion recognition and judgment system is studied. A data glove with multi-channel data transmission based on hand gesture recognition and emotion recognition is constructed. The system of virtual hand control and manipulator driven by emotion is established using Unity3D as a construction tool of computer system. In the test of virtual hand control driven by emotion, the data glove is used to simultaneously control the virtual hand on the PC side and external mechanical claw, while the system of emotion recognition and judgment is only used in the virtual hand control. In the test of the manipulator driven by emotion, the data glove is used to directly control the manipulator, and the arm angle control is replaced by the optimal features of physiological signal. The test results show that the virtual hand and manipulator can be simultaneously controlled by the data glove. The main innovation lies in the discovery that, in the case that the subjects do not make any hand gesture change, the system can directly control the gesture of the virtual hand by reading the physiological signal of the subject, and the gesture control and emotion control can be carried out at the same time. In the test of the manipulator driven by emotion, only the results driven by two emotional trends achieve the desired purpose. Although the system of the manipulator driven by emotion is feasible, it needs to be improved.

Author Contributions: Conceptualization, Y.Z. and W.L.; methodology, W.L. and C.L.; software, W.L. and C.L.; validation, C.L. and W.L.; writing, W.L. and C.L.; resources, W.L. and Y.Z.; review, Y.Z. All authors have read and agreed to the published version of the manuscript.

Funding: This work was supported by the National Natural Science Foundation of China (Grant no. 12132015).

Institutional Review Board Statement: Not applicable.

Informed Consent Statement: Not applicable.

Data Availability Statement: Data sharing not applicable.

Conflicts of Interest: There are no conflict of interest regarding the publication of this paper.

References

1. Kundu, A.S.; Mazumder, O.; Lenka, P.K.; Bhaumik, S. Hand gesture recognition based omnidirectional wheelchair control using IMU and EMG sensors. *J. Intell. Robot. Syst.* **2018**, *91*, 529–541. [CrossRef]
2. Zhang, X.L.; Yang, Z.Q.; Chen, T.Y.; Chen, D.L.; Huang, M.C. Cooperative sensing and wearable computing for sequential hand gesture recognition. *IEEE Sens. J.* **2019**, *9*, 5775–5783. [CrossRef]
3. Song, X.Y.; van de Ven, S.S.; Chen, S.G.; Kang, P.Q.; Gao, Q.H.; Jia, J.; Shull, P.B. Proposal of a wearable multimodal sensing-based serious games approach for hand movement training after stroke. *Front. Physiol.* **2022**, *13*, 811950. [CrossRef] [PubMed]
4. Jiang, S.; Lv, B.; Guo, W.C.; Zhang, C.; Wang, H.T.; Sheng, X.J.; Shull, P.B. Feasibility of wrist-worn, real-time hand, and surface gesture recognition via sEMG and IMU sensing. *IEEE Trans. Ind. Inform.* **2018**, *14*, 3376–3385. [CrossRef]
5. Yang, K.; Xu, M.J.; Yang, X.T.; Yang, R.H.; Chen, Y.M. A novel EMG-based hand gesture recognition framework based on multivariate variational mode decomposition. *Sensors* **2021**, *21*, 7002. [CrossRef]

6. Alfaro, J.G.C.; Trejos, A.L. User-independent hand gesture recognition classification models using sensor fusion. *Sensors* **2022**, *4*, 1321. [CrossRef] [PubMed]
7. Wu, J.; Sun, L.; Jafari, R. A wearable system for recognizing american sign language in real-time using IMU and surface EMG sensors. *IEEE J. Biomed. Health Inform.* **2016**, *20*, 2598302. [CrossRef]
8. Shin, S.; Tafreshi, R.; Langari, R. EMG and IMU based real-time HCI using dynamic hand gestures for a multiple-DoF robot arm. *J. Intell. Fuzzy Syst.* **2018**, *35*, 861–876. [CrossRef]
9. Shahzad, W.; Ayaz, Y.; Khan, M.J.; Naseer, N.; Khan, M. Enhanced performance for multi-forearm movement decoding using hybrid IMU sEMG interface. *Front. Neurorobot.* **2019**, *13*, 43. [CrossRef]
10. Ordóñez Flores, J.A.; Álvarez, R.; Benalcázar, M.; Barona, L.; Leonardo, Á.; Cruz, P.; Vásconez, J.P. A new methodology for pattern recognition applied to hand gestures recognition using emg. Analysis of intrapersonal and interpersonal variability. In Proceedings of the 2021 IEEE Fifth Ecuador Technical Chapters Meeting (ETCM), Cuenca, Ecuador, 12–15 October 2021. [CrossRef]
11. Romero, R.; Cruz, P.; Vasconez, J.P.; Benalcazar, M.; Alvarez, R.; Barona, L.; Valdivieso, A.L. Hand gesture and arm movement recognition for multimodal control of a 3-DOF helicopter. *Lect. Notes Netw. Syst.* **2022**, *429*, 363–377.
12. Chico, A.; Cruz, P.; Vásconez, J.P.; Benalcázar, M.; Álvarez, R.; Barona, L.; Valdivieso, A.L. Hand gesture recognition and tracking control for a virtual UR5 robot manipulator. In Proceedings of the 2021 IEEE Fifth Ecuador Technical Chapters Meeting (ETCM), Cuenca, Ecuador, 12–15 October 2021. [CrossRef]
13. Tarchanidis, K.N.; Lygouras, J.N. Data glove with a force sensor. *IEEE Trans. Instrum. Meas.* **2003**, *52*, 984–989. [CrossRef]
14. Kamel, N.S.; Sayeed, S. SVD-based signature verification technique using data glove. *Int. J. Pattern Recognit. Artif. Intell.* **2008**, *22*, 431–443. [CrossRef]
15. Yoon, J.W.; Yang, S.I.; Cho, S.B. Adaptive mixture-of-experts models for data glove interface with multiple users. *Expert Syst. Appl.* **2012**, *39*, 4898–4907. [CrossRef]
16. Kim, K.W.; Lee, M.S.; Soon, B.R.; Ryu, M.H.; Kim, J.N. Recognition of sign language with an inertial sensor-based data glove. *Technol. Health Care* **2016**, *24*, S223–S230. [CrossRef]
17. Chen, S.; Lou, Z.; Chen, D.; Jiang, K.; Shen, G.Z. Polymer-Enhanced Highly Stretchable Conductive Fiber strain sensor used for electronic data gloves. *Adv. Mater. Technol.* **2016**, *1*, 1600136. [CrossRef]
18. Fang, B.; Sun, F.C.; Liu, H.P.; Liu, C.F. 3D human gesture capturing and recognition by the IMMU-based data glove. *Neurocomputing* **2018**, *277*, 198–207. [CrossRef]
19. Lin, B.S.; Lee, I.J.; Yang, S.Y.; Lo, Y.C.; Lee, J.; Chen, J.L. Design of an inertial-sensor-based data glove for hand function evaluation. *Sensors* **2018**, *18*, 1545. [CrossRef]
20. Wang, D.M.; Wang, Y.K.; Pang, J.W.; Wang, Z.Y.; Zi, B. Development and control of an MR brake-based passive force feedback data glove. *IEEE Access* **2019**, *7*, 172477–172488. [CrossRef]
21. Li, Y.T.; Di, H.T.; Xin, Y.; Jiang, X.S. Optical fiber data glove for hand posture capture. *OPTIK* **2021**, *233*, 166603. [CrossRef]
22. Wu, C.C.; Wang, K.E.; Cao, Q.Q.; Fei, F.; Yang, D.H.; Lu, X.; Xu, B.G.; Zeng, H.; Song, A.G. Development of a low-cost wearable data glove for capturing finger joint angles. *Micromachines* **2021**, *12*, 771. [CrossRef]
23. Sarwat, H.; Sarwat, H.; Maged, S.A.; Emara, T.H.; Elbokl, A.M.; Awad, M.I. Design of a data glove for assessment of hand performance using supervised machine learning. *Sensors* **2021**, *21*, 6948. [CrossRef] [PubMed]
24. Takigawa, S.; Mimura, H. Development of contralaterally controlled functional electrical stimulation to realize multiple grasping postures with data glove. *Sens. Mater.* **2021**, *33*, 3645–3656. [CrossRef]
25. Ueda, Y. Understanding mood of the crowd with facial expressions: Majority judgment for evaluation of statistical summary perception. *Atten. Percept. Psychophys.* **2022**, *84*, 843–860. [CrossRef] [PubMed]
26. Zhou, M.X. Interactive Visual Analysis of Human Emotions from Text. In *The Workshop on Emovis*; ACM: New York, NY, USA, 2016; p. 3.

sensors

MDPI

Article

Motion Analysis of Football Kick Based on an IMU Sensor

Chun Yu [1], Ting-Yuan Huang [1] and Hsi-Pin Ma [2,3,*]

[1] Interdisciplinary Program of Engineering, National Tsing Hua University, Hsinchu 300044, Taiwan
[2] Department of Electrical Engineering, National Tsing Hua University, Hsinchu 300044, Taiwan
[3] Center for Sport Science and Technology, National Tsing Hua University, Hsinchu 300044, Taiwan
* Correspondence: hp@ee.nthu.edu.tw

Abstract: A greater variety of technologies are being applied in sports and health with the advancement of technology, but most optoelectronic systems have strict environmental restrictions and are usually costly. To visualize and perform quantitative analysis on the football kick, we introduce a 3D motion analysis system based on a six-axis inertial measurement unit (IMU) to reconstruct the motion trajectory, in the meantime analyzing the velocity and the highest point of the foot during the backswing. We build a signal processing system in MATLAB and standardize the experimental process, allowing users to reconstruct the foot trajectory and obtain information about the motion within a short time. This paper presents a system that directly analyzes the instep kicking motion rather than recognizing different motions or obtaining biomechanical parameters. For the instep kicking motion of path length around 3.63 m, the root mean square error (RMSE) is about 0.07 m. The RMSE of the foot velocity is 0.034 m/s, which is around 0.45% of the maximum velocity. For the maximum velocity of the foot and the highest point of the backswing, the error is approximately 4% and 2.8%, respectively. With less complex hardware, our experimental results achieve excellent velocity accuracy.

Keywords: sports technology; football; motion analysis; IMU; trajectory reconstruction

Citation: Yu, C.; Huang, T.-Y.; Ma, H.-P. Motion Analysis of Football Kick Based on an IMU Sensor. *Sensors* **2022**, *22*, 6244. https://doi.org/10.3390/s22166244

Academic Editor: Giovanni Saggio

Received: 17 July 2022
Accepted: 17 August 2022
Published: 19 August 2022

1. Introduction

For any sports, repeated practice is required to improve performance and techniques. In addition to the amount of training, it is more important to use the correct method to enhance the quality of training. Practicing with improper methods is not only ineffective but also more likely to cause sports injuries. While performing a shot, players maximize speed and power, trying to make the shot more effective. However, for amateurs, exerting excessive force can easily lead to stiffness of the kicking leg. This results in insufficient knee bending which leads to momentum reduction during the foot swing before contacting with the ball. This problem is difficult to realize by athletes themselves. One way to analyze the motion is by applying multiple high-speed cameras combined with image analysis software to reconstruct the human body model and the state of motion. However, such equipment is relatively expensive and has environmental restrictions since image-related equipment needs to be set in a specific space or venue. On the other hand, IMU sensors have features such as light weight, low power, low cost and small size. An IMU can consist of a three-axis accelerometer, a three-axis gyroscope and a three-axis magnetometer. With proper filtering and data fusion, the information can be used for attitude and position estimation. Applications of IMU include military, automobile and sports.

1.1. Related Work

1.1.1. IMU in Sports

Wearable sensors with IMUs have been utilized in pedestrian dead-reckoning systems by detecting the stationary stance phase and applying zero-velocity updates (ZUPTs) for position tracking [1]. Inertial sensors were placed on the side of the shoe in [2] to obtain information about foot clearance and mean step velocity, which helps assess foot kinematics in steady-state running. Another study [3] developed a system for field-based performance analysis based on IMUs which are attached to both ankles. The system detects stance duration, providing users with real-time feedback. In [4], the study used eight IMU sensors with velocity-based localization to capture the human spatial behavior and velocity during motions such as walking, jumping and running. The system was reduced to three IMU sensors and utilized the velocity-based localization with acceleration fine tuning [5].

To help prevent shoulder injuries, ref. [6] presented a classification approach by tracking and discriminating shoulder motions using an IMU. The wearable motion capture platform proposed in [7] provides physical quantities during the high-speed motion of baseball pitchers. With an array of inertial and magnetic sensors, the method allows for the analysis of various biomechanical parameters. A wearable device was developed by incorporating IMU sensors with flow sensors. The device in [8] measures human limbs velocity, acceleration and attitude angles. Experiments include boxing motion capture with the device on the forearm and kicking motion capture with the device on the shank. Ref. [9] presented a wearable sensing system consisting of multiple IMU sensors for basketball activity recognition. The system is able to identify walking, jogging, running, sprinting and shooting. Another basketball-related study built a wrist-worn sensor consisting of an IMU, five environmental sensors, a processor and a microcontroller. The activity recognition part was conducted by machine learning [10]. The algorithm proposed in [11] detects four key temporal events and three temporal phases in skateboarding. It can provide quantitative assessment for injury prevention.

1.1.2. Football-Related Motion Analysis

Lower extremity and pelvis kinematics such as linear velocities and angular velocities were measured by an off-the-shelf product of 17 inertial sensors during kicking. The measurements were then compared with those obtained from an optoelectronic motion analysis system [12]. The hip joint motion of football players during practice was recorded directly on a sports field by a three IMU system [13]. The motion was characterized by hip acceleration and orientation. To quantify movement intensity and improve training load estimation, the system in [14] obtained knee and hip joint kinematics for football-specific movements performed at different intensities. A pressure-sensitive material was placed on the kicking foot in [15]. The device measured the force and center of pressure during the impact phase for players to further improve their technique. Biomechanical differences were observed during kicking with the preferred and the non-preferred leg [16]. Both kinetics and kinematics were derived from the filmed movements. By the full-body modeling and three-dimensional motion capture system, quantitative evaluations of kick quality were provided [17]. Using a single IMU and the acceleration data, the system in [18] distinguished between running and dribbling, passing and shooting. The study also compared three sensor locations (inside ankle, lower back and upper back) for better accuracy. Detection and segmentation of a soccer kick were performed by a system of wearable sensors and video cameras for sports motion analysis [19].

From the above paragraphs, most IMU-related motion analysis research focuses on activity classification or motion recognition during training or in a match. With the environmental limitations possessed by camera-based optoelectronic systems, the size and weight of IMU has a clear advantage. It is a popular choice when performing motion analysis. Although some research studies look at the motion itself, most of them dive into the information related to training load or biomechanical parameters on a specific joint or body part. In particular, no previous research reconstructed and analyzed the instep kicking motion with a single IMU. This paper aims to present a motion analysis system with increased accessibility, providing football players of all levels with instant feedback and an auxiliary training method to improve the instep kicking technique.

The field application of this study is expected to help players or general football lovers to adjust their movement posture before actual kicking. Preliminarily changing the posture in the empty kick stage will make the players develop good kicking habits more effectively, resulting in a better performance when actually kicking the ball. Therefore, this paper is mainly focusing on dealing with the trajectory of the foot during the kicking motion. The sensors are calibrated and the threshold setting is tailored for the kicking motion to avoid some tiny impact.

To validate the reliability of the system, we utilize the high-speed cameras to obtain the golden pattern for the trajectory. According to the systematic review study [20], one of the most commonly used measures of agreement is the Bland–Altman plot. It is a scatter plot which shows the relationship between two methods. The metric is used in this study to evaluate the accuracy of the trajectory reconstructed from the IMU data.

The system architecture is shown in Figure 1. We collect acceleration and angular velocity data during movement through the accelerometer and gyroscope in the six-axis inertial measurement unit (IMU). After the steps of deviation calibration, attitude estimation with quaternions, the transformation of coordinates, and gravity compensation, we analyze the maximum velocity and highest point of the foot before contacting with the ball while reconstructing the 2D and 3D trajectory of the kicking motion.

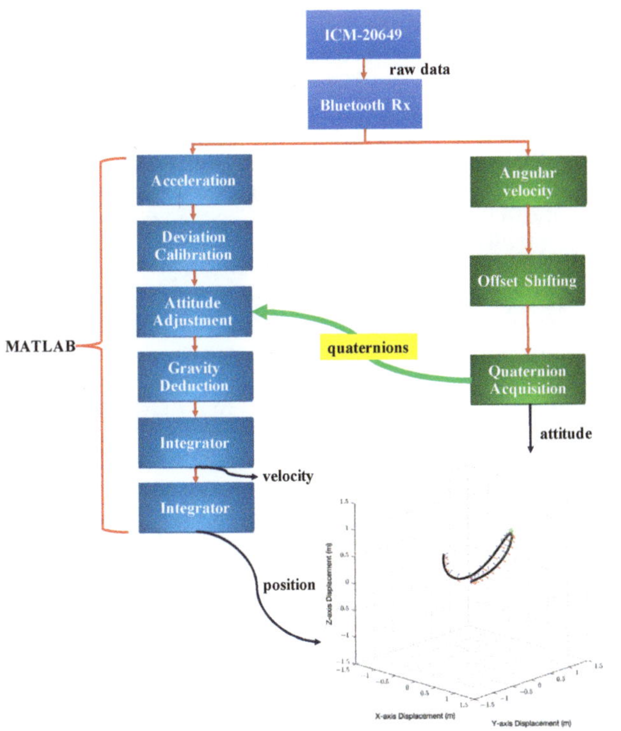

Figure 1. An illustration of the proposed system. The three colored small axes on the trajectory represent the coordinates of the sensor.

This paper aims to present a 3D motion analysis system which allows users to observe the kicking motion and acquire significant motion information, with only a single IMU sensor attached to the kicking foot, avoiding complex accessories which might affect training and eluding the hassle of setting up optoelectronic devices. The main contributions of this study are: (1) the synthesis of a simple motion trajectory reconstruction system for the data collected by a single six-axis IMU during an instep kicking motion, which employs the quaternion representation of orientation to describe the attitude change; (2) the customized adjustments to various parameters for the football kicking action during the signal processing, and the elimination of various possible noises to ensure that the accumulation of integral errors is minimized; (3) the extraction of specific motion data from the reconstructed trajectory to provide motion parameters that affect the quality of the kick during the process from backswing to kicking.

2. Methodology

The proposed sensing system includes data collection and several data processing procedures. More detailed steps will be given later in this chapter.

2.1. Data Collection and Deviation Calibration

The sensor selected in this research is ICM-20649 [21]. It is a wide-range six-axis motion tracking device which contains a three-axis accelerometer and a three-axis gyroscope, each with a 16-bit ADC, and the sampling frequency is set to 100 Hz. In the previous measurement, we found that the upper limit of the kicking motion is about 12 g, so we set the full signal range to ± 30 g and ± 4000 °/s for the application in this research. The precision measured from this range is acceptable because there are subsequent mechanisms for threshold and stationary judgement to distinguish the state of motion.

This experiment uses Bluetooth to transmit real-time data. After pairing the sensor with the Bluetooth receiver, the acceleration data and angular velocity data will be transferred to the computer and stored as text files. After the data are converted to decimal, it is necessary to perform the two's complement to obtain the negative number.

A modified sphere model is applied in the calibration process for sensor deviation. First, we assume the calibration equation to be G = L(g + b), G is the acceleration before calibration, g is the real acceleration, L is the linear proportional deviation of the sensor itself and b is the deviation of the center value of the sensor. In an ideal static state, the sum of the squares of the three-axis acceleration should be equal to one, so the gravitational acceleration values at various angles will form a sphere with a radius of one. When calculating, one must first assume that the linear proportional deviation is one, and one must use the least square method to obtain the center of the three axes. The same method can be used to find the linear proportional deviation, but the actual test found that the three-axis acceleration square sum will be less than one when the sensor is stationary. Therefore, normalization is performed in the end to complete the accelerometer calibration.

2.2. Attitude Estimation with Quaternion

In the common state of motion, rotation is bound to participate, and the acceleration received by the three axes of the sensor is actually the acceleration of the sensor's coordinates, not the acceleration of the earth coordinates. Data can only be applied and analyzed through attitude processing. The six-axis sensor chosen for this research only includes an accelerometer and a gyroscope. Without a magnetometer, we can only obtain the sensor attitude by obtaining the respective angle changes of the sensor and comparing them with that of the initial coordinates.

Quaternion representation of rotation is derived from the characteristics of inner and outer products between vectors. It can be considered to be the extension of two-dimensional real and imaginary numbers to four-dimensional to show the rotation in three-dimensional space. Similar to complex numbers, quaternions are composed of real numbers and three elements i, j and k. Each quaternion q can be represented by a linear combination of them, generally expressed as $q = a + bi + cj + dk$, and they follow the following relationship:

$$i^2 = j^2 = k^2 = ijk = -1 \tag{1}$$

The attitude quaternion (q) is a column vector of four parameters to describe a rotation along a specific axis, which can be written as:

$$q = \begin{bmatrix} q_0 \\ q_x \\ q_y \\ q_z \end{bmatrix} \triangleq \begin{bmatrix} \cos\left(\frac{\theta}{2}\right) \\ E_x \sin\left(\frac{\theta}{2}\right) \\ E_y \sin\left(\frac{\theta}{2}\right) \\ E_z \sin\left(\frac{\theta}{2}\right) \end{bmatrix} \quad (2)$$

However, in a general movement, it is difficult to know the rotation axis of each sampling point, and the angle information is of the sensor axes instead of the axis with which the sensor rotates along. Since the angle information is obtained through the gyroscope, we decide to directly update the quaternion by using the angular velocity data. The vector $\boldsymbol{S_\omega}$ which contains the angular velocities is defined as:

$$\boldsymbol{S_\omega} = \begin{bmatrix} 0 & \omega_x & \omega_y & \omega_z \end{bmatrix} \quad (3)$$

Then, we consider the quaternion derivative that describes the rate of change in orientation:

$$\frac{dQ_k}{dt} = \frac{1}{2} \cdot \hat{Q}_{k-1} \bigotimes S_\omega \quad (4)$$

The first parameter, $\frac{dQ_k}{dt}$, is the derivative at time step k expressed in quaternion, $\hat{Q}_{k-1}$ is the estimated orientation at time step k, and $\otimes$ is the quaternion product operator. By integrating the quaternion derivative, it would be possible to estimate the orientation over time:

$$\hat{Q}_k = \hat{Q}_{k-1} + \frac{dQ_k}{dt} \cdot \Delta t \quad (5)$$

Finally, we can use the following equation to complete the quaternion update:

$$Q_k + = 0.5 \times Q_{k-1} \times \boldsymbol{angVel} \times dt \quad (6)$$

In addition, after each update of the quaternion, the quaternion must be normalized to obtain the true quaternion, so as to avoid the phenomenon of scaling while the vector is rotating. When a new quaternion is obtained, the acceleration data of the sensor can be converted into the acceleration data of the initial coordinates through the following formula:

$$\boldsymbol{accl_{transformed}} = \boldsymbol{Q} \times \boldsymbol{accl} \times \boldsymbol{Q_{conj}} \quad (7)$$

where $\boldsymbol{accl_{transformed}}$ is the acceleration data in initial coordinates, $\boldsymbol{accl}$ is the acceleration data before attitude processing, $\boldsymbol{Q}$ and $\boldsymbol{Q_{conj}}$ represent the quaternion and the conjugate quaternion, respectively.

2.3. Gravity Compensation

This subsection will introduce the method of compensating the gravity components and the transformation of coordinates. Since the sensor data during the entire motion have been converted into the initial sensor coordinates, we can subtract the average acceleration of the first 500 sampling points obtained in the static state offset_accl from the raw acceleration data. Through this process, we can obtain the movement data of the sensor without the influence of gravity.

After gravity compensation, the misalignment between the initial coordinates and the earth coordinate still needs to be dealt with. If this problem remains unsolved, the 2D and 3D motion trajectory will be tilted. Different from the previous processing of attitude changes, since the initial coordinates are those at rest and cannot be processed with angular velocity information, we implement the rotation matrix of the initial coordinates to the earth coordinates to calculate the inclination of the gravity component.

First, we divide the rotation into three parts: roll, pitch and yaw. The tilt of a three-dimensional space can be achieved with two axial rotations.

$$\text{roll} = \arctan\left(\frac{\text{offset}_y}{\text{offset}_z}\right), \quad \text{pitch} = -\arctan\left(\frac{\text{offset}_x}{\text{offset}_z}\right), \quad \text{yaw} = 0 \tag{8}$$

After obtaining the rotation angles around each axis, we find the rotation matrix, and combine the three with matrix multiplication to obtain the complete rotation matrix, written as the following matrices:

$$\begin{gathered} \boldsymbol{R_x} = \begin{bmatrix} 1 & 0 & 0 \\ 0 & \cos(\text{roll}) & -\sin(\text{roll}) \\ 0 & \sin(\text{roll}) & \cos(\text{roll}) \end{bmatrix} \\ \boldsymbol{R_y} = \begin{bmatrix} \cos(\text{pitch}) & 0 & \sin(\text{pitch}) \\ 0 & 1 & 0 \\ -\sin(\text{pitch}) & 0 & \cos(\text{pitch}) \end{bmatrix} \\ \boldsymbol{R_z} = \begin{bmatrix} \cos(\text{yaw}) & -\sin(\text{yaw}) & 0 \\ \sin(\text{yaw}) & \cos(\text{yaw}) & 0 \\ 0 & 0 & 1 \end{bmatrix} \\ \boldsymbol{T_{rotate}} = \boldsymbol{R_z} \times \boldsymbol{R_y} \times \boldsymbol{R_x} \end{gathered} \tag{9}$$

Lastly, we multiply it by the three-axis acceleration after compensating the gravity to complete the transformation of the coordinates, written as:

$$\boldsymbol{accl_{corrected}} = \boldsymbol{T_{rotate}} \times \boldsymbol{accl_{corrected}} \tag{10}$$

2.4. Quadratic Integration and Threshold Setting

After completing the transformation of the coordinates and the gravity compensation, we proceed to the trajectory construction part. The velocity can be obtained by integrating the acceleration once, and the displacement can be obtained after the second integration. The displacement between every two sampling points can be used to reconstruct the trajectory of the sensor movement.

In this research, we slightly modified the integration method by averaging the acceleration value between two sampling points to calculate the acceleration value belonging to the time interval. The formula can be written as:

$$v_i = v_{i-1} + \frac{a_i + a_{i-1}}{2}\Delta t \tag{11}$$

The result calculated by this integration method is more accurate than that calculated by the original formula $v_i = v_{i-1} + a_i\Delta t$. The velocity change, which is the area calculated by this method, is shown by the area $a'\Delta t$ in Figure 2, and a' is the average acceleration of a_1 and the acceleration from the previous sampling point. It can be found that the purple area on the left can be roughly compensated to the original missing area, so the integral error will be smaller than the original formula. We perform the integration separately on the three-axis data collected by the sensor to obtain the velocity of each axis, and then we use a similar integration technique to obtain the displacement.

Threshold setting is a crucial aspect when integrating. During the experiment, the sensor will inevitably be affected by some external factors, such as vibration, wind and incomplete compensation of gravity components. The slight fluctuation of acceleration has a considerable influence on the error of the integration. Therefore, after repeating several experiments, we found that the acceleration of the target motion is mostly above 3.92 m/s^2. We set 0.392 m/s^2 as the acceleration threshold to filter the acceleration value of the target movement before integration.

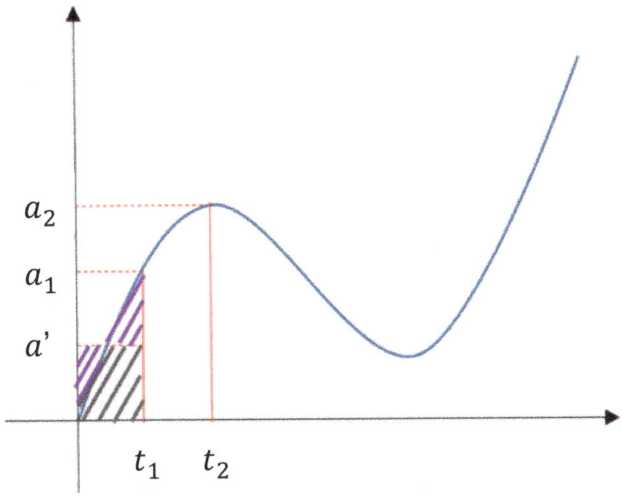

Figure 2. Illustration of integration error cancellation. The average acceleration of two adjacent sampling points is taken for calculation.

In addition, there will be a physical blind spot in the actual acceleration integration. When the sensor is stationary after a motion, the acceleration integration area during acceleration and deceleration cannot completely offset each other. Even if the sensor is at rest and the acceleration has become exactly zero, the velocity remains at the same value of the previous sampling point. In this case, when the velocity is integrated to obtain the displacement, the sensor will seem to continue its motion at a constant velocity instead of being in a static state. Therefore, a new judgment condition is added here. When the acceleration of fifteen consecutive sampling points is zero, it is determined to be a static state, and the velocity is returned to zero. A reasonable velocity threshold is also obtained through multiple experiments, and is set to 0.196 m/s to ensure that the above-mentioned accumulation of errors will not occur.

3. Results

3.1. Experimental Setup

Two high-speed cameras are used to capture the image from the front and side view to provide golden patterns for the experiment; we use tripods to secure the camera to avoid shaking, and place multiple scale bars within the capture range as a reference for depth correction. After setting up the cameras, we tie the sensor (ICM-20649) on the top of the athlete's foot with a rubber band, and perform an instep kicking motion without hitting a ball. The data received from the IMU will be collected and imported to MATLAB for data processing, then we draw trajectory diagrams and analyze different motion data.

The theoretical value of the experiment is provided by the video of the cameras. We import the video into Tracker for mapping and export the 2D data of each angle of view, align the peaks through the front view and the side view, and then perform the depth correction separately. The 3D data can be combined and the data can also be imported into MATLAB as the theoretical values. The results will then be used to calculate the error of each analysis.

3.2. Experimental Results

Motion Trajectory Analysis

After completing the data processing introduced in the previous chapter, the 3D position information of each sampling point of the IMU will be obtained, and the 3D trajectory diagram will be drawn with MATLAB. The average path length in several repeating experiments and the root mean square error (RMSE) with the theoretical value of the entire path will be calculated to verify the accuracy of the system. The two trajectories are aligned from the beginning of the motion, and then we utilize the relative sampling rate according to the different sample rates of the IMU and the frame rate of the camera. We calculate the distance between the corresponding sample points and calculate the RMSE of the position and the velocity in the direction of the kick. Figure 3 is a 3D motion trajectory diagram, the blue solid line in the figure is the theoretical trajectory obtained by Tracker, and the line composed of the red dots is the trajectory obtained after IMU data are processed.

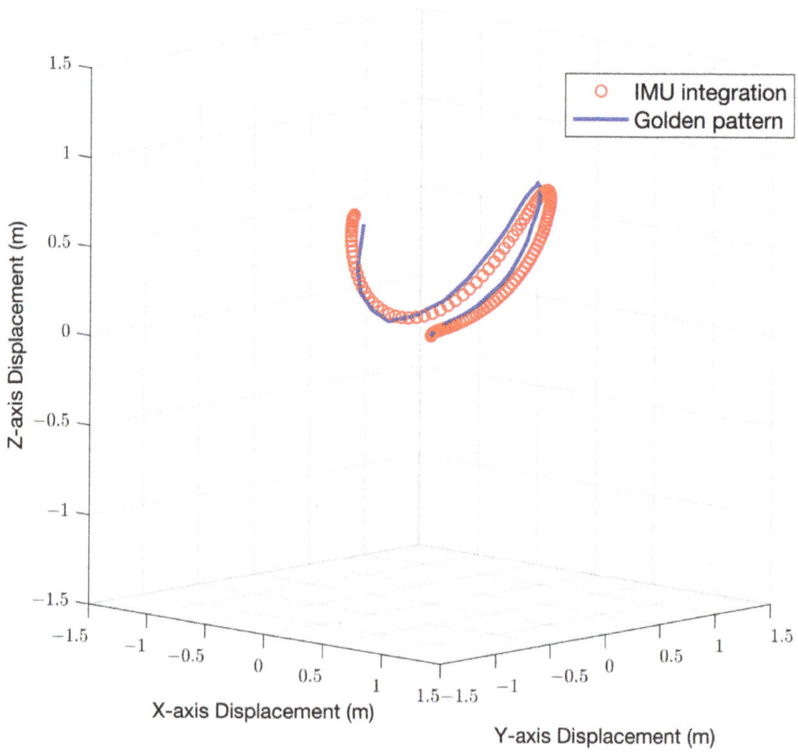

Figure 3. Three-dimensional motion trajectory diagram. For an instep kicking motion of path length around 3.63 m, the position RMSE and the velocity RMSE of the two trajectories are 0.07 m and 0.034 m/s, respectively.

3.3. Foot Velocity Analysis

On the football field, whether it is passing or shooting, the velocity of the ball is a crucial factor. We hope to observe the maximum velocity of the athlete's foot swing and where the maximum value occurs so that we can help athletes transmit the most kinetic energy to the ball. With the golden pattern obtained by Tracker, we can compare the velocity of the sensor with the velocity from the video. Figure 4 is a 2D motion trajectory diagram, the blue cross is the position where the maximum velocity appears in the theoretical trajectory, and the red circle is the position where the maximum velocity appears in the IMU motion trajectory.

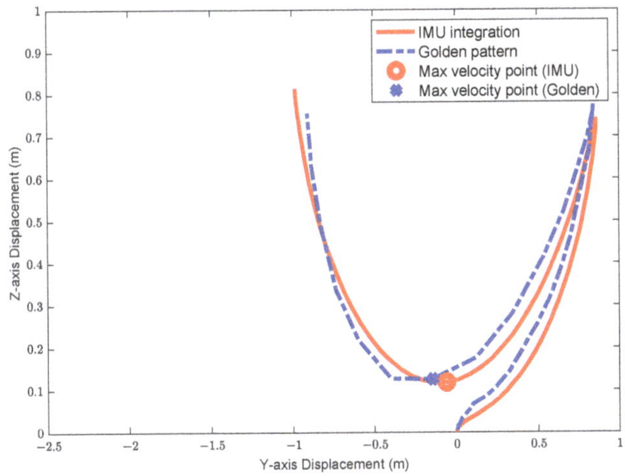

Figure 4. Two-dimensional trajectory diagram with maximum velocity position. The maximum velocity occurs when the foot reaches the bottom of the motion trajectory. An average value of the maximum instantaneous velocity in repeated experiments is around 7.4 m/s, and an error of 4% is achieved.

3.4. Backswing Height Analysis

When shooting or hitting a long ball, if the knee of the kicking foot is not bent enough to increase the height of the foot, the power of the ball will be significantly affected. Therefore, we would like to observe the height of the highest point of the foot during the pull-back motion on the reconstructed trajectory. With the golden patterns obtained by Tracker, we can discuss the accuracy of the system by comparing the highest points during the backswing. We can also use the 3D trajectory graph to obtain the position of the highest point for visualization. Figure 5 is the 3D motion trajectory diagram and the highest point of the backswing.

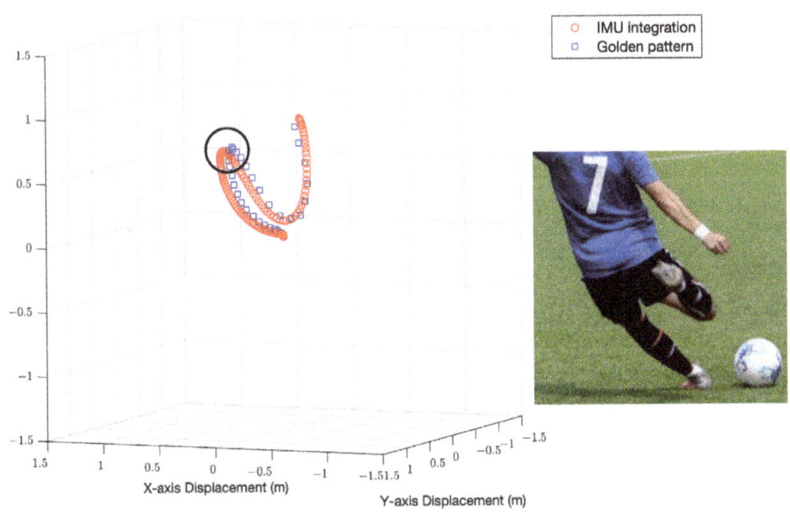

Figure 5. Three-dimensional trajectory diagram with backswing height illustrated. An error of 2.8% is achieved for an average backswing height of 0.756 m. The image on the right shows the highest point during the backswing.

Table 1 shows the quantified results generated from IMU data and also the results from high-speed cameras. From the results below, we can observe that in the motion with an average path of about 3.6 m, the entire trajectory obtained by IMU's data processing with the theoretical trajectory only has an absolute error of about 0.07 m. It is considered a very accurate result when constructing a motion trajectory, thus it proves that our signal processing system has a certain degree of credibility. As for the instantaneous velocity of the foot and the backswing height, the error is approximately 4% and 2.8%, respectively.

Table 1. Comparison of reconstructed trajectory, instantaneous velocity and backswing height generated by IMU data with high-speed cameras' results.

Instep Kicking Test of Sample Size 10	Reconstructed Trajectory			Foot Velocity Analysis (Instantaneous Velocity)			Backswing Height Analysis		
	Average Length (m)	Position RMSE (m)	Velocity RMSE (m/s)	IMU (m/s)	Image Analysis (m/s)	Error	IMU (m)	Image Analysis (m)	Error
	3.63	0.07	0.034	7.468	7.409	4.0%	0.741	0.756	2.8%

Figure 6 shows the validation of position (three axes) during the motion by comparing the IMU algorithm results with high-speed camera results. From the Bland–Altman plot, it can be seen that only 4.17% (10 out of 240) of the points are outside the 95% limits of agreement, the extent of the difference is clinically acceptable, so the two methods can be considered to be in good agreement, inferring that this IMU algorithm can be clinically substituted for high-speed camera.

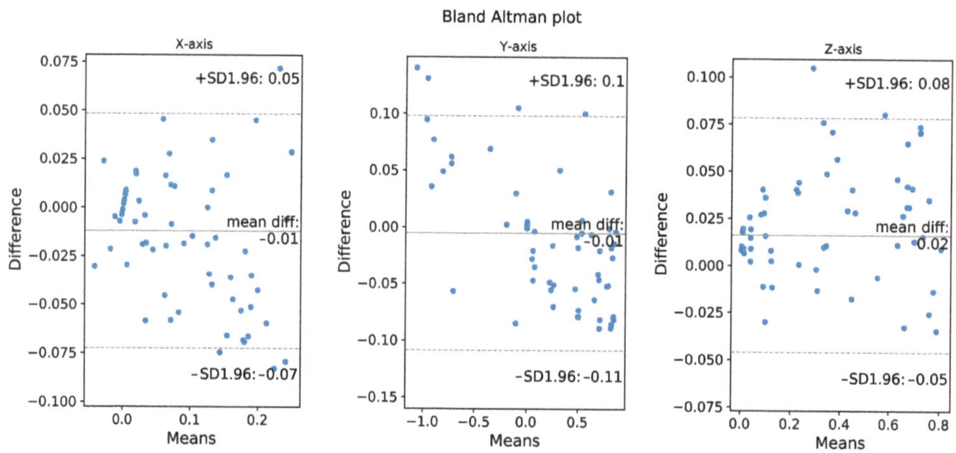

Figure 6. Validation of position during the motion by comparing the IMU algorithm results with high-speed camera results. The Bland–Altman plots for the three axes show that the data obtained by these two methods have high similarity.

4. Discussion

While camera-based optoelectronic systems can provide high accuracy for motion capturing, it has environmental restrictions and has limitations in capture rate. When calculating derivatives greater than or equal to second order using the measurement data, it has a high level of noise, often resulting in limited or no physical meaning unless the raw data are filtered to 10–20 Hz [22]. When these optoelectronic systems are applied to targets moving in high speed, although the position will be accurate, the velocity and acceleration might not be of adequate accuracy. At the same time, the device settings of these image analysis systems are cumbersome and can only be used in a specific environment. The

IMU sensor is undoubtedly a perfect substitute in this case. It provides the information of inertial data such as acceleration and angular velocity directly. The sensor can be easily mounted on the person without interfering with their performance, and since the sensor is light, players can easily adapt to the existence of the new device.

Focusing on the football kicking motion, we constructed a motion analysis system based on an IMU sensor, trying to analyze the physical quantities related to improving the football kicking performance. To preliminarily evaluate and assess a kicking motion, the foot velocity and backswing phase are both key factors related to the quality of the kick. In [23], the results showed that the foot velocity at the initial instant at the initial impact phase affects the ball velocity more than any other factors. The quality of foot–ball contact is crucial to the spin and speed of the ball. Higher foot velocity is related to more powerful kicks [24].

For the reconstructed trajectory, our system has achieved results with high accuracy and low RMSE in both position and velocity. Since the types of target motions are different, it sometimes cannot fully explain whether a method outplays another simply by comparing the RMSE without considering the length of the motion and the dimensions evaluated. For the gait analysis algorithm that Zhou et al. performed in [25] on the action of striding forward, they achieved an RMSE of about 0.05 m in a stride of about 1.5 m. As for the acceleration-based simultaneous localization and capture method (A-SLAC) proposed in [5], the RMSE in the main walking direction is 0.038 m for a length of 3.6 m for each trial. The RMSE is 0.032 m for the vertical direction and 0.057 m in the sideways direction, which is about 2% of the trial length. While they focus on performing the error calculation on the direction of the stride, we conduct the error calculation of the 3D motion. For an instep kicking motion with the average path length of around 3.63 m, our system achieved the position RMSE of 0.07 m.

For velocity, we extracted the maximum instantaneous velocity from the kick; the results showed a 4% error compared to the image captured by the high-speed cameras. Moreover, the RMSE of the foot velocity is about 0.034 m/s, which is around 0.45% of the maximum velocity (7.47 m/s). For the velocity in the main walking direction in [5], the RMSE is 0.051 m/s, which is around 3% of the maximum velocity (1.5 m/s). The results indicate our system performs better in the accuracy of velocity. Table 2 shows the accuracy evaluation results obtained for different types of motion using different IMU-based systems.

Table 2. Accuracy evaluation results obtained for different types of motion using different IMU-based systems. For the position RMSE in the gait-related system and A-SLAC system, the error is calculated according to the direction of movement, while our system performs it with the 3D trajectory.

	Motion Type	Motion Length	Position RMSE	Maximum Velocity	Velocity RMSE	Velocity RMSE %	IMU Used
Gait-related	stride	1.5 m	0.05 m	N/A	N/A	N/A	2
A-SLAC	walking	3.6 m	0.038 m	1.5 m/s	0.051 m/s	3%	3
Our system	instep kicking	3.63 m	0.07 m	7.47 m/s	0.034 m/s	0.45%	1

With the steady evolvement of wearable IMUs, inertial components are now commonly integrated onto a single die, allowing users to receive various motion-related data. The development of high-resolution and wide-range devices would be ideal for measuring motion poses in high-intensity motion. Moreover, the stretchable electronics would enable devices with multiple sensors to be embedded into forms that are more suitable for mounting on the body [26–28]. Multiple inertial sensor nodes would even provide better motion tracking; since there are more data, we can use the gradient descent method to fuse data and obtain a more accurate trajectory [29]. Moreover, by fusing the position and orientation data from the optoelectronic systems with the inertial data obtained from the IMU, we might be able to obtain the best set of kinematics data. By applying sensor fusion techniques based on a multiple-model linear Kalman filter for deflection estimation, the data can be fused with low processing cost, compatible with real-time embedded applications [30].

5. Conclusions

For the motion analysis, we develop a data processing procedure to fuse data from the accelerometer and gyroscope of the IMU. According to the experiment results, for the instep kicking motion of trajectory length around 3.63 m, the root mean square error of the position and the velocity compared with the golden patterns obtained from the high-speed cameras and image analysis software is about 0.07 m and 0.034 m/s, respectively. For the maximum velocity of the foot, the error is approximately 4%. This metric is related to the contact point with the ball and the timing of acceleration. The error for the highest point of the foot before hitting the ball is 2.8%.

This system can be applied to players of all ages and levels, whether it is to observe movement changes by trajectory, or simply to measure the height or velocity of the feet. The motion information provided in the quantified form allows players or coaches to have a more specific and clear method to analyze the action. The experiment in this research does not require a large amount of equipment, nor does it need to be carried out in a specific place or room, hence the convenience of practical application is greatly improved.

Author Contributions: Conceptualization, C.Y. and T.-Y.H.; Formal analysis, C.Y. and T.-Y.H.; Methodology, C.Y., T.-Y.H. and H.-P.M.; Writing—original draft, C.Y. and T.-Y.H.; Writing—review and editing, H.-P.M. All authors have read and agreed to the published version of the manuscript.

Funding: This research was supported in part by the Ministry of Science and Technology in Taiwan, R.O.C. (Grant No. 110-2221-E-007-126-), and in part by National Tsing Hua University (NTHU 110Q2703E1).

Informed Consent Statement: Informed consent was obtained from all subjects involved in the study.

Data Availability Statement: The data presented in this study are available on request from the corresponding author.

Conflicts of Interest: The authors declare no conflict of interest.

References

1. Foxlin, E. Pedestrian tracking with shoe-mounted inertial sensors. *IEEE Comput. Graph. Appl.* **2005**, *25*, 38–46. [CrossRef] [PubMed]
2. Bailey, G.P.; Harle, R. Assessment of foot kinematics during steady state running using a foot-mounted IMU. *Procedia Eng.* **2014**, *72*, 32–37. [CrossRef]
3. Schmidt, M.; Rheinländer, C.; Nolte, K.F.; Wille, S.; Wehn, N.; Jaitner, T. IMU-based determination of stance duration during sprinting. *Procedia Eng.* **2016**, *147*, 747–752. [CrossRef]
4. Yuan, Q.; Chen, I.M. Human velocity and dynamic behavior tracking method for inertial capture system. *Sens. Actuators A Phys.* **2012**, *183*, 123–131. [CrossRef]
5. Yuan, Q.; Chen, I.M. Localization and velocity tracking of human via 3 IMU sensors. *Sens. Actuators A Phys.* **2014**, *212*, 25–33. [CrossRef]
6. Rawashdeh, S.A.; Rafeldt, D.A.; Uhl, T.L. Wearable IMU for shoulder injury prevention in overhead sports. *Sensors* **2016**, *16*, 1847. [CrossRef]
7. Lapinski, M.; Medeiros, C.B.; Scarborough, D.M.; Berkson, E.; Gill, T.J.; Kepple, T.; Paradiso, J.A. A wide-range, wireless wearable inertial motion sensing system for capturing fast athletic biomechanics in overhead pitching. *Sensors* **2019**, *19*, 3637. [CrossRef]
8. Liu, S.; Zhang, J.; Zhang, Y.; Zhu, R. A wearable motion capture device able to detect dynamic motion of human limbs. *Nat Commun.* **2020**, *11*, 5615. [CrossRef]
9. Nguyen, L.N.N.; Rodríguez-Martín, D.; Català, A.; Pérez-López, C.; Samà, A.; Cavallaro, A. Basketball Activity Recognition using Wearable Inertial Measurement Units. In *Proceedings of the XVI International Conference on Human Computer Interaction (Interacción '15)*; Association for Computing Machinery: New York, NY, USA, 2015; Article 60; pp. 1–6. [CrossRef]
10. Hölzemann, A.; Laerhoven, K.V. Using Wrist-Worn Activity Recognition for Basketball Game Analysis. In Proceedings of the 5th International Workshop on Sensor-Based Activity Recognition and Interaction (iWOAR '18), Berlin, Germany, 20–21 September 2018; Association for Computing Machinery: New York, NY, USA, 2018. Article 13. pp. 1–6. [CrossRef]
11. Hu, X.; Liang, F.; Fang, Z.; Qu, X.; Zhao, Z.; Ren, Z.; Cai, W. Automatic temporal event detection of the Ollie movement during skateboarding using wearable IMUs. *Sports Biomech.* **2021**. [CrossRef]

12. Blair, S.; Duthie, G.; Robertson, S.; Hopkins, W.; Ball, K. Concurrent validation of an inertial measurement system to quantify kicking biomechanics in four football codes. *J. Biomech.* **2019**, *73*, 24–32. [CrossRef]
13. Horenstein, R.E.; Goudeau, Y.R.; Lewis, C.L.; Shefelbine, S.J. Using magneto-inertial measurement units to pervasively measure hip joint motion during sports. *Sensors* **2020**, *20*, 4970. [CrossRef] [PubMed]
14. Wilmes, E.; De Ruiter, C.J.; Bastiaansen, B.J.C.; Van Zon, J.F.J.A.; Vegter, R.J.K.; Brink, M.S.; Goedhart, E.A.; Lemmink, K.A.P.M.; Savelsbergh, G.J.P. Inertial sensor-based motion tracking in football with movement intensity quantification. *Sensors* **2020**, *20*, 2527. [CrossRef] [PubMed]
15. Fuss, F.K.; Düking, P.; Weizman, Y. Discovery of a sweet spot on the foot with a smart wearable soccer boot sensor that maximizes the chances of scoring a curved kick in soccer. *Front. Physiol.* **2018**, *9*, 63. [CrossRef] [PubMed]
16. Dörge, H.C.; Andersen, T.B.; Sørensen, H.; Simonsen, E.B. Biomechanical differences in soccer kicking with the preferred and the non-preferred leg. *J. Sports Sci.* **2002**, *20*, 293–299. [CrossRef]
17. Shan, G.; Westerhoff, P. Full-body kinematic characteristics of the maximal instep soccer kick by male soccer players and parameters related to kick quality. *Sports Biomech.* **2005**, *4*, 59–72. [CrossRef]
18. Kondo, Y.; Ishii, S.; Aoyagi, H.; Hossain, T.; Yokokubo, A.; Lopez, G. FootbSense: Soccer Moves Identification Using a Single IMU. In *Sensor- and Video-Based Activity and Behavior Computing*; Ahad, M.A.R., Inoue, S., Roggen, D., Fujinami, K., Eds.; Smart Innovation, Systems and Technologies; Springer: Singapore, 2022; Volume 291. [CrossRef]
19. Kim, W.; Kim, M. Sports motion analysis system using wearable sensors and video cameras. In Proceedings of the 2017 International Conference on Information and Communication Technology Convergence (ICTC), Jeju, Korea, 18–20 October 2017; IEEE: Piscataway, NJ, USA, 2017; pp. 1089–1091. [CrossRef]
20. Zaki, R.; Bulgiba, A.; Ismail, R.; Ismail, N.A. Statistical methods used to test for agreement of medical instruments measuring continuous variables in method comparison studies: A systematic review. *PLoS ONE* **2012**, *7*, e37908. [CrossRef]
21. TDK InvenSense. World's First Wide-Range 6-Axis MEMS MotionTracking™ Device for Sports and High Impact Applications, DS-000192 Datasheet. March 2016 [Revised July 2021]. Available online: https://invensense.tdk.com/smartmotion/6-axis/ (accessed on 1 May 2021).
22. Lapinski, M. A Platform for High-Speed Biomechanical Data Analysis Using Wearable Wireless Sensors. Ph.D. Thesis, MIT, Cambridge, MA, USA, 2013.
23. Ishii, H.; Yanagiya, T.; Naito, H.; Katamoto, S.; Maruyama, T. Theoretical Study of Factors Affecting Ball Velocity in Instep Soccer Kicking. *J. Appl. Biomech.* **2012**, *28*, 258–270. [CrossRef]
24. Kellis, E.; Katis, A. Biomechanical characteristics and determinants of instep soccer kick. *J. Sports Sci. Med.* **2007**, *6*, 154–165.
25. Zhou, L.; Tunca, C.; Fischer, E.; Brahms, M.; Ersoy, C.; Granacher, U.; Arnrich, B. Validation of an IMU Gait Analysis Algorithm for Gait Monitoring in Daily Life Situations. In Proceedings of the 2020 42nd Annual International Conference of the IEEE Engineering in Medicine & Biology Society (EMBC), Montreal, QC, Canada, 20–24 July 2020; IEEE: Piscataway, NJ, USA, 2020; pp. 4229–4232. [CrossRef]
26. Servati, A.; Zou, L.; Wang, Z.J.; Ko, F.; Servati, P. Novel Flexible Wearable Sensor Materials and Signal Processing for Vital Sign and Human Activity Monitoring. *Sensors* **2017**, *17*, 1622. [CrossRef]
27. Garlant, J.A.; Ammann, K.R.; Slepian, M.J. Stretchable Electronic Wearable Motion Sensors Delineate Signatures of Human Motion Tasks. *ASAIO J.* **2018**, *64*, 351–359. [CrossRef]
28. Rogers, J.A.; Someya, T.; Huang, Y. Materials and Mechanics for Stretchable Electronics. *Science* **2010**, *327*, 1603–1607. [CrossRef] [PubMed]
29. Liu, L.; Qiu, S.; Wang, Z.L.; Li, J.; Wang, J.X. Canoeing Motion Tracking and Analysis via Multi-Sensors Fusion. *Sensors* **2020**, *20*, 2110. [CrossRef] [PubMed]
30. Medeiros, C.B.; Wanderley, M.M. Multiple-Model Linear Kalman Filter Framework for Unpredictable Signals. *IEEE Sens. J.* **2014**, *14*, 979–991. [CrossRef]

sensors

Article

Detection of Horse Locomotion Modifications Due to Training with Inertial Measurement Units: A Proof-of-Concept

Benoît Pasquiet [1,*], Sophie Biau [1], Quentin Trébot [2], Jean-François Debril [3], François Durand [3] and Laetitia Fradet [2]

1 Plateau technique «Equitation et performance sportive», Institut français du cheval et de l'équitation, Avenue de l'École Nationale d'Équitation, 49411 Saumur, France; sophie.biau@ifce.fr

2 Equipe Robotique, Biomécanique, Sport, Santé, Institut PPRIME, UPR3346 CNRS Université de Poitiers ENSMA, 86073 Poitiers, France; quentin.trebot@univ-poitiers.fr (Q.T.); laetitia.fradet@univ-poitiers.fr (L.F.)

3 Centre d'Analyse d'Image et Performance Sportive, CREPS de Poitiers, 86580 Vouneuil sous Biard, France; jean-francois.debril@creps-poitiers.sports.gouv.fr (J.-F.D.); francois.durand@creps-poitiers.sports.gouv.fr (F.D.)

* Correspondence: benoit.pasquiet@ifce.fr

Abstract: Detecting fatigue during training sessions would help riders and trainers to optimize their training. It has been shown that fatigue could affect movement patterns. Inertial measurement units (IMUs) are wearable sensors that measure linear accelerations and angular velocities, and can also provide orientation estimates. These sensors offer the possibility of a non-invasive and continuous monitoring of locomotion during training sessions. However, the indicators extracted from IMUs and their ability to show these locomotion changes are not known. The present study aims at defining which kinematic variables and indicators could highlight locomotion changes during a training session expected to be particularly demanding for the horses. Heart rate and lactatemia were measured to attest for the horse's fatigue following the training session. Indicators derived from acceleration, angular velocities, and orientation estimates obtained from nine IMUs placed on 10 high-level dressage horses were compared before and after a training session using a non-parametric Wilcoxon paired test. These indicators were correlation coefficients (CC) and root mean square deviations (RMSD) comparing gait cycle kinematics measured before and after the training session and also movement smoothness estimates (SPARC, LDLJ). Heart rate and lactatemia measures did not attest to a significant physiological fatigue. However, the statistics show an effect of the training session ($p < 0.05$) on many CC and RMSD computed on the kinematic variables, indicating a change in the locomotion with the training session as well as on SPARCs indicators ($p < 0.05$), and revealing here a change in the movement smoothness both in canter and trot. IMUs seem then to be able to track locomotion pattern modifications due to training. Future research should be conducted to be able to fully attribute the modifications of these indicators to fatigue.

Keywords: horse locomotion; training effect; inertial measurement units

Citation: Pasquiet, B.; Biau, S.; Trébot, Q.; Debril, J.-F.; Durand, F.; Fradet, L. Detection of Horse Locomotion Modifications Due to Training with Inertial Measurement Units: A Proof-of-Concept. *Sensors* **2022**, *22*, 4981. https://doi.org/10.3390/s22134981

Academic Editor: Raffaele Gravina

Received: 24 May 2022
Accepted: 28 June 2022
Published: 1 July 2022

1. Introduction

Assessing the athlete, including a sport horse gives meaning to his training. Fatigue and, in particular, muscular fatigue, is a normal outcome of physical exercise. However, delaying the onset of fatigue is a key aim to any training program and is essential for athletic success. [1]. The challenge within a training program is not to cross the limit from which muscular fatigue causes detrimental effects to the musculoskeletal system [2], as fatigue is associated with injury and underperformance [3,4]. Identifying the onset of fatigue and understanding its effects are then essential to optimize training.

Physiological indicators such as heart rate, body temperature, and lactatemia [5] or muscle damage biomarkers [6] provide indications of fatigue during physical exercise. As a consequence of these muscular and physiological effects, movement and locomotion are also modified, which can be seen in the kinematics and spatio-temporal indicators.

This has been shown in humans [7,8] and also in horses [9,10]. For horses, their length and stride frequency are thus affected by fatigue. More specifically, difficulty keeping a stable stride frequency [9] and a decrease in stride length [10] have been reported. During endurance events, trot asymmetry was seen to increase as a response to an increase in physical demands [11]. A more recent study also highlighted a decrease in the diagonal step length for thoroughbred horses during canter races. According to the authors, this decrease suggests that horses could not extend their body when fatigued [12].

By combining electromyographic and kinematic analyses, some studies have linked these spatio-temporal modifications to muscular activity changes. Takahashi et al. [13] found that fatigue induced a decrease in the activity of the splenius and brachiocephalicus muscles during canter and trotting exercises, whereas the infraspinatus and deltoid muscles' activity did not change. This decrease in muscular activity affected the horse's speed and stride frequency. A change in splenius muscular activity related to fatigue has also been shown to impact head and neck movements [14].

The methods used in these studies were based on optoelectronic systems or electromyography, which are difficult to use in field conditions. Inertial measurement units (IMUs) are wearable sensors that measure 3-dimensional acceleration, angular velocity, and, most of the time, sensor orientation using a combination of accelerometers, gyroscopes, and, most of the time, magnetometers. IMUs have been used effectively in the field to characterize anomalies in horse locomotion since they offer a non-invasive wearable measurement of locomotor indicators. These sensors have been proposed as an aid for lameness detection [15–17], for gait classification [18], for horse speed estimation [19], for evaluation of the effect of shoes on break over [20], or for the evaluation of different rehabilitation methods [20]. They also offer an alternative to traditional systems such as optoelectronic systems to identify the phases of locomotion [21–25] or estimate the horse's protraction and retraction angles [16].

In addition to the question relative to the indicators that could be used to detect fatigue, another question concerns the location of the sensors. A recent study measured the trot movement symmetry of reining horses with three IMUs fixed at the head, wither, and sacrum [15]. Several lameness detection systems are now available on the market (Lameness Locator® (Equinosis, LLC, Columbia, MO, USA), Equigait® (Equigait UK, London, UK)) that are based on three locations, the head, croup, and withers, whereas the Equimoves system® is based on measures located at the four cannons [26]. Thus, no specific sensor location seems to be more favourable to attest for changes in locomotion.

Locomotion indicators modified by fatigue are probably not exhaustive and, as such, it is difficult to determine which of them are the most pertinent for the detection of locomotion changes associated with fatigue. In fact, since IMUs measure acceleration and angular velocities they offer a wide spectrum of kinematic variables. In human studies, they have also been used to assess movement smoothness, which enable the appreciation of defects in motor control [27]. A smoother movement is indeed associated with a skilled behaviour, which represents less effort [28]. Indicators have been developed for the analysis of the smoothness of periodic movements [29]. Among these indicators, the SPARC (for spectral arc length) and the LDLJ (log dimensionless jerk) can be computed from the acceleration and angular velocities measured by IMUs [30]. SPARC has been applied, for example, to IMUs monitoring patients with Parkinson's disease [31].

The aim of this study is to define which kinematic variables and indicators could be the best to highlight locomotion changes during a training session. To meet these objectives, the evolution of indicators characterising gait derived from acceleration, angular velocities, and orientation estimates extracted from IMUs were compared before and after a training session. The findings from this study could be used to help develop an on-board system to detect in horses the locomotion changes that could be associated with fatigue.

2. Materials and Methods

2.1. Horses

Ten high level dressage horse/rider combinations competing at the advanced level were involved in this study (1 male, 5 geldings, 4 females; average age of 9.2 ± 2.6 years). These horses were free of clinical signs of lameness.

2.2. Material

Nine time-synchronised IMUs (Opal, APDM Inc., Portland, OR, USA) were placed on the horse on the forehead and the pool, at the withers, the sternum and the sacrum, and on the distal forelimbs and hindlimbs (Figure 1). They were positioned such that the x-axis of the sensor case was aligned with the segment longitudinal axis and the y-axis or the z-axis was roughly parallel to the segment medio-lateral axis.

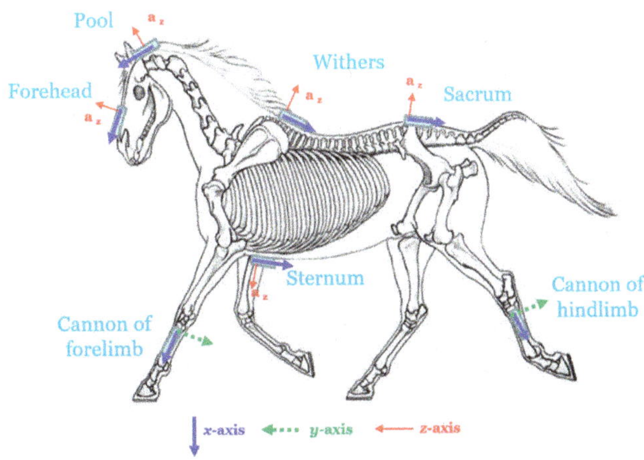

Figure 1. IMUs' localizations and orientations.

The recordings took place in a horse-riding arena. A straight corridor was marked with bars on one side of the arena. The length of this corridor allowed for at least two canter strides to be taken on each hand.

Two monitoring methods were applied to obtain the physiological state of each horse: measurement of the heart rate with a Polar® monitor during the session; measurement of the lactatemia at the end of the work using an Akray® Lactate Pro 2 portable analyser.

2.3. Protocol

After a warm-up, the horses were equipped with the IMUs. A first series of locomotion measurements including 2 passages at each gait (trot, canter) was performed in the corridor, one for each direction. Then, the working session was carried out. Immediately at the end of the training session, lactate was taken. Afterwards, a new series of 2 passages at each gait was undertaken in the corridor. Per run, a total of 4 recordings were obtained.

High-level horse training requires individualized planning. Warm-up and working session were decided by the rider, the instruction being that it had to be the most demanding of the sessions scheduled for that period. The volume of the session (gait duration, gait order, and figure repetition) was then entirely managed by the rider. It was not possible to impose the same fatigue protocol to horses of that level with an individualized planning of training.

2.4. Data Processing

Figure 2 presents data processing.

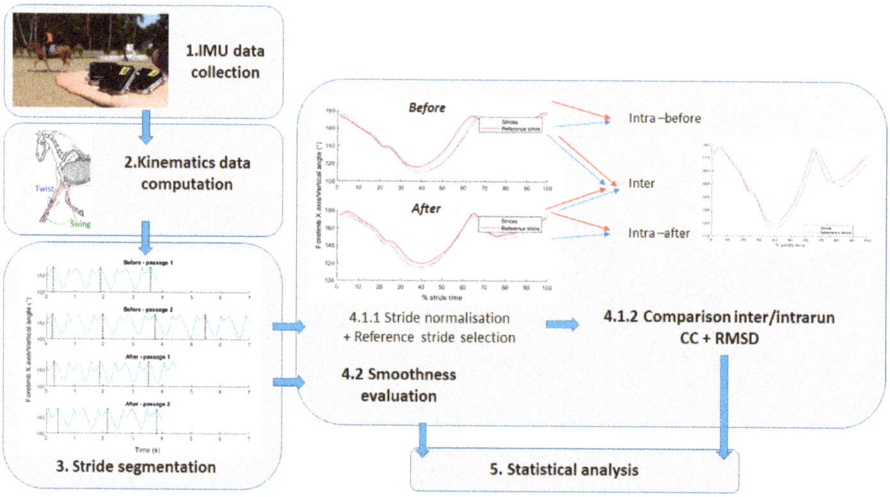

Figure 2. Data processing.

2.4.1. Kinematics Data

First, the accelerations and angular velocities expressed along the three axes of the IMU reference frame were extracted. With the IMU orientation estimation, we expressed the acceleration in a global coordinate system in order to calculate the components of the acceleration along the vertical axis and in the horizontal plane. In addition, the angle of each axis of the inertial unit was calculated relative to the earth's vertical. Finally, for the sensors placed on the limbs, a swing–twist decomposition [26,32] was applied. This decomposition provides the rotation angle around the limb axis (twist) and the angle around the lateral axis of the horse (swing).

In total, 13 kinematic signals for the cannons and 11 for the other sensors placed on the head or the sacrum and sternum were recorded.

2.4.2. Stride Segmentation

From the vertical acceleration of the sternum, a stride segmentation of the collected data was taken. For that, we applied a 4th order Butterworth filter to the raw signal. We chose a low-pass filter with a cut-off frequency of 3 Hz, which is a sufficient value to include stride frequency in passband for each gait [33]. A search for consecutive peaks in the filtered signal was performed, imposing a delay greater than 75% of the period between two successive peaks. A manual check on the signal graph was conducted to ensure the correct stride segmentation.

2.4.3. Stride Kinematics' Comparison

Strides were recorded before and after the training session. A Pchip interpolation provided the same signal length for each stride (100 points). For each one of the kinematics previously described and each gait, strides of a same run (before work or after work) were then grouped. For each group, the stride that minimized the root mean square deviation (RMSD) from this average signal was chosen as the stride of reference.

We then compared each stride to the reference stride of the same run (intra before work or intra after work) or the other run (inter before work vs inter after work), in order to highlight possible modifications in coordination and/or movement amplitude. For this, Pearson correlation coefficient (CC) and root mean square deviation (RMSD) were used.

2.4.4. Smoothness

Two movement smoothness indicators were qualified according to Melendez-Calderon [30]. More specifically, for each stride and each signal, the spectral arc length (SPARC) was calculated, with a cut-off frequency of 10 Hz. Spectrum magnitudes were normalized by their maximum values, to avoid division by zero. The SPARC on angular velocity norm was calculated. In addition, the log dimensionless jerk on acceleration vector (LDLJ-A) after removing gravity was also calculated.

SPARC uses spectrum to quantify sub-movements' dispersion. It uses magnitude spectrum curve length, until a cut-off frequency. For a signal $v(t)$, with normalized Fourier magnitude spectrum $V\hat{(}f)$ and a cut-off frequency f_c, SPARC is defined as

$$\text{SPARC} = -\int_0^{f_c} \sqrt{\left(\frac{1}{f_c}\right)^2 + \left(\frac{dV\hat{(}f)}{df}\right)^2}\, df \quad (1)$$

LDLJ is an indicator built from the minimum jerk model. For acceleration vector $\vec{a}(t)$, over time segment $[t_1, t_2]$, LDLJ-A is computed as

$$\text{LDLJ}_a = -ln\left(\frac{t_2 - t_1}{{a_{peak}}^2}\right) \int_{t_1}^{t_2} \left\| \frac{d\vec{a}(t)}{dt} \right\|_2^2 dt \quad (2)$$

where a_{peak} is acceleration magnitude peak after removing the mean acceleration vector.

All calculations were made using MATLAB 2019b software (The MathWorks Inc., Natick, MA, USA).

2.5. Statistical Analysis

To define the effect of the training session, the indicators computed before and after the training session were compared as well as the CC and the RMSD intra versus inter-run. Intra-run values give a null hypothesis as the training session does not cause differences. For these comparisons, we used a non-parametric Wilcoxon paired test. Significance was set to $p = 0.05$. R 4.0.2 software was used to perform statistical analysis.

3. Results

3.1. Energetic Solicitation during the Training Session

The durations of the training sessions were variable (28 ± 10 min), while the results of HR and lactatemia were homogeneous. The average heart rate of the 10 horses was 112 ± 9 bpm (96 < Fcmav (bpm) < 123) and the blood lactate at the end of the session reached 1 ± 0 mmol/L on average (0.7 < lactate (mmol/L) < 1.1).

3.2. Kinematics' Modifications

The processing of the IMU data made it possible to calculate 834 indicators. The Wilcoxon test provides a p-value < 0.01 for 72 indicators and a p-value between 0.01 and 0.050 for 168 indicators (see Table S1).

There was no significant difference for the stride durations before training vs after.

Some coefficient correlations computed on variables on the pre-training gait cycles were small. Only the variables for which CC on the pre-training gait cycles were greater than 0.80 were kept. Table 1 presents these significant indicators for CC, Table 2 for SPARC, and Table 3 for RMSD. Appendix A proposes the complete significant results for CC.

Thirteen indicators related to the CC were significantly affected by training for the canter and 12 for the trot. Six indicators related to the SPARC were significantly affected by training for the canter and four for the trot. LDLJ did not provide any significant difference. Finally, 43 indicators related to the RMSD were significantly affected by training for the canter and 58 for the trot.

Eight of them were common to trot and canter for the CC, 39 for RMSD, and none for the SPARC. Among them, both the CC and the RMSD were significantly affected by the training session when computed for: the acceleration along the x- and z-axes of the forehead sensor, along the x-axis of the pool sensor, as well as for the vertical acceleration of the sternum sensor; the angle between the x-axis of the sternum sensor and vertical axis; the angular velocities obtained around the y-axis of the sacrum and sternum sensor.

Sixty-six indicators related to the acceleration (obtained in the global or local coordinate system) were modified as a result of the training, whereas 34 were related to angles and 36 to angular velocities.

Regarding the location of the sensor where the significant indicators were seen (Figure 3), when including all the CC significantly affected by the training session, 41 of them were obtained at the sensor located at the forelimb, 38 at the hindlimb cannon, 37 at the sternum, 27 at the pool, 26 at the forehead, and 21 at the sacrum, and none at the withers.

From these results, it emerges, for the trot and the canter, that the CC computed between intra-run strides were superior to the CC computed between inter-run strides.

For the RMSDs, the results go in the same direction, since the RMSD applied to two cycles of a same run was on average lower than the RMSD calculated between two cycles of two different runs.

Regarding the SPARC, at canter, most of the SPARC indicators tend to increase after training, while they tend to decrease at trot.

Table 1. CC significantly affected by the training session. Only the variables for which CC were greater than 0.8 were kept. In grey are highlighted indicators common to canter and trot.

CANTER						TROT					
Variable	Axis	Position	*p*-Value	Before/Intra	Post/Inter	Variable	Axis	Position	*p*-Value	Before/Intra	Post/Inter
Acceleration	h	FH	0.031	0.8 (0.06)	0.76 (0.07)	Acceleration	x	St	0.008	0.83 (0.06)	0.77 (0.06)
	x	FH	0.016	0.94 (0.03)	0.92 (0.04)		x	FH	0.047	0.93 (0.03)	0.91 (0.03)
	x	Po	0.008	0.92 (0.04)	0.84(0.11)		x	Po	0.047	0.93 (0.03)	0.91 (0.03)
	z	FH	0.016	0.88 (0.06)	0.81 (0.12)		z	Po	0.004	0.94 (0.02)	0.91 (0.03)
	z	St	0.039	0.92 (0.04)	0.89 (0.03)		z	Sa	0.031	0.96 (0.03)	0.93 (0.05)
	z	Po	0.008	0.92 (0.05)	0.87 (0.09)		v	St	0.039	0.94 (0.04)	0.93 (0.05)
	v	St	0.004	0.91 (0.04)	0.87 (0.04)		v	Sa	0.031	0.95 (0.03)	0.93 (0.05)
Angle	x/v	Sa	0.031	0.88 (0.25)	0.86 (0.26)	Angle	x/v	Sa	0.031	0.82 (0.13)	0.69 (0.24)
	x/v	St	0.027	0.96 (0.03)	0.95 (0.04)		x/v	FC	0.039	0.84 (0.17)	0.51 (0.5)
Angular Velocity	y	FH	0.016	0.87 (0.09)	0.83 (0.11)		x/v	St	0.004	0.9 (0.07)	0.82 (0.15)
	y	Sa	0.031	0.82 (0.27)	0.76 (0.28)	Angular Velocity	y	Sa	0.031	0.84 (0.08)	0.76 (0.12)
	y	St	0.02	0.89 (0.07)	0.85 (0.08)		y	St	0.008	0.89 (0.07)	0.83 (0.13)
	y	Po	0.004	0.82 (0.12)	0.75 (0.2)						

FH: forehead, Po: pool, St: sternum, Sa: sacrum, FC: forelimb cannon, HC: hindlimb cannon. v: vertical, h: horizontal according to the global reference frame. x, y, z: axis in IMU reference frame. x/v, y/v, z/v: angle between IMU axis and global reference frame vertical. (cf. Figure 1).

Table 2. SPARC significantly affected by the training session. No indicators were common to canter and trot.

CANTER						TROT					
Variable	Axis	Position	*p*-Value	Before/Intra	Post/Inter	Variable	Axis	Position	*p*-Value	Before/Intra	Post/Inter
Acceleration	v	Po	0.02	−6.8 (1.6)	−6.2 (1.2)	Acceleration	v	FC	0.039	−7.5 (1.8)	−8.5 (1.6)
Angle	Swing	HC	0.004	−2.4 (0)	−2.4 (0)	Angle	Twist	HC	0.004	−2.1 (0.1)	−2.4 (0.1)
	y/v	FH	0.031	−2.4 (0)	−2.4 (0)	Angular Velocity	y	Po	0.008	−2.7 (0.2)	−2.5 (0.2)
	z/v	HC	0.004	−2.4 (0)	−2.4 (0)		z	St	0.012	−2.5 (0.2)	−2.7 (0.3)
Angular Velocity	y	Sa	0.031	−2.3 (0.2)	−2.4 (0.3)						
	Norm	FH	0.016	−2.3 (0.1)	−2.2 (0.1)						

FH: forehead, Po: pool, St: sternum, Sa: sacrum, FC: forelimb cannon, HC: hindlimb cannon. v: vertical, h: horizontal according to the global reference frame. x, y, z: axis in IMU reference frame. x/v, y/v, z/v: angle between IMU axis and global reference frame vertical. (cf. Figure 1).

Table 3. RMSD significantly affected by the training session. In grey are highlighted indicators common to canter and trot.

CANTER

Variable	Axis	Position	*p*-Value	Before/Intra	Post/Inter
Acceleration (m/s^{-2})	*h*	FC	0.047	24.7 (1.8)	29.1 (3.9)
	h	HC	0.02	22 (2.3)	24.7 (2.5)
	h	FH	0.031	3.5 (0.9)	4.3 (0.7)
	h	St	0.004	3.1 (0.4)	3.7 (0.8)
	x	FC	0.031	24.4 (3)	30.3 (5.1)
	x	HC	0.027	21.8 (2.3)	25.5 (4.1)
	x	FH	0.016	3.2 (0.8)	3.8 (0.9)
	x	Sa	0.031	3.7 (1.1)	4.9 (1)
	x	St	0.008	3.5 (0.5)	4.3 (0.7)
	x	Po	0.004	4.1 (1.2)	5.4 (1.8)
	y	FC	0.031	36.2 (4.2)	42.6 (4.6)
	y	HC	0.02	28.2 (2.7)	32.9 (4.7)
	y	St	0.012	3.9 (0.5)	4.4 (0.7)
	z	FC	0.039	15.2 (3.1)	16.7 (2.5)
	z	HC	0.039	13.9 (1.3)	15.4 (1.8)
	z	FH	0.016	2.3 (1)	2.8 (1)
	z	St	0.039	2.7 (0.5)	3.1 (0.3)
	z	Po	0.004	2.6 (0.8)	3.5 (1.2)
	v	FC	0.031	29.6 (1.8)	35.8 (5.2)
	v	HC	0.039	28 (2.6)	32.7 (4.6)
	v	St	0.004	2.8 (0.5)	3.5 (0.5)
	v	Po	0.039	6.6 (3)	7.3 (2.7)
Angle (°)	Swing	FC	0.031	8.1 (4.7)	10 (5.8)
	Twist	FC	0.008	65.6 (7.7)	149.1 (77.9)
	x/v	FC	0.016	15.3 (6)	17.5 (6.6)
	x/v	HC	0.004	10.9 (1.6)	12.8 (2.1)
	x/v	St	0.008	3.2 (1.2)	4.7 (1.6)
	y/v	HC	0.039	16.1 (2.8)	17.9 (2.7)
	y/v	Po	0.02	4.7 (1.1)	6 (2)
	z/v	FC	0.031	8.1 (4.7)	10 (5.8)
	z/v	Sa	0.031	2.9 (0.9)	3.3 (0.9)
	z/v	St	0.02	3.4 (1.2)	4.5 (1.6)
Angular Velocity (rad/s)	*x*	FC	0.031	2.5 (0.4)	3.1 (0.6)
	x	HC	0.02	2.2 (0.4)	2.6 (0.6)
	x	FH	0.047	0.6 (0.1)	0.7 (0.1)
	x	St	0.004	0.6(0.1)	0.7(0.1)
	y	FC	0.02	1.3 (0.2)	1.5 (0.3)
	y	HC	0.02	1.2 (0.3)	1.4 (0.4)
	y	FH	0.016	0.5 (0.2)	0.6 (0.2)
	y	Sa	0.031	0.5 (0.3)	0.6 (0.3)
	y	St	0.008	0.4 (0.1)	0.4 (0.1)
	y	Po	0.02	0.5 (0.2)	0.6 (0.2)
	z	Sa	0.031	0.4 (0.1)	0.5 (0.1)

TROT

Variable	Axis	Position	*p*-Value	Before/Intra	Post/Inter
Acceleration (m/s^{-2})	*h*	FC	0.016	12.8 (2.7)	17.4 (4)
	h	HC	0.016	10.1 (2)	13.6 (4.6)
	h	FH	0.016	2.7 (0.9)	3.2 (1)
	h	Sa	0.031	2.8 (0.2)	3.4 (0.3)
	h	St	0.008	2.7 (0.4)	3.1 (0.5)
	h	Po	0.008	4.5 (1.1)	5.1 (1)
	x	FC	0.02	12.6 (3.3)	17.3 (5.2)
	x	HC	0.016	10.9 (2.8)	15.1 (5.3)
	x	FH	0.016	3.1 (0.5)	3.6 (0.7)
	x	Sa	0.031	2.2 (0.3)	3 (0.8)
	x	St	0.008	2.7 (0.5)	3.2 (0.6)
	x	Po	0.004	3.3 (0.5)	4.2 (0.8)
	y	FC	0.027	21.3 (5.1)	29.6 (9)
	y	HC	0.039	15.2 (4.1)	21.3 (8.1)
	y	FH	0.016	2.2 (0.6)	2.8 (1)
	y	St	0.004	2.3 (0.5)	3.1 (1)
	y	Po	0.004	2.3 (0.4)	3 (0.8)
	z	FC	0.016	8.7 (1.4)	10.5 (1.6)
	z	HC	0.039	8.3 (1.4)	10.3 (3.2)
	z	FH	0.016	2 (0.5)	2.7 (0.8)
	z	Sa	0.031	2 (0.5)	2.7 (0.6)
	z	St	0.012	2.1 (0.4)	2.4 (0.5)
	z	Po	0.004	2.1 (0.4)	2.7 (0.4)
	v	FC	0.031	25.7 (3.1)	29.4 (3.5)
	v	HC	0.039	21.7 (1.4)	25.2 (3.4)
	v	Sa	0.031	2.2 (0.4)	2.8 (0.5)
	v	St	0.02	2.2 (0.3)	2.5 (0.4)
	v	Po	0.02	4.4 (2.1)	5 (2)
Angle (°)	Swing	HC	0.02	4.5 (0.9)	7 (2.9)
	Twist	FC	0.016	88.3 (10.7)	122 (40.7)
	Twist	HC	0.039	51.3 (18.8)	112.5 (67.3)
	x/v	FC	0.016	5.6 (2.8)	16.9 (16.4)
	x/v	HC	0.008	5 (2.8)	9.4 (5.9)
	x/v	St	0.004	2.3 (0.5)	3.7 (1.6)
	y/v	FC	0.016	8.4 (7.7)	24.6 (21.4)
	y/v	Sa	0.031	4.9 (1.3)	6 (1.8)
	y/v	Po	0.02	3.4 (0.8)	4.7 (2.6)
	z/v	FC	0.012	4.9 (2)	6.8 (2.4)
	z/v	HC	0.02	4.5 (0.9)	7 (2.9)
	z/v	FH	0.031	5.5 (1.2)	8.2 (5)
	z/v	Sa	0.031	2.5 (0.7)	3.7 (1.5)
	z/v	St	0.004	2.6 (0.5)	4.4 (1.7)
	z/v	Po	0.008	4.9 (0.9)	7.7 (3.6)
Angular Velocity (rad/s)	*x*	FC	0.016	1.7 (0.4)	2.4 (0.9)
	x	HC	0.008	1.3 (0.4)	1.9 (0.9)
	x	FH	0.016	0.5 (0.1)	0.6 (0.3)
	x	St	0.004	0.4 (0.1)	0.6 (0.3)
	x	Po	0.004	0.4 (0.1)	0.5 (0.1)
	y	FC	0.031	0.8 (0.3)	1.2 (0.4)
	y	HC	0.023	0.7 (0.2)	0.9 (0.5)
	y	FH	0.031	0.4 (0.1)	0.5 (0.2)
	y	Sa	0.031	0.3 (0)	0.3 (0)
	y	St	0.008	0.3 (0.1)	0.4 (0.1)
	z	FC	0.031	1.6 (1)	3.6 (2.8)
	z	HC	0.02	1.4 (0.7)	2.4 (1.7)
	z	FH	0.016	0.3 (0.1)	0.4 (0.1)
	z	St	0.02	0.3 (0.1)	0.6 (0.4)
	z	Po	0.004	0.4 (0.1)	0.5 (0.2)

FH: forehead, Po: pool, St: sternum, Sa: sacrum, FC: forelimb cannon, HC: hindlimb cannon. *v*: vertical, *h*: horizontal according to the global reference frame. *x*, *y*, *z*: axis in IMU reference frame. *x/v*, *y/v*, *z/v*: angle between IMU axis and global reference frame vertical. (cf. Figure 1).

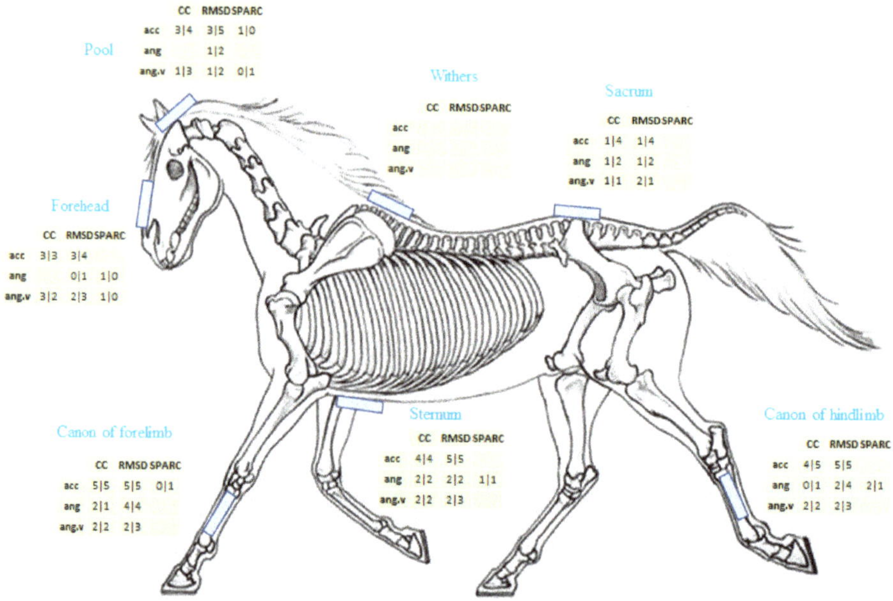

Figure 3. Number of indicators significantly affected by the training session for each sensor location for canter/trot. Here, for the CC, all the significant results are considered.

4. Discussion

The aim of the present paper was to define which kinematic indicators extracted from the IMUs' data could identify locomotion changes following a training session. These indicators could be used in the future to build an on-board system that could help to anticipate detrimental fatigue during training. For this, various locomotion indicators were considered to assess for changes in the kinematics following a training session.

It was expected for the training session to induce fatigue. The riders all interpreted the instruction in the same way and worked with the same objective; however, it was well below the physiological reality. Given the heart rate and blood lactate values, the energy demand (anaerobic and aerobic) was indeed not elevated. The anaerobic threshold expected for the measurements [34] was then not reached, even if the horses were felt to be fatigued according to their riders. In the present study, the maximal blood lactate value measured did not exceed 1.1 mmol/L, which is below the values presented in [35] following two standardised exercise tests with dressage horses and well below the 4 mm/L that can be found following exercises implying running at a specific intensity level [35]. Regarding the heart rate, the maximal heart rate did not exceed 123 bpm in the present study when values over 170 bpm were obtained following the two standardised exercise tests with dressage horses presented in [35].

Despite this, some locomotor indicators were statistically different after vs before the training session. CC and RMSD show a stronger similarity between strides of a same measurement run than between strides of two different runs, which indicates a change in the locomotion following the training session. The change in CC can be interpreted as a modification in the kinematics pattern, whereas a change in the RMSD, in the absence of a change in the pattern, can be seen as a modification in the movement amplitude. We could have expected the forehead and pool kinematics to be more affected than the other kinematics by the training session since modifications in the horse's head movement has previously been associated with fatigue [14]. In human athletes, it has been shown that the trunk kinematics changed notably and prematurely with fatigue during cycling [36] or running [7]. The absence of a larger effect on the forehead and pool kinematics might be due to the rider, who might influence and reduce the head movements, especially in dressage during which the horse's posture is evaluated.

The variation in SPARC also reveals a change in the fluidity of movement. The SPARC values were greater in the present study when computed on the angular velocity than in elderly humans during assessments of their gait [31], with values between −2 and −3 obtained in the present study against values around −5 obtained by Beck et al. [31]. The greater smoothness found in the present study might come from many factors such as the fact that horses walk/run on four limbs whereas humans walk/run on two, the "age" of the subjects (adults in the present study vs elderly in [31]), or the potential difference in the IMUs' noise. At trot, the sternum sensor was impacted regarding this SPARC indicator. It is assumed that this location reflects the body movement because of its rather constant proximity to the centre of gravity at trot, whereas this constant proximity is rather questionable in canter, in particular, because of the pitch of the gait [37]. Thus, the SPARC coefficient calculated from the angular velocity of the sternum seems interesting, and the results show a degradation of the fluidity of the trot after the effort. In canter, SPARC presents an increase in the fluidity, except for the sacrum's angular velocity. It is known that gluteal muscles and biceps femoris, actors of flexion/extension and abduction/adduction of the hindlimb, can be fatigued at canter [12]. This suggest that fatigue could affect the pelvis movements.

It appears that the indicators could attest to changes in the horse's locomotion patterns and smoothness following a training session. To the best of our knowledge, no such study has been conducted in horses. In humans, changes in the kinematic range of motion or spatio-temporal indicators have been investigated [38] as well as indicators associated with performance [39].

However, some of the proposed indicators, the CC and RMSD, characterising the evolution in the kinematics might not be the most appropriate to use to develop an on-board device for fatigue detection. This requires the detection on-line of the strides, to save the corresponding stride kinematics, and to compute the CC and RMSD on consecutive kinematic strides and regularly between non-consecutive strides. Approaches based on artificial intelligence could be developed to identify and compare gait cycles during a training session [40].

Because the traditional physiological indicators used to evidence fatigue were moderately affected, a doubt remains: were the observed modifications due to a slight fatigue or were the modifications due to something else such as the technical exercises? If the results are due to the effort itself, then the indicators are very sensitive, which would be promising for the development of a system helping to monitor fatigue.

In fact, the training session was expected to induce a fatigue so important that it would affect these physiological indicators, which was not the case. However, the small impact of the training session on the heart rate and lactate does not necessarily mean that the horses did not experience muscular fatigue. It has been shown that, during sub-maximal exercises, lactatemia does not necessarily significantly increase despite a decrease in the maximal force production [41]. Unfortunately, more elaborated protocols that could attest for a decrease in force production, which is the recognized method to attest for muscular fatigue [42], are not that easy to implement, especially with horses. Nevertheless, other studies measuring, for instance, EMG, as proposed by Takahashi et al. [13] but for locomotion on a treadmill, should be performed to confirm that the modifications seen in the present study can really be attributed to muscular fatigue.

The greatest number of significant variations were obtained in descending order from the fore and hind cannons, then for the sternum, the pool, the forehead, and lastly the sacrum sensors. The sensor located at the withers did not show significant modification, but this is probably due to parasite movements resulting from an inadequate fixation chosen for this sensor. In fact, with the number of indicators being relatively close for the other sensors, it is not possible to really discard a sensor location from the present study.

As a limitation of the present study, the absence of clear fatigue limited the scope of the results. A second limitation is related to the heterogeneous training session chosen by the riders. Another limitation of the study is the lack of information about the warm-up, in terms of duration, intensity, and type (active, passive, specific). It is indeed assumed that these elements improve performance as well as negatively impact it if the content is poorly managed [43]. The warm-up must also be adapted to the needs of the horse according to its lifestyle, its pains, etc. [44]. This is why it is difficult to impose the warm-up on these competitive horses for a protocol of research. However, it would have been relevant to equip the horses with sensors as soon as they had left the stables.

Regarding the kinematics affected by the training session, acceleration and angular velocity-based indicators were numerous even when the kinematics were not expressed in a global reference frame. This can be viewed as positive since these indicators do not require much computation time, which makes it possible to consider the integration of these indicators in an on-board system, to follow the changes in locomotion due to training.

5. Conclusions

The aim of the present paper was to define indicators extracted from IMUs' data that could reveal locomotion changes associated with fatigue in horses. For this, various locomotion indicators were considered to assess for changes in the kinematics following a training session supposed to provoke fatigue in the horses.

If the training session did not seem to induce an important physiological fatigue, some locomotor indicators were statistically different after vs before the training session, indicating that IMUs seem appropriate to track locomotion pattern modifications due to training. Future research should be conducted to be able to fully attribute the modifications of these indicators to fatigue.

Supplementary Materials: The following supporting information can be downloaded at: https://www.mdpi.com/article/10.3390/s22134981/s1, Table S1: Complete statistical results.

Author Contributions: Conceptualization, L.F. and S.B.; methodology, L.F., S.B. and B.P.; software, B.P.; formal analysis, L.F., S.B. and B.P.; investigation, L.F., S.B., Q.T., J.-F.D. and B.P.; resources, S.B.; writing—original draft preparation, L.F., S.B. and B.P.; supervision, L.F. and S.B.; project administration, F.D. All authors have read and agreed to the published version of the manuscript.

Funding: This research was funded by the "Conseil scientifique de la filière équine" (CS_2019_44).

Institutional Review Board Statement: Ethical review and approval were waived for this study because the measurements were taken during normal training sessions without any disturbance.

Informed Consent Statement: Informed consent was obtained from all subjects (riders) involved in the study.

Data Availability Statement: Not applicable.

Acknowledgments: Authors thank the riders and horses who took part in the experimentation. They also thank Boichot, L., Pycik, E. and Pacher, L. who contributed to the measurements. They are grateful to Lewis, V. who reviewed their manuscript.

Conflicts of Interest: The authors declare no conflict of interest. The funders had no role in the design of the study; in the collection, analyses, or interpretation of data; in the writing of the manuscript, or in the decision to publish the results.

Appendix A

Other CC values significantly different between the comparison of cycles from the same run (intra) versus the comparison of cycles measured before and after training (inter) according to the statistical tests. They were not presented in the corpse text of the manuscript because the CC values were smaller than 0.80 for the intra-run.

Table A1. CC values significantly different at canter.

CANTER					
Variable	**Axis**	**Position**	***p*-Value**	**Before/Intra**	**Post/Inter**
Acceleration	*h*	FC	0.027	0.28 (0.09)	0.18 (0.14)
	h	St	0.004	0.61 (0.14)	0.5 (0.19)
	x	FC	0.031	0.31 (0.11)	0.1 (0.1)
	x	HC	0.039	0.39 (0.08)	0.26 (0.12)
	x	Sa	0.031	0.77 (0.23)	0.67 (0.21)
	x	St	0.008	0.72 (0.09)	0.59 (0.07)
	y	FC	0.031	0.32 (0.07)	0.16 (0.1)
	y	HC	0.039	0.41 (0.07)	0.29 (0.11)
	z	FC	0.020	0.24 (0.07)	0.12 (0.07)
	z	HC	0.023	0.33 (0.1)	0.23 (0.12)
	v	FC	0.016	0.45 (0.12)	0.28 (0.14)
	v	HC	0.027	0.36 (0.08)	0.21 (0.1)
	v	Po	0.039	0.63 (0.26)	0.56 (0.27)
Angle	Twist	FC	0.008	0.65 (0.09)	0.15 (0.57)
	x/v	FC	0.031	0.24 (0.13)	0.13 (0.07)
	z/v	St	0.012	0.74 (0.29)	0.66 (0.37)
Angular velocity	*x*	FC	0.027	0.36 (0.08)	0.25 (0.13)
	x	HC	0.020	0.38 (0.1)	0.25 (0.1)
	x	FH	0.016	0.27 (0.07)	0.1 (0.12)
	x	St	0.039	0.2 (0.06)	0.07 (0.13)
	y	FC	0.031	0.35 (0.12)	0.23 (0.11)
	y	HC	0.016	0.29 (0.11)	0.17 (0.1)
	z	FH	0.039	0.28 (0.14)	0.14 (0.24)

FH: forehead, Po: pool, St: sternum, Sa: sacrum, FC: forelimb cannon, HC: hindlimb cannon. *v*: vertical, *h*: horizontal according to the global reference frame. *x*, *y*, *z*: axis in IMU reference frame. *x/v*, *y/v*, *z/v*: angle between IMU axis and global reference frame vertical. (cf. Figure 1).

Table A2. CC values significantly different at trot.

TROT					
Variable	**Axis**	**Position**	***p*-Value**	**Intra**	**Inter**
Acceleration	*h*	FC	0.02	0.62 (0.16)	0.37 (0.25)
	h	HC	0.039	0.54 (0.18)	0.22 (0.4)
	h	Sa	0.031	0.53 (0.14)	0.33 (0.28)
	h	St	0.008	0.6 (0.13)	0.5 (0.15)
	h	Po	0.039	0.61 (0.14)	0.51 (0.19)
	x	FC	0.016	0.6 (0.2)	0.32 (0.34)
	x	HC	0.016	0.62 (0.15)	0.34 (0.33)
	x	Sa	0.031	0.76 (0.09)	0.53 (0.28)
	y	FC	0.039	0.62 (0.2)	0.32 (0.36)
	y	HC	0.004	0.62 (0.18)	0.29 (0.35)
	y	FH	0.016	0.55 (0.13)	0.22 (0.37)
	y	St	0.008	0.57 (0.21)	0.28 (0.37)
	y	Po	0.004	0.63 (0.13)	0.35 (0.39)
	z	FC	0.008	0.42 (0.14)	0.18 (0.18)
	z	HC	0.004	0.56 (0.13)	0.24 (0.29)
	z	FH	0.016	0.75 (0.09)	0.6 (0.21)
	v	FC	0.012	0.26 (0.13)	0.04 (0.21)
	v	HC	0.02	0.22 (0.13)	0.09 (0.14)
Angle	Twist	HC	0.039	0.66 (0.4)	0.33 (0.67)
	z/v	Sa	0.031	0.7 (0.19)	0.55 (0.26)
	z/v	St	0.004	0.72 (0.26)	0.66 (0.28)

Table A2. *Cont.*

		TROT			
Variable	Axis	Position	*p*-Value	Intra	Inter
Angular velocity	*x*	FC	0.004	0.7 (0.12)	0.43 (0.21)
	x	HC	0.039	0.74 (0.12)	0.49 (0.38)
	x	FH	0.016	0.5 (0.14)	0.11 (0.34)
	x	St	0.039	0.67 (0.22)	0.27 (0.57)
	x	Po	0.008	0.51 (0.17)	0.22 (0.31)
	y	FC	0.02	0.67 (0.23)	0.26 (0.44)
	y	HC	0.039	0.63 (0.17)	0.35 (0.43)
	y	FH	0.031	0.64 (0.13)	0.48 (0.29)
	y	Po	0.02	0.56 (0.18)	0.4 (0.32)
	z	Po	0.008	0.51 (0.16)	0.13 (0.37)

FH: forehead, Po: pool, St: sternum, Sa: sacrum, FC: forelimb cannon, HC: hindlimb cannon. *v*: vertical, *h*: horizontal according to the global reference frame. *x*, *y*, *z*: axis in IMU reference frame. *x/v*, *y/v*, *z/v*: angle between IMU axis and global reference frame vertical. (cf. Figure 1).

References

1. Di Domenico, F.; Raiola, G. Effects of Training Fatigue on Performance. *J. Human Sport Exer.* **2021**, *16*, S769–S780. [CrossRef]
2. Jones, C.M.; Griffiths, P.C.; Mellalieu, S.D. Training Load and Fatigue Marker Associations with Injury and Illness: A Systematic Review of Longitudinal Studies. *Sports Med.* **2017**, *47*, 943–974. [CrossRef] [PubMed]
3. McGowan, C.M.; Whitworth, D.J. Overtraining Syndrome in Horses. *CEP* **2008**, *5*, 57. [CrossRef]
4. Kellmann, M. Preventing Overtraining in Athletes in High-Intensity Sports and Stress/Recovery Monitoring: Preventing Overtraining. *Scand. J. Med. Sci. Sports* **2010**, *20*, 95–102. [CrossRef]
5. Lindner, A.; Mosen, H.; Kissenbeck, S.; Fuhrmann, H.; Sallmann, H.P. Effect of Blood Lactate-Guided Conditioning of Horses with Exercises of Differing Durations and Intensities on Heart Rate and Biochemical Blood Variables. *J. Anim. Sci.* **2009**, *87*, 3211–3217. [CrossRef]
6. Mami, S.; Khaje, G.; Shahriari, A.; Gooraninejad, S. Evaluation of Biological Indicators of Fatigue and Muscle Damage in Arabian Horses After Race. *J. Equine Veterinary Sci.* **2019**, *78*, 74–78. [CrossRef]
7. Winter, S.; Gordon, S.; Watt, K. Effects of Fatigue on Kinematics and Kinetics during Overground Running: A Systematic Review. *J. Sports Med. Phys. Fitness* **2017**, *57*, 887–899. [CrossRef]
8. Barbieri, F.A.; dos Santos, P.C.R.; Lirani-Silva, E.; Vitório, R.; Gobbi, L.T.B.; van Diëen, J.H. Systematic Review of the Effects of Fatigue on Spatiotemporal Gait Parameters. *BMR* **2013**, *26*, 125–131. [CrossRef]
9. Johnston, C.; Gottlieb-Vedi, M.; Drevemo, S.; Roepstorff, L. The Kinematics of Loading and Fatigue in the Standardbred Trotter. *Equine Veterinary J.* **1999**, *31*, 249–253. [CrossRef]
10. Wickler, S.J.; Greene, H.M.; Egan, K.; Astudillo, A.; Dutto, D.J.; Hoyt, D.F. Stride Parameters and Hindlimb Length in Horses Fatigued on a Treadmill and at an Endurance Ride. *Equine Veterinary J.* **2006**, *38*, 60–64. [CrossRef]
11. Muñoz, A.; Cuesta, I.; Riber, C.; Gata, J.; Trigo, P.; Castejón, F.M. Trot Asymmetry in Relation to Physical Performance and Metabolism in Equine Endurance Rides. *Equine Veterinary J.* **2006**, *38*, 50–54. [CrossRef] [PubMed]
12. Takahashi, Y.; Takahashi, T.; Mukai, K.; Ohmura, H. Effects of Fatigue on Stride Parameters in Thoroughbred Racehorses during Races. *J. Equine Veterinary Sci.* **2021**, *101*, 103447. [CrossRef] [PubMed]
13. Takahashi, Y.; Mukai, K.; Matsui, A.; Ohmura, H.; Takahashi, T. Electromyographic Changes in Hind Limbs of Thoroughbreds with Fatigue Induced by Treadmill Exercise. *Am. J. Veterinary Res.* **2018**, *79*, 828–835. [CrossRef]
14. Kienapfel, K. The Effect of Three Different Head-Neck Positions on the Average EMG Activity of Three Important Neck Muscles in the Horse. *J. Anim. Physiol. Anim. Nutr.* **2015**, *99*, 132–138. [CrossRef]
15. Pfau, T.; Scott, W.M.; Sternberg Allen, T. Upper Body Movement Symmetry in Reining Quarter Horses during Trot In-Hand, on the Lunge and during Ridden Exercise. *Animals* **2022**, *12*, 596. [CrossRef] [PubMed]
16. Sapone, M.; Martin, P.; Ben Mansour, K.; Chateau, H.; Marin, F. The Protraction and Retraction Angles of Horse Limbs: An Estimation during Trotting Using Inertial Sensors. *Sensors* **2021**, *21*, 3792. [CrossRef]
17. Lopes, M.A.F.; Nichols, J.T.; Dearo, A.C.O.; Nelson, S.R. Effects of Forelimb Instrumentation on Lameness Detection in Horses Using a Portable Inertial Sensor-Based System. *J. Am. Vet. Med. Assoc.* **2021**, *259*, 892–898. [CrossRef]
18. Serra Bragança, F.M.; Broomé, S.; Rhodin, M.; Björnsdóttir, S.; Gunnarsson, V.; Voskamp, J.P.; Persson-Sjodin, E.; Back, W.; Lindgren, G.; Novoa-Bravo, M.; et al. Improving Gait Classification in Horses by Using Inertial Measurement Unit (IMU) Generated Data and Machine Learning. *Sci. Rep.* **2020**, *10*, 17785. [CrossRef]
19. Darbandi, H.; Serra Bragança, F.; van der Zwaag, B.J.; Voskamp, J.; Gmel, A.I.; Haraldsdóttir, E.H.; Havinga, P. Using Different Combinations of Body-Mounted IMU Sensors to Estimate Speed of Horses—A Machine Learning Approach. *Sensors* **2021**, *21*, 798. [CrossRef]

20. Walker, V.A.; Tranquillle, C.A.; MacKechnie-Guire, R.; Spear, J.; Newton, R.; Murray, R.C. Effect of Ground and Raised Poles on Kinematics of the Walk. *J. Equine Vet. Sci.* **2022**, *115*, 104005. [CrossRef]
21. Sapone, M.; Martin, P.; Mansour, K.B.; Château, H.; Marin, F. Comparison of Trotting Stance Detection Methods from an Inertial Measurement Unit Mounted on the Horse's Limb. *Sensors* **2020**, *20*, 2983. [CrossRef] [PubMed]
22. Hatrisse, C.; Macaire, C.; Sapone, M.; Hebert, C.; Hanne-Poujade, S.; De Azevedo, E.; Marin, F.; Martin, P.; Chateau, H. Stance Phase Detection by Inertial Measurement Unit Placed on the Metacarpus of Horses Trotting on Hard and Soft Straight Lines and Circles. *Sensors* **2022**, *22*, 703. [CrossRef] [PubMed]
23. Tijssen, M.; Hernlund, E.; Rhodin, M.; Bosch, S.; Voskamp, J.P.; Nielen, M.; Serra Bragança, F.M. Automatic Detection of Break-over Phase Onset in Horses Using Hoof-Mounted Inertial Measurement Unit Sensors. *PLoS ONE* **2020**, *15*, e0233649. [CrossRef]
24. Tijssen, M.; Hernlund, E.; Rhodin, M.; Bosch, S.; Voskamp, J.P.; Nielen, M.; Serra Bragança, F.M. Automatic Hoof-on and -off Detection in Horses Using Hoof-Mounted Inertial Measurement Unit Sensors. *PLoS ONE* **2020**, *15*, e0233266. [CrossRef]
25. Hagen, J.; Jung, F.T.; Brouwer, J.; Bos, R. Detection of Equine Hoof Motion by Using a Hoof-Mounted Inertial Measurement Unit Sensor in Comparison to Examinations with an Optoelectronic Technique—A Pilot Study. *J. Equine Vet. Sci.* **2021**, *101*, 103454. [CrossRef]
26. Bosch, S.; Serra Bragança, F.; Marin-Perianu, M.; Marin-Perianu, R.; van der Zwaag, B.J.; Voskamp, J.; Back, W.; Van Weeren, R.; Havinga, P. Equimoves: A Wireless Networked Inertial Measurement System for Objective Examination of Horse Gait. *Sensors* **2018**, *18*, 850. [CrossRef]
27. Balasubramanian, S.; Melendez-Calderon, A.; Burdet, E. A Robust and Sensitive Metric for Quantifying Movement Smoothness. *IEEE Trans. Biomed. Eng.* **2012**, *59*, 2126–2136. [CrossRef]
28. Harris, C.M.; Wolpert, D.M. Signal-Dependent Noise Determines Motor Planning. *Nature* **1998**, *394*, 780–784. [CrossRef]
29. Balasubramanian, S.; Melendez-Calderon, A.; Roby-Brami, A.; Burdet, E. On the Analysis of Movement Smoothness. *J. NeuroEng. Rehabil.* **2015**, *12*, 112. [CrossRef]
30. Melendez-Calderon, A.; Shirota, C.; Balasubramanian, S. Estimating Movement Smoothness From Inertial Measurement Units. *Front. Bioeng. Biotechnol.* **2020**, *8*, 558771. [CrossRef]
31. Beck, Y.; Herman, T.; Brozgol, M.; Giladi, N.; Mirelman, A.; Hausdorff, J.M. SPARC: A New Approach to Quantifying Gait Smoothness in Patients with Parkinson's Disease. *J. NeuroEng. Rehabil.* **2018**, *15*, 49. [CrossRef] [PubMed]
32. Back, W.; Clayton, H.M. *Equine Locomotion*; Elsevier Health Sciences: Amsterdam, Netherlands, 2013; ISBN 978-0-7020-2950-9.
33. Dobrowolski, P. Swing-Twist Decomposition in Clifford Algebra. *arXiv* **2015**, arXiv:1506.05481.
34. Gondim, F.J.; Zoppi, C.C.; Pereira-da-Silva, L.; de Macedo, D.V. Determination of the Anaerobic Threshold and Maximal Lactate Steady State Speed in Equines Using the Lactate Minimum Speed Protocol. *Compar. Biochem. Physiol. Part A Mol. Integr. Physiol.* **2007**, *146*, 375–380. [CrossRef] [PubMed]
35. Lindner, A.; von Wittke, P.; Schmald, M.; Kusserow, J.; Sommer, H. Maximal Lactate Concentrations in Horses after Exercise of Different Duration and Intensity. *J. Equine Veterinary Sci.* **1992**, *12*, 36–39. [CrossRef]
36. Galindo-Martínez, A.; López-Valenciano, A.; Albaladejo-García, C.; Vallés-González, J.M.; Elvira, J.L.L. Changes in the Trunk and Lower Extremity Kinematics Due to Fatigue Can Predispose to Chronic Injuries in Cycling. *Int. J. Environ. Res. Public Health* **2021**, *18*, 3719. [CrossRef]
37. Galloux, P.; Richard, N.; Dronka, T.; Léard, M.; Perrot, A.; Jouffroy, J.L.; Cholet, A. Analysis of Equine Gait Using Three-dimensional Accelerometers Fixed on the Saddle. *Equine Veterinary J.* **1994**, *26*, 44–47. [CrossRef]
38. Giandolini, M.; Gimenez, P.; Temesi, J.; Arnal, P.J.; Martin, V.; Rupp, T.; Morin, J.-B.; Samozino, P.; Millet, G.Y. Effect of the Fatigue Induced by a 110-Km Ultramarathon on Tibial Impact Acceleration and Lower Leg Kinematics. *PLoS ONE* **2016**, *11*, e0151687. [CrossRef]
39. Marotta, L.; Scheltinga, B.L.; van Middelaar, R.; Bramer, W.M.; van Beijnum, B.-J.F.; Reenalda, J.; Buurke, J.H. Accelerometer-Based Identification of Fatigue in the Lower Limbs during Cyclical Physical Exercise: A Systematic Review. *Sensors* **2022**, *22*, 3008. [CrossRef]
40. Jiang, Y.; Hernandez, V.; Venture, G.; Kulić, D.; Chen, B.K. A Data-Driven Approach to Predict Fatigue in Exercise Based on Motion Data from Wearable Sensors or Force Plate. *Sensors* **2021**, *21*, 1499. [CrossRef]
41. Strojnik, V.; Komi, P.V. Fatigue after Submaximal Intensive Stretch-Shortening Cycle Exercise. *Med. Sci. Sports Exerc.* **2000**, *32*, 1314–1319. [CrossRef]
42. Enoka, R.M.; Duchateau, J. Muscle Fatigue: What, Why and How It Influences Muscle Function. *J. Physiol.* **2008**, *586*, 11–23. [CrossRef] [PubMed]
43. Bishop, D. Warm Up II: Performance Changes Following Active Warm Up and How to Structure the Warm Up. *Sports Med.* **2003**, *33*, 483–498. [CrossRef] [PubMed]
44. Clayton, H.M. Warming Up Horses When Riding: Why and How. Available online: https://thehorse.com/190044/warming-up-horses-when-riding-why-and-how/ (accessed on 2 May 2022).

sensors

MDPI

Article

A Lightweight Pose Sensing Scheme for Contactless Abnormal Gait Behavior Measurement

Yuliang Zhao [1,2,*], Jian Li [1,2], Xiaoai Wang [1,2], Fan Liu [3], Peng Shan [1,2], Lianjiang Li [1,2] and Qiang Fu [4]

1 Sensor and Big Data Laboratory, Northeastern University, Qinhuangdao 066000, China; 2172080@stu.neu.edu.cn (J.L.); 2101929@stu.neu.edu.cn (X.W.); peng.shan@neuq.edu.cn (P.S.); lilianjiang@live.cn (L.L.)

2 Hebei Key Laboratory of Micro-Nano Precision Optical Sensing and Measurement Technology, Qinhuangdao 066000, China

3 College of Life Science and Technology, Xi'an Jiaotong University, Xi'an 710049, China; 2213722832@stu.xjtu.edu.cn

4 Shijiazhuang School, People Liberation Army Engineering University—Shijiazhuang, Shijiazhuang 050003, China; fuq2124@gmail.com

* Correspondence: zhaoyuliang@neuq.edu.cn

Abstract: The recognition of abnormal gait behavior is important in the field of motion assessment and disease diagnosis. Currently, abnormal gait behavior is primarily recognized by pressure and inertial data obtained from wearable sensors. However, the data drift and wearing difficulties for patients have impeded the application of these wearable sensors. Here, we propose a contactless abnormal gait behavior recognition method that captures human pose data using a monocular camera. A lightweight OpenPose (OP) model is generated with Depthwise Separable Convolution to recognize joint points and extract their coordinates during walking in real time. For the walking data errors extracted in the 2D plane, a 3D reconstruction is performed on the walking data, and a total of 11 types of abnormal gait features are extracted by the OP model. Finally, the XGBoost algorithm is used for feature screening. The final experimental results show that the Random Forest (RF) algorithm in combination with 3D features delivers the highest precision (92.13%) for abnormal gait behavior recognition. The proposed scheme overcomes the data drift of inertial sensors and sensor wearing challenges in the elderly while reducing the hardware requirements for model deployment. With excellent real-time and contactless capabilities, the scheme is expected to enjoy a wide range of applications in the field of abnormal gait measurement.

Keywords: abnormal gait behavior; OpenPose; machine learning; XGBoost; random forest

Citation: Zhao, Y.; Li, J.; Wang, X.; Liu, F.; Shan, P.; Li, L.; Fu, Q. A Lightweight Pose Sensing Scheme for Contactless Abnormal Gait Behavior Measurement. *Sensors* **2022**, *22*, 4070. https://doi.org/10.3390/s22114070

Academic Editor: Giovanni Saggio

Received: 3 May 2022
Accepted: 25 May 2022
Published: 27 May 2022

1. Introduction

Abnormal gait behavior is highly-related to many neurodegenerative diseases, such as Parkinson's disease, cerebral palsy, lumbar disc herniation, cerebral infarction and osteoarthritis. Therefore, the recognition and measurement of abnormal gait behavior has been an important topic of research in the field of diagnosis and treatment [1]. Abnormal gait behavior is highly prevalent, especially in the elderly. According to the statistics of the China Parkinson's Disease Registry (CPDR), more than 3 million patients suffer from Parkinson's symptoms in China. This indicates an urgent need for recognition systems for behavioral disorders [2]. Preliminary diagnosis of a patient's disease based on their abnormal gait behavior is needed in many everyday life settings, such as houses, nursing homes, and other public places, which saves much cost and time for the patient. At present, the abnormal gait behavior of patients can be recognized by two main categories of methods: by using inertial measurement units (IMUs) and by using contactless models and machine learning algorithms.

A preliminary diagnosis of abnormal gait behavior can be achieved by wearing micro sensors. SIJOBERT et al. [3] extracted features from frozen gait by placing a wireless inertial sensor on the patient's lower leg to acquire changes in gait parameters. Zhao et al. [3,4] developed a gait analysis system consisting of a bipedal IMU. By using an inequality-constrained zero-velocity update (ZUPT) aided INS algorithm, this system provides an efficient method for estimating gait parameters and characterizing gait performance to assess the rehabilitation process of patients with gait disorders. Wang et al. [5] developed a new IMU-based clinical gait assessment method. Their experiment extracted nine variables from two calf-mounted IMUs and used them to quantify the patient's gait deviation. Based on these parameters, an IMU-based gait normal index (INI) was derived to assess the overall gait performance. However, the use of sensors to recognize abnormal gait behavior in patients with mobility impairments suffers from data drift problems and wearing difficulties [5].

In recent years, inertial and pressure sensors have been widely used in hospitals and nursing homes for analyzing patients' gait [6]. However, such methods are suitable for patients who have difficulty wearing sensors for data acquisition. Therefore, it is necessary to explore contactless systems for diagnosing different behavioral disorders [7]. Kursun et al. [2] proposed a method that combines the support vector machine (SVM) algorithm and a recognition model for the preliminary diagnosis of patients with Parkinson's symptoms. Using acoustic data with the smallest deviation, the method can distinguish patients with Parkinson's disease from healthy people at an accuracy of 92.75%. Yaman et al. [1] found through experiments that patients with Parkinson's disease have poor verbal ability, so they proposed a method in which SVM and k-nearest neighbors (KNN) algorithms are used to obtain features from the Parkinson's acoustic data set for the recognition of Parkinson's disease. The accuracy was calculated to be 91.25% and 92.33%, respectively, by using the two algorithms. Sato et al. [8] obtained frozen gait and Magnetic step data of Parkinson's patients by using OP forward gait features. By analyzing the data curve, they found patients with Parkinson's disease have a different movement curve from healthy people. Liu et al. [9] proposed a locally weighted discriminant-preserving projection embedding ensemble algorithm to solve the problems of high noise and small sample size with Parkinson's disease data. The algorithm achieved improved accuracy in Parkinson's disease recognition. Studies have found that contactless methods can better differentiate patients with Parkinson's disease and healthy people.

However, there is still a lack of studies on the recognition of gait behavior differences caused by diseases such as cerebral infarction, cervical compression, cerebellar lesions, and lumbar disc herniation. When it comes to diagnosing a patient, it is necessary for the doctor to first make a preliminary diagnosis of the type of disease that causes the abnormal gait behavior.

Guo et al. [10] went a further step by using an OP model to assess six abnormal toe types with a mobile 3D gait analysis system. Later, D'Antonio et al. [11] solved the problem of information concealment in videos with a corrected OP model. They also used an IMU sensor to calibrate the collected data, which verified the authenticity of features extracted by the OP model. At present, SVM and KNN are among the mainstream algorithms for the recognition of behavioral disorders. Chen et al. [12] used a new FKNN model to classify the Parkinson's data set and achieved an experimental accuracy of 96.07%. Hariharan et al. [13] adopted a feature reduction/selection technique and a recognition algorithm to detect Parkinson's symptoms. The recognition process was performed using least squares SVM (LS-SVM), probabilistic neural network (PNN), and general regression neural network (GRNN), and the recognition accuracy was as high as 100%.

To sum up, most of these techniques emphasized the recognition accuracy of abnormal gait behavior over the recognition efficiency. This means challenges remain to deploy these techniques in devices in daily applications. When a traditional OP model is used, in particular, it can be difficult to achieve real-time disease identification without the support of powerful hardware. Therefore, we developed a novel method that can recognize abnormal gait behavior accurately and efficiently. First, an ultra-lightweight OP model was developed to enable much-increased efficiency at the price of a little bit lower accuracy.

Then, based on the gait features obtained using the OP model, a simple 3D reconstruction model was developed to supplement more accurate features. At last, some highly efficient machine learning algorithms were used to recognize abnormal gait behavior. Our scheme achieves a contactless recognition of abnormal gait behavior due to multiple types of diseases compared to previous work.

Here is a summary of what we did and accomplished in this work:

1. We constructed a lightweight OP model with Depthwise Separable Convolution for real-time extraction of abnormal gait features. This significantly reduced the computing workload required for hardware-intensive devices.
2. We performed a 3D reconstruction on the 2D lower limb data extracted from subjects and obtained a total of 11 abnormal gait features from that data. Then, we further processed the extracted data to obtain step length features. These steps improved the data structure and diversified feature types.
3. We used machine learning algorithms to filter and classify abnormal gait features to the measurement of abnormal gait behavior caused by different diseases.

2. Experimental Method

2.1. Establishment of Experimental Models

Usually, the lower limb behavior of the human body is captured by a lightweight OP model, which offers a quick solution to process video and image data in real time [11,14].

Our work used this model to identify the 2D joint coordinates of patients during walking and to obtain their walking pose data by extracting the coordinates of their lower limb joints [15,16]. This vision-based model eliminates the inertial drift problem with traditional sensors, and its structure is illustrated in Figure 1. With the image stream data to be processed by the OP model, the feature map F was obtained through the VGG19 network. Then, the data entered the dual-branch convolutional neural network in multiple stages through the feature map F. The upper branch was used to predict the heat map of the joint, which was obtained as the heat map S. The lower branch was used to predict the affinity field of the joint. Each stage was further predicted, and finally, the joint heat map and affinity field of the entire network were obtained after t times of recognition [14].

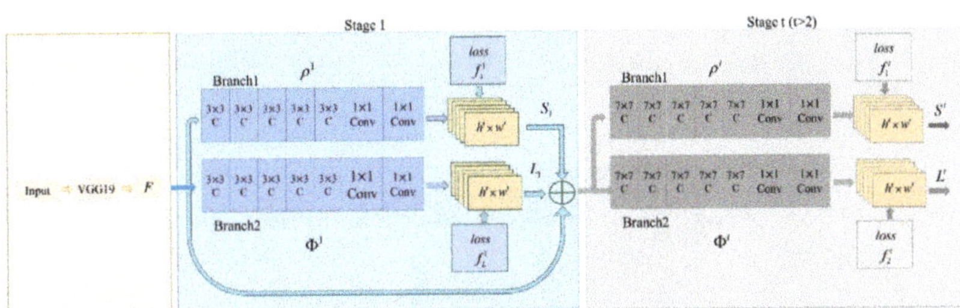

Figure 1. Structure of the OP model.

In Figure 1, ρ^t and φ^t are convolutional neural networks used to read features in stage t to generate a joint heat map $S^1 = \rho^1(F)$ and joint affinity field $L^1 = \varphi^1(F)$; $\rho^1, \rho^t, \varphi^1, \varphi^t$ were composed of five convolution blocks and two 1×1 ones. The input for each stage was the image feature F and the recognition result of the previous stage. S^t, L^t are the heat map and affinity field of joint at stage t, respectively. Then, the convolutional network of this stage was used to predict the joint heat map and joint affinity field of this stage. The recognition process can be expressed as follows:

$$S^t = \rho^t(F, S^{t-1}, L^{t-1}), \forall t \geq 2 \quad (1)$$

$$L^t = \varphi^t(F, S^{t-1}, L^{t-1}) \quad (2)$$

To obtain the coordinates of the lower limbs for real-time gait recognition, a Depthwise Separable Convolution structure, instead of the conventional convolution in VGG19, was used in our experiment. This can significantly reduce the number of model parameters required [17]. The size of all convolution kernels was set to 3 × 3, and the number of convolution kernels increased with the number of layers. The Depthwise Separable Convolution used different convolution kernels to convolve different channels, and decomposed the ordinary convolution into two processes: Depthwise Convolution and Pointwise Convolution, so as to decouple channel correlation and spatial correlation [17]. The Depthwise Convolution process split the convolution kernel into single channels and convolved each channel without changing the depth of the input feature image. The Pointwise Convolution process was used to up- and down-dimension the feature map with 1 × 1 convolution. The combination of these two processes made the model more lightweight. N conventional convolution kernels of size $D_K \times D_K \times M$ were equivalent to one Depthwise Convolution and N Pointwise Convolutions. Therefore, the FLOPS and Params of the Depthwise Separable Convolution were reduced to $(1/N) + (1/ D_K{}^2)$ conventional convolutions. Since there were 16 convolutions of size 3 × 3 in VGG-19, the FLOPS and Params of the lightweight OP model dropped to 17.36% of the original model.

A convolutional neural network with a smaller size and less computation was formed, which was well-suited for mobile devices and enabled faster and more efficient extraction of features from video stream data and reduced hardware requirements for model deployment.

2.2. 3D Construction of Lower Limbs

During data acquisition, the camera was located in the middle of the walking distance of the person, at a distance of 3 m from the vertical position of the person. There was a smaller angle between the video of the person during walking and the position of the camera, as shown in Figure 2a. The computer displayed the knee angle in motion, the velocity of the knee angle variation, and the acceleration of the knee angle variation. The positions of the thighs, calves and feet in the video were mapped to a two-dimensional (2D) plane. Therefore, errors were present in the length and angle data mapped in the video, and traditional IMU sensors have demonstrated errors in the knee angle measured by the OP [11,16,17]. As shown in Figure 2b,c, in the 3D space, there was an angle θ_1 error between the mapped thigh and the real thigh. The real knee angle is represented by θ_2, and the false knee angle of the mapped surface is represented by θ_3.

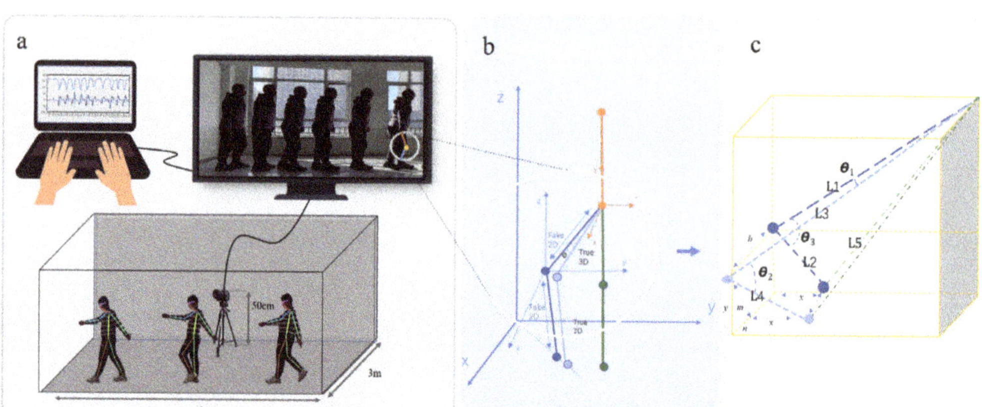

Figure 2. (**a**) Data acquisition process. (**b**) The 3D spatial relationship between the real knee and the mapped knee. (**c**) The lower limb reconstructed through the 2D data.

Since the data output from the OP model was 2D data, the angle data output by the leg needed to be reconstructed, and the reconstruction process is shown in Figure 2c. We obtained all the position coordinates of the leg joints and generated length and angle data by connecting the positions of the joint points. In the 2D image, L_1, L_2 and L_5 denote length data directly output by the OP model as extractable quantities, while L_3 and L_4 denote the real leg lengths in space. In the experiment, the data of the person standing in the video was used as the real leg length data. Finally, the 3D knee angle θ_2 was obtained as follows:

$$\cos\theta_2 = \frac{L_1^2 - L_5^2 + L_2^2 + 2\sqrt{L_3^2 - L_1^2} \times \sqrt{L_4^2 - L_2^2}}{2L_3 \times L_4} \quad (3)$$

3. Extraction of Step Length Features

Traditional gait features include multidimensional features such as step length, average stride time, average pace time, average stride length, and the lowest knee angle [6,18]. With the OP extraction model, we obtained abnormal gait features directly by intercepting each gait cycle in the program. In addition, the left and right step length features needed to be obtained by further processing the extracted data. Therefore, we designed an experiment for step length data extraction by observing the walking pose of the subject, as shown in Figure 3.

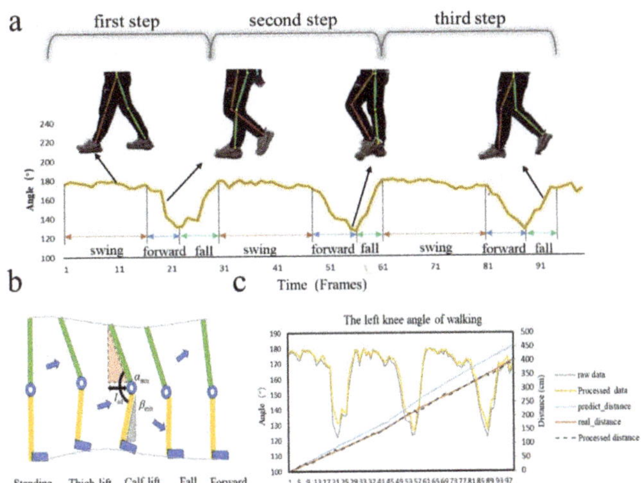

Figure 3. (**a**) The lightweight OP model captures the phases of the knee angle change during walking. (**b**) Step length calculation process. (**c**) Step length correction process.

The human walking process mainly consists of forward, swing and fall, as shown in Figure 3a. Patients with gait behavioral disorders generally walk with left and right swings and rapid changes in step length. Therefore, step length data was extracted to serve as the predictive features for the subsequent experiments.

The step length feature extraction process is shown in Figure 3b. The distance for which a person walks one step with one foot is determined by the person's leg length and knee angle during walking [19]. When a person leans forward, the raised foot is affected by the bending angle of the knee and moves forward. Taking the fixed angle of the step length as the lowest knee angle, we designed an experiment to measure the walking distance of a single person. We denoted the length of the thigh as Γ_1, the length of the calf as Γ_2, the lowest knee angle as α_{min}, and the angle between the calf and the vertical direction of the knee at the lowest knee angle as β_{min}.

When the step length was completely determined by thigh length Γ_1 and lowest knee angle α_{min}, we obtained:

$$l_{\alpha 1} = \Gamma_1 \times \cos(\alpha_{min} - 90) \quad (4)$$

$l_{\alpha 1}$ represents the predicted real step. The predicted distances in Figure 3c represent the time-varying movement distance curve fitted by this step method. As time went by, the difference between the predicted real distance and the real value became larger. Therefore, it was necessary to process the experimental data. It was found that the position coordinate of the knee at the lowest angle did not accurately reflect the distance moved by a single step during the real walking process.

When the step length was determined by thigh and calf lengths Γ_1, Γ_2 and knee angles α_{min} and β_{min}, we obtained:

$$l_{\alpha 2} = \Gamma_1 \times \cos(\alpha_{min} - 90) - \Gamma_2 \times \sin(\beta_{min}) \quad (5)$$

The processed single-step step length data is also presented in Figure 3c. This data was close to the real data. This demonstrated that this contactless step length measurement method is scientifically feasible. In the experiment, left and right step lengths were used as the features for classifying different types of abnormal gait behavior.

4. Analysis of Abnormal Gait Behavior

4.1. Analysis of Gait Characteristics for Different Diseases

In the medical field, behavioral disorders are mostly diagnosed in patients with Parkinson's disease, lumbar disc herniation, cerebral infarction, diabetes mellitus, and cerebellar lesions. Stimulation of electrical muscle signals can cause abnormal gait when walking. Therefore, there is a need to classify and assess these patients' disorders in a quantitative and contactless manner. In our experiments, we extracted gait data by asking the subject to walk for a distance under an indoor camera. Then, using machine learning algorithms, we achieved a preliminary diagnosis of these diseases.

Table 1 lists five different abnormal gaits that may be caused by behavioral disorders and their characteristics. These characteristics can be used as the motor characteristics of subjects who showed such abnormal behavioral symptoms in the experiment. Five types of abnormal gait behaviors are caused by different types of diseases. When the walking stride is small and the movement is stiff and slow, it is a manifestation of Parkinson's disease. Therefore, this scheme is convenient for doctors to pre-diagnose patients by classifying abnormal gait.

Table 1. Characteristics of abnormal gait behavior for different diseases.

Gait	Gait Characteristics	Corresponding Types of Diseases
Magnetic step (or Freezing gait)	The walking steps are small and the movements are stiff and slow.	This gait may indicate Parkinson's disease. The patient has symptoms of tremor, stiff limbs, and slow movement [20]
Mop step	The patient moves their left and right legs at inconsistent paces, and tends to walk by dragging their feet.	This gait may indicate lumbar disc herniation or cervical spondylitis myelopathy. Due to nerve compression, the patient has weak muscle on one leg, and generally drags one foot during walking [18]
Scissor Step	The patient tends to walk with their toes facing inward and their legs crossed.	This gait may indicate cerebral palsy or spinal cord injury, which can lead to impaired neurological function and affect physical activity [21]

Table 1. *Cont.*

Gait	Gait Characteristics	Corresponding Types of Diseases
Intermittent fragmentation	The patient experiences lameness and often feels the need to stop and rest due to pain and numbness in legs.	This gait may indicate osteoarthritis, lumbar spinal stenosis, vasculitis, or diabetes [22]
Drunk step	The patient cannot walk in a straight line and tend to stagger.	This gait may indicate cerebral hemorrhage, cerebral infarction, brain tumor, or cerebellar lesions. These diseases can cause cerebellar damage or cerebellar dysfunction [23].

4.2. Collection of Experimental Data

Due to sensitive neuromuscular changes, patient gait can serve as an important tool for patient state prediction and classification, widely affecting most gait features such as knee angle, step size, and stride length [24,25]. Exploring the gait changes caused by the muscles caused by lesions can help to understand the gait changes and the rehabilitation process of various diseases [26].

A total of eight subjects of different heights and weights (five males and three females) were involved in this experiment for data collection. Based on the behavioral characteristics of the disease in Table 1, the subjects were asked to imitate walking with a normal gait and five abnormal gaits. All the subjects walked back and forth along a five-meter-long experimental route. Five sets of experimental data were collected from each subject for each gait. Through the method of data undersampling, the imbalanced data set becomes balanced, and 40 experimental data are saved for each type of feature. The knee angle changes as the subjects walked with different gaits are shown in Figure 4. During the experiment, the subjects showed significant differences in knee angle changes among the six gaits. The changes in the left and right knee angles during walking with abnormal gaits suggested that the body was unbalanced.

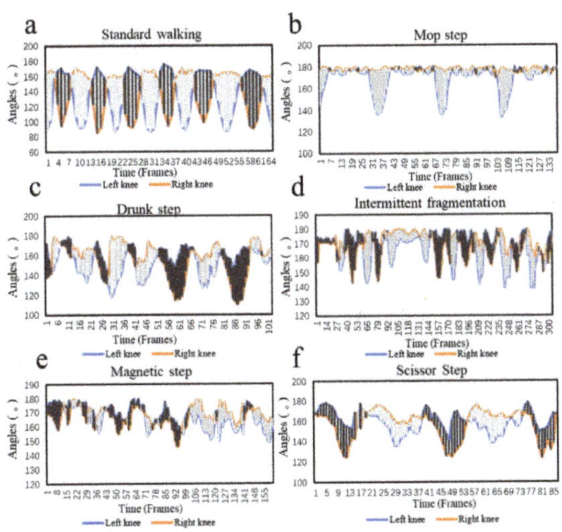

Figure 4. Variation curves of left and right knee angles under different gait. (**a**) Standard walking. (**b**) Mop step. (**c**) Drunk step. (**d**) Intermittent fragmentation. (**e**) Magnetic step. (**f**) Scissor step.

4.3. Feature Screening

We set one step for each left and right leg as a motion cycle. By collecting and processing the raw data, we obtained a total of 11 gait features, including left step length (LSS), right step length (RSS), lowest left knee angle (LLK), lowest right knee angle (LRK), average stride length (AS), average pace time (APT), average stride time (AST), variance of right knee angle variation (VOR), variance of left knee angle variation (VOL), average value of knee angular velocity (KAV), and average value of knee angular acceleration (KAA). We performed a 3D reconstruction on four of these features: LSS, RSS, LLK, and LRK. The four feature statistics of the six gaits with large differences are shown in Figure 5. Through statistical data, it is found that there are large differences in the features of different gait types, which contributes to higher accuracy of classification.

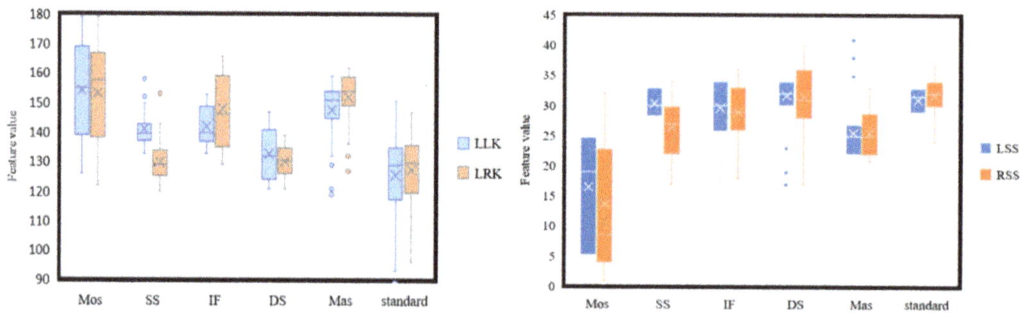

Figure 5. Difference distribution statistics of six gaits. (Features included are LLK, LRK, LSS, RSS).

Admittedly, accidental errors and interfering characteristics were present in this experimental data. Therefore, before classifying abnormal gait behavior, we screened the feature data and obtained the importance scores of each feature using the XGBoost algorithm [27], as shown in Figure 6. (After GridSearch, it is determined that the parameter combination is booster is gbtree, the learning rate is 0.3, tree depth is 6, and the maximum number of iterations is 100).

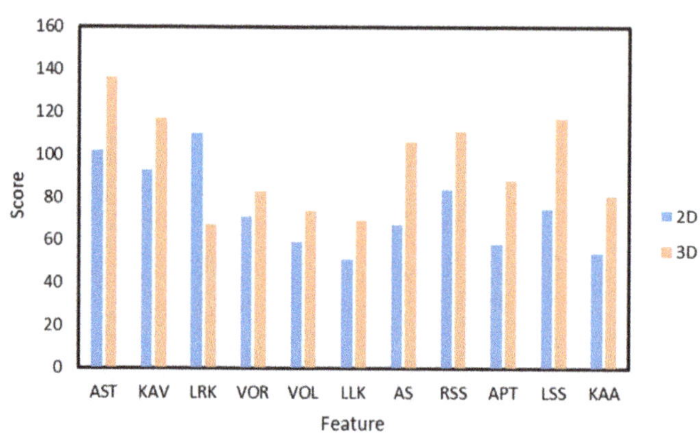

Figure 6. 2D and 3D feature importance scores after XGBoost screening.

From Figure 5, five of the eleven features, i.e., AST, KAV, APT, RSS, and LSS, have relatively high importance scores in both 2D and 3D conditions. For the four features processed with 3D reconstruction, the models showed increased importance scores on LSS, RSS, LLK and a decreased importance score on LRK. For models containing 3D features,

AST achieved the highest importance score. For models containing 2D features, LRK achieved the highest importance score. The experimental results showed that the overall model performance varied greatly with the type of feature used in the experiment. Some features produced low importance scores, indicating that these features did not contribute much to the overall model performance due to the masking problem and the randomness of different subjects during data collection. Better results were achieved for AST and KAV than for the other features, indicating that the average stride time and knee angular velocity played a bigger role in assessing body balance during walking.

XGBoost is used as a key machine learning algorithm for feature importance ranking, which can eliminate unfavorable features of machine learning models [27–29]. As shown in Table 2, by reducing the low-scoring features in order of importance scores, we obtained the acceptance scores for different numbers of features. With 3D features, the acceptance scores decreased as the number of features decreased in the range of 1~8. The best score of 0.9306 was achieved with 11 or 8 features. For the 2D features, the best score was achieved with 11 features, and the score basically decreased as the number of features decreased. Since the same score was obtained with 8 or 11 3D gait features, the abnormal gait behavior is recognized with 8 and 11 features, respectively.

Table 2. Acceptance scores for different numbers of features.

Number of Features	Score-2D	Score-3D
11	0.9167	0.9306
10	0.9028	0.8889
9	0.8889	0.9167
8	0.8472	0.9306
7	0.8611	0.8611
6	0.8472	0.8750
5	0.8056	0.8333
4	0.7639	0.6944
3	0.7083	0.7083
2	0.5556	0.5833
1	0.3333	0.4028

5. Discussion

The five abnormal gaits were mainly determined by different types of gait behavioral disorders. In our experiment, we used five recognition methods, i.e., Gradient Boosting (GB), KNeighbors (KN), Multilayer Perception (MLP), Random Forest (RF), and SVM, to classify the six gait features [28–33]. The parameter settings of the machine learning model obtained by GridSearch are shown in Table 3. Finally, 2D adopts 11 features for classification, and 3D adopts 8 and 11 features for classification, respectively, as shown in Table 4. For the multi-classification problem of abnormal gait behavior, we introduce the evaluation index Macro-average method and use Recall and Precision to express the classification results.

Table 3. Parameter combinations for machine learning models.

Machine Learning Algorithms	Parameters
GB	α = 10, loss function = deviance, subsample = 1.0.
KN	Weights = distance, n = 4, distance measure = 1.
MLP	Activation = ReLU, χ = (50,50), optimizer = Adam, α = 800, γ = 1.
RF	Number of decision trees = 57.
SVM	Kernel = 'linear', Kernel coefficient = 1.

n is the number of neighbors, α is the maximum number of iterations, γ is the state of the random number generator.

Table 4. Recognition results obtained for 8 and 11 features using five machine learning algorithms.

Machine Learning Algorithms	2D—11 Features		3D—8 Features		3D—11 Features	
	Recall	Precision	Recall	Precision	Recall	Precision
GB	0.7661	0.7778	0.8333	0.8611	0.8194	0.8472
KN	0.7211	0.7361	0.7500	0.7778	0.7533	0.7638
MLP	0.7557	0.7778	0.7944	0.8055	0.8344	0.8472
RF	0.8888	0.8918	0.9167	0.9213	0.9032	0.9048
SVM	0.7881	0.7918	0.8917	0.9027	0.8571	0.8611

TP represents the number of samples that predict the correct gait as the correct gait; FN represents the number of samples that predict the correct gait as the incorrect gait; FP represents the number of samples that predict the incorrect gait as the correct gait; TN represents the number of samples that predict incorrect gait as incorrect gait;

$$Precision = \frac{TP}{TP + FP} \tag{6}$$

$$Recall = \frac{TP}{TP + FN} \tag{7}$$

$$Precision_{\text{macro}} = \frac{1}{n}\sum_{i=1}^{n} Precision_{i} \tag{8}$$

$$Recall_{\text{macro}} = \frac{1}{n}\sum_{i=1}^{n} Recall_{i} \tag{9}$$

In this experiment, we used five machine learning algorithms to classify the feature data of abnormal gait behavior. As seen in Table 3, the recognition accuracies were improved after the 3D reconstruction of some features extracted from the OP model, with the highest precisions being 89.18% for 2D features and 92.13% for 3D features. As shown in Figure 7, high recognition accuracy was achieved for all gaits using random forest (RF). The lower recognition accuracy of abnormal gait types also reached 83%, and the highest recognition can reach 100%. The different gait recognition accuracy of 3D features has been improved to varying degrees, indicating that abnormal gait features show more obvious differences after 3D reconstruction. The highest recognition accuracy for abnormal gait (Magnetic step) caused by Parkinson's disease is 92%. Under the interference of a large number of different abnormal gaits, Parkinson's gait achieved the same level of accuracy as previous work [1,2]. The overall experimental results were as expected, and high recognition accuracy was achieved for different types of abnormal gaits.

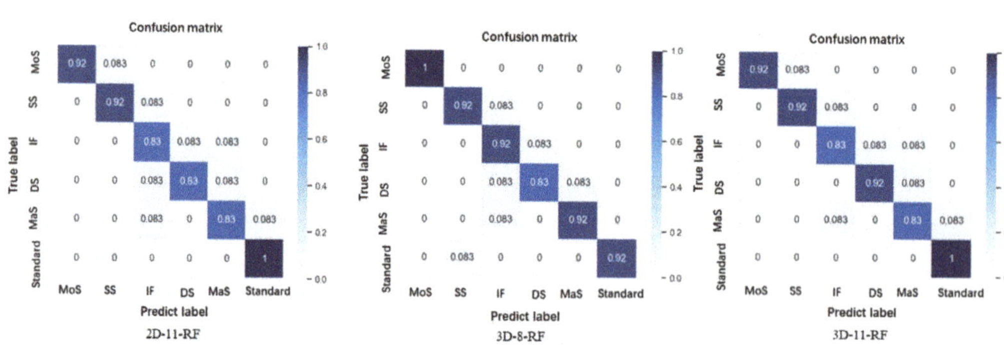

Figure 7. Best recognition precisions for 2D and 3D features.

This work realizes the lightweight of the model and quickly completes the gait recognition of volunteers, which overcomes the problems caused by wearing sensors and include multiple types of abnormal gait diseases and is no longer limited to Parkinson's [1]. However, with the introduction of a more abnormal gait, there may be some impact on Parkinson's recognition.

6. Conclusions

In this paper, we presented a lightweight contactless pose sensing scheme for abnormal gait behavior recognition. With this scheme, a lightweight OP model was used to extract abnormal gait features in experiments and satisfactory results were achieved for the recognition of diseases with abnormal gait behavior. The scheme offered a more lightweight and less hardware-intensive alternative to traditional approaches for the recognition of abnormal behavior in the elderly. Specifically, we used Depthwise Separable Convolution to make the OP model more lightweight, with its FLOPs and Params reduced to 17.36% of the original model. This design reduced the hardware requirements for the model and allowed for real-time contactless recognition of abnormal gait behavior by cameras.

For the data collected by the OP model, we first performed 3D reconstruction on the lower limb data to obtain the real walking data. Then, we screened out the invalid features from the acquired features, completed feature importance analysis and filtered out gait features with poor results. Finally, we used five machine learning algorithms to classify the gait data and performed disease type recognition based on abnormal gait features. In the experiments, the RF algorithm achieved the best recognition precisions, which was 92.13%. The experiments verified that our proposed scheme can classify diseases with abnormal gait behavior accurately and efficiently. This scheme can assist doctors to recognize patient lesions by different abnormal gait behavior caused by different diseases. With this scheme, we can continue to study high-precision quantitative evaluation of such diseases in the future.

Author Contributions: Conceptualization, Y.Z. and J.L.; methodology, J.L., F.L. and Q.F.; software, J.L.; validation, Y.Z. and P.S.; formal analysis, P.S. and L.L.; data curation, X.W.; writing—original draft preparation, J.L.; writing—review and editing, P.S. and Y.Z.; visualization, J.L.; supervision, Y.Z., L.L. and Q.F.; project administration, Y.Z.; funding acquisition, Y.Z. All authors have read and agreed to the published version of the manuscript.

Funding: This work was supported in part by the National Natural Science Foundation of China under Grant 61873307, in part by the Natural Science Foundation of Hebei Province of China under Grant F2021203070 and F2021501021, in part by the Scientific Research Project of Colleges and Universities in Hebei Province under Grant ZD2019305, in part by the Fundamental Research Funds for the Central Universities under Grant N2123004, in part by the Qinhuangdao Science and Technology Planning Project under Grant 201901B013, in part by the Administration of Central Funds Guiding the Local Science and Technology Development under Grant 206Z1702G and in part by the Chinese Academy of Sciences (CAS)-Research Grants Council (RGC) Joint Laboratory Funding Scheme under Project JLFS/E-104/18.

Institutional Review Board Statement: Not applicable.

Informed Consent Statement: Not applicable.

Data Availability Statement: The data presented in this study are available in the link: https://github.com/dlj0214/GAIT (accessed on 4 May 2022).

Conflicts of Interest: The authors declare no conflict of interest.

References

1. Yaman, O.; Ertam, F.; Tuncer, T. Automated Parkinson's Disease Recognition Based on Statistical Pooling Method Using Acoustic Features. *Med. Hypotheses* **2020**, *135*, 109483. [CrossRef]
2. Sakar, C.O.; Kursun, O. Telediagnosis of Parkinson's Disease Using Measurements of Dysphonia. *J. Med. Syst.* **2010**, *34*, 591–599. [CrossRef]
3. Sijobert, B.; Azevedo Coste, C.; Denys, J.; Geny, C. IMU Based Detection of Freezing of Gait and Festination in Parkinson's Disease. In Proceedings of the 2014 IEEE 19th International Functional Electrical Stimulation Society Annual Conference (IFESS), Kuala Lumpur, Malaysia, 17–19 September 2014; pp. 1–3.
4. Zhao, H.; Wang, Z.; Qiu, S.; Shen, Y.; Wang, J. IMU-Based Gait Analysis for Rehabilitation Assessment of Patients with Gait Disorders. In Proceedings of the 2017 4th International Conference on Systems and Informatics, Hangzhou, China, 11–13 November 2017; pp. 622–626.
5. Wang, L.; Sun, Y.; Li, Q.; Liu, T.; Yi, J. IMU-Based Gait Normalcy Index Calculation for Clinical Evaluation of Impaired Gait. *IEEE J. Biomed. Health Inform.* **2021**, *25*, 3–12. [CrossRef]
6. Li, W.; Lu, W.; Sha, X.; Xing, H.; Lou, J.; Sun, H.; Zhao, Y. Wearable Gait Recognition Systems Based on MEMS Pressure and Inertial Sensors: A Review. *IEEE Sens. J.* **2022**, *22*, 1092–1104. [CrossRef]
7. Guan, Z.; Li, S.; Cheng, Y.; Man, C.; Mao, W. A Video-Based Fall Detection Network by Spatio-Temporal Joint-Point Model on Edge Devices. In Proceedings of the 2021 Design, Automation & Test in Europe, Virtual, 1–5 February 2021; pp. 422–427.
8. Sato, K.; Nagashima, Y.; Mano, T.; Iwata, A.; Toda, T. Quantifying Normal and Parkinsonian Gait Features from Home Movies: Practical Application of a Deep Learning–Based 2D Pose Estimator. *PLoS ONE* **2019**, *14*, e0223549. [CrossRef] [PubMed]
9. Liu, C.; Tan, X.; Wang, P.; Zhang, Y.; Li, Y. Recognition Algorithm of Parkinson's Disease Based on Weighted Local Discriminant Preservation Projection Embedded Ensemble Algorithm. In Proceedings of the BIBE 2019 the Third International Conference on Biological Information and Biomedical Engineering, Hangzhou, China, 20–22 July 2019; pp. 100–105.
10. Guo, Y.; Deligianni, F.; Gu, X.; Yang, G.Z. 3-D Canonical Pose Estimation and Abnormal Gait Recognition with a Single RGB-D Camera. *IEEE Robot. Autom. Lett.* **2019**, *4*, 3617–3624. [CrossRef]
11. D'Antonio, E.; Taborri, J.; Mileti, I.; Rossi, S.; Patane, F. Validation of a 3D Markerless System for Gait Analysis Based on OpenPose and Two RGB Webcams. *IEEE Sens. J.* **2021**, *21*, 17064–17075. [CrossRef]
12. Chen, H.L.; Huang, C.C.; Yu, X.G.; Xu, X.; Sun, X.; Wang, G.; Wang, S.J. An Efficient Diagnosis System for Detection of Parkinson's Disease Using Fuzzy k-Nearest Neighbor Approach. *Expert Syst. Appl.* **2013**, *40*, 263–271. [CrossRef]
13. Hariharan, M.; Polat, K.; Sindhu, R. A New Hybrid Intelligent System for Accurate Detection of Parkinson's Disease. *Comput. Methods Programs Biomed.* **2014**, *113*, 904–913. [CrossRef]
14. Cao, Z.; Hidalgo, G.; Simon, T.; Wei, S.E.; Sheikh, Y. OpenPose: Realtime Multi-Person 2D Pose Estimation Using Part Affinity Fields. *IEEE Trans. Pattern Anal. Mach. Intell.* **2021**, *43*, 172–186. [CrossRef]
15. Abe, K.; Tabei, K.-I.; Matsuura, K.; Kobayashi, K.; Ohkubo, T. OpenPose-Based Gait Analysis System For Parkinson's Disease Patients From Arm Swing Data. In Proceedings of the 2021 International Conference on Advanced Mechatronic Systems, Tokyo, Japan, 9–12 December 2021; pp. 61–65.
16. Yamamoto, M.; Shimatani, K.; Hasegawa, M.; Kurita, Y.; Ishige, Y.; Takemura, H. Accuracy of Temporo-Spatial and Lower Limb Joint Kinematics Parameters Using OpenPose for Various Gait Patterns with Orthosis. *IEEE Trans. Neural Syst. Rehabil. Eng.* **2021**, *29*, 2666–2675. [CrossRef]
17. Mai, W.; Wu, F.; Guo, Z.; Xiang, Y.; Liu, G.; Chen, X. A Fall Detection Alert System Based on Lightweight Openpose and Spatial-Temporal Graph Convolution Network. *J. Phys. Conf. Ser.* **2021**, *2035*, 012036. [CrossRef]
18. Nagai, T.; Takahashi, Y.; Endo, K.; Ikegami, R.; Ueno, R.; Yamamoto, K. Analysis of Spastic Gait in Cervical Myelopathy: Linking Compression Ratio to Spatiotemporal and Pedobarographic Parameters. *Gait Posture* **2018**, *59*, 152–156. [CrossRef]
19. Zeng, H.; Chen, W. An Evaluation Approach of Multi-Person Movement Synchronization Level Using OpenPose. In Proceedings of the 40th Chinese Control Conference, Shanghai, China, 26–28 July 2021; pp. 3900–3905.
20. Alcock, L.; Galna, B.; Perkins, R.; Lord, S.; Rochester, L. Step Length Determines Minimum Toe Clearance in Older Adults and People with Parkinson's Disease. *J. Biomech.* **2018**, *71*, 30–36. [CrossRef]
21. Armand, S.; Decoulon, G.; Bonnefoy-Mazure, A. Gait Analysis in Children with Cerebral Palsy. *EFORT Open Rev.* **2016**, *1*, 448–460. [CrossRef]
22. Schmitt, D.; Vap, A.; Queen, R.M. Effect of End-Stage Hip, Knee, and Ankle Osteoarthritis on Walking Mechanics. *Gait Posture* **2015**, *42*, 373–379. [CrossRef]
23. Hoogkamer, W.; Potocanac, Z.; van Calenbergh, F.; Duysens, J. Quick Foot Placement Adjustments during Gait Are Less Accurate in Individuals with Focal Cerebellar Lesions. *Gait Posture* **2017**, *58*, 390–393. [CrossRef]
24. Hussain, I.; Park, S.J. Prediction of Myoelectric Biomarkers in Post-Stroke Gait. *Sensors* **2021**, *21*, 5334. [CrossRef]
25. Hussain, I.; Park, S.J. Quantitative Evaluation of Task-Induced Neurological Outcome after Stroke. *Brain Sci.* **2021**, *11*, 900. [CrossRef]
26. Hussain, I.; Park, S.J. HealthSOS: Real-Time Health Monitoring System for Stroke Prognostics. *IEEE Access* **2020**, *8*, 213574–213586. [CrossRef]

27. Tang, Q.; Xia, G.; Zhang, X.; Long, F. A Customer Churn Prediction Model Based on XGBoost and MLP. In Proceedings of the 2020 International Conference on Computer Engineering and Application, ICCEA 2020, Guangzhou, China, 18–20 March 2020; pp. 608–612.
28. Rahman, B.; Hendric Spits Warnars, H.L.; Subirosa Sabarguna, B.; Budiharto, W. Heart Disease Classification Model Using K-Nearest Neighbor Algorithm. In Proceedings of the 2021 6th International Conference on Informatics and Computing, ICIC 2021, Hyderabad, India, 16–17 September 2021.
29. Zhang, S.; Li, X.; Zong, M.; Zhu, X.; Wang, R. Efficient KNN Classification with Different Numbers of Nearest Neighbors. *IEEE Trans. Neural Netw. Learn. Syst.* **2018**, *29*, 1774–1785. [CrossRef]
30. Priyadarshini, R.K.; Bazila, A.B.; Nagamani, T. Gradient Boosted Decision Tree Based Classification for Recognizing Human Behavior. In Proceedings of the 2019 International Conference on Advances in Computing & Communication Engineering (ICACCE-2019), Sathyamangalam, India, 4–6 April 2019.
31. Zheng, X.; Wang, Z.; Chung, W. Efficient Parameter Selection for SVM: TheCase of Business Intelligence Categorization. In Proceedings of the IEEE International Conference on Intelligence and Security Informatics: Security and Big Data, Beijing, China, 22–24 July 2017; pp. 158–160.
32. Sheykhmousa, M.; Mahdianpari, M.; Ghanbari, H.; Mohammadimanesh, F.; Ghamisi, P.; Homayouni, S. Support Vector Machine Versus Random Forest for Remote Sensing Image Classification: A Meta-Analysis and Systematic Review. *IEEE J. Sel. Top. Appl. Earth Obs. Remote Sens.* **2020**, *13*, 6308–6325. [CrossRef]
33. Özel, E.; Tekin, R.; Kaya, Y. Implementation of Artifact Removal Algorithms in Gait Signals for Diagnosis of Parkinson Disease. *Traitement Signal* **2021**, *38*, 587–597. [CrossRef]

MDPI
St. Alban-Anlage 66
4052 Basel
Switzerland
Tel. +41 61 683 77 34
Fax +41 61 302 89 18
www.mdpi.com

Sensors Editorial Office
E-mail: sensors@mdpi.com
www.mdpi.com/journal/sensors

www.ingramcontent.com/pod-product-compliance
Lightning Source LLC
Chambersburg PA
CBHW040804250926
51850CB00007BA/1079
* 9 7 8 3 0 3 6 5 8 5 4 7 5 *